Land and Disaster Management Strategies in Asia

Huong Ha
Editor

Land and Disaster Management Strategies in Asia

Springer

Editor
Huong Ha
UON Singapore (University of Newcastle,
Singapore Campus)
Singapore
Singapore

ISBN 978-81-322-1975-0 ISBN 978-81-322-1976-7 (eBook)
DOI 10.1007/978-81-322-1976-7

Library of Congress Control Number: 2014943249

Springer New Delhi Heidelberg New York Dordrecht London

Printed on acid-free paper

Springer is part of Springer Science+Business Media (www.springer.com)

Foreword

Even before the world entered an era of globalisation, the term "spaceship earth" emphasised the interconnectedness of our planet's peoples, resources, climates, ecological, economic and natural systems, and future prospects. Given man's huge dependence on non-renewable resources from time immemorial and systemic tendencies towards entropy, the increasingly pertinent need to manage the planet became clear. Perhaps the challenge and task were best captured by the title of R. Buckminster Fuller's 1969 book, *Operating Manual for Spaceship Earth*. Along the same lines, but more darkly, Garrett Hardin introduced the concept of "Lifeboat Ethics" (1974). Fortunately, as Julian Simon contended in *The Ultimate Resource* and *Ultimate Resource 2* (1981 and 1996), the earth's human resources may be up to the task of securing the earth's sustainability.

Land and Disaster Management Strategies in Asia is an excellent contribution to thinking about what key aspects of managing earth will demand. Its chapters deftly analyse the complex relationships among land distribution and use, economic and social development, industrialization, ecological and climate change, food security, governance, public administration, and peace. Part I on "Land Management and Food Security" emphasises the social as well as the political, economic, and agricultural importance of land distribution and reform. Land is not simply a factor of production. It is central to rural society, of course, but also to urban life and climate change when its use impels large migrations to cities. In some countries, land distribution is also central to governmental stability.

The importance of Part II on "Disaster Governance" requires no emphasis. Weather-related disasters, often attributed to permanent climate change, abound. Disasters involving earth, wind, fire, and water have been pronounced in recent years and many contend they are likely to become increasingly severe in the years to come. One learns almost monthly, if not daily, about landslides, sinkholes and earthquakes; storms, tornadoes and cyclones; extensive forest fires; and floods and droughts. As pointed out in Chap. 10, many disasters, such as terror attacks, toxic spills and mishandling of hazardous materials are "man-made" rather than natural. How can governance best reduce, mitigate and prepare for disasters? This section of the book emphasises the importance of inclusive planning that addresses the needs of populations that are especially vulnerable, such as persons with disabilities and in some societies, women.

Returning to Simon's conviction that humans are the earth's ultimate resource, Part III on "Community Education and Other Issues Related to Disaster Management" reminds us that disaster management goes well beyond dealing with natural systems. Fuller's "Operating Manual" must be written by humans for humans. Governments, civil society and each of us have a critical role in promoting sustainability. This section of the book addresses education, peace building, civil society and community as core components of mobilising human resources to forestall, mitigate and respond effectively to disasters.

I urge readers to give deep thought to each of the chapters in this outstanding book. It contains many essential perspectives, ideas and frameworks for developing effective disaster management strategies for Asia and the world as a whole. I certainly learnt a great deal from it and I hope the reader will do so as well.

Prof. David Rosenbloom
Distinguished Professor
American University
USA

References

Fuller RB (1969) Operating manual for spaceship earth. Simon and Schuster, New York
Hardin G (1974) "Lifeboat Ethics: The Case against Helping the Poor." Psychology Today. September 1974 http://www.garretthardinsociety.org/articles/art_lifeboat_ethics_case_against_helping_poor.html
Simon JL (1981) The ultimate resource. University Press, Princeton, New Jersey: Princeton
Simon JL (1996) The ultimate resource 2. University Press Princeton, New Jersey: Princeton

Preface

Management of land records, protection of public lands, assignment of land to the landless and economically weaker sections and development based on adequate land-use planning, form the basis of Land Governance. A country classified as 'developed' invariably has a reasonably fool-proof land management plan in place. In developing countries this issue assumes greater importance as the pressure on land and natural resources is often severe. The issue is also linked to the presence of rural poverty, agricultural land not in the name of tiller and skewed land holdings. The situation is no better in urban areas where while the economically powerful control holdings over most of the lands, the working class is forced to occupy miniscule proportion of the lands mostly in slum pockets and tenements. This situation is evident in most, if not all, urban settlements in the developing countries.

Over the years, the State has devised policies to ensure that the gap between the landed and the landless narrowed as quickly and as effectively as possible. However, there are many impediments to achieve this goal: vested interests that are forced to give up right over the land, spiralling land prices in urban areas and land being used as a source of authority and empowerment are some of them. The crucial role played by land in overall governance is brought out by the fact that even in countries where free distribution of land to the landless has been part of the State policy with a view to achieving agrarian reforms, in many cases the gap between the top 10 % and the bottom 10 % remains un-bridged. Most of the land distribution policies end up with the lands returning into the hands of the landed.

Apart from being a necessary condition for the healthy growth of a country's economy, land and land-based resources is also the commonest modes of exploitation. The power-holders get more land and resources assigned in their names, while the less powerful tend to lose whatever little they have. So notwithstanding land reforms measures in certain countries or regions, the probability of a neo-assignee alienating that land within his/her lifetime is three times higher than the average.

Given its importance in a country's economic growth, it is not surprising that land is also the source of many corrupt practices. There are a number of examples that emerge on a daily basis of how land and land-linked resources are grabbed by the political and administrative power-holders in the name of their family members or for their cronies.

Land governance is linked to the management of land as well as natural resources, water bodies, forests and hilly areas. The poor enforcement of land laws becomes a cause for disasters. In other words, a well-designed Disaster Management Policy should have good land protection policies as its base. Whether it is in the context of landslides, floods, droughts or coastal erosion, it is often noticed that many of these disasters are the result of man-made illegal and forced interventions. The example of some countries or regions is often quoted in the context of climate change and how this impinges on the economic growth of the country or region. It is feared that with the rise in sea levels, some of the low-lying agricultural lands could get inundated that would result in higher rural to urban migration and larger pressure on the urban areas.

It is unfortunate that notwithstanding the numerous studies, Committee Reports and Court judgments, the extent of landlessness amongst the economically weaker section, does not seen to be decreasing. The lack of accountability among the law-makers and law-enforcers, the high level of human greed and the lack of transparency in land records are the main reasons. The seemingly futile battle of the average citizen against the powerful law enforcers can be won only when the positive forces combine. Such exploitation will continue unless and until the adversely-affected learn to ask questions; and they can do this only when they are empowered. Digitizing all land records, textual and spatial, ensuring such records are geo-referenced and thence made tamper-proof and placing all land-holding data in the public domain, would go a long way in empowering the citizens.

It is in this context that all the aforementioned factors need to be taken seriously, especially by the policy makers in these countries. A lot of research and surveys are carried out by academicians and researchers on these areas. However, most of these remain as publications on the shelves of libraries. Correlation and intermingling of ideas between academicians and researchers, on the one hand, and administrators and policy-makers on the other hand, rarely happen. The policy-makers, therefore, often act in the interest of the undeserving or for benefiting certain vested interests with nobody to seriously question them. Adverse agricultural policies, change in cropping patterns, lack of land use planning, encroachments permitted on river banks and within rivers and distribution of forest lands, hillocks, etc. are the result of such lack of co-ordination. It is amazing to think how much more can be achieved if both sets of persons can synergise their ideas and their strengths so that both could work for the benefit of not any single class or group but for welfare of the entire society. It is this message that comes across from each of the papers contained in this Anthology.

Dr. Nivedita P. Haran
Additional Chief Secretary
Home & Vigilance
Personnel & Administrative Reforms
Department Government of Kerala
India

Acknowledgments

This book would not have been completed without the support of members of the advisory board, the reviewers, family members, colleagues, the authors and the research assistants. The editor wishes to thank (i) the advisors and reviewers for their professional and constructive feedback which is valuable to the authors, and (ii) the authors for their cooperation during the revision stages. Finally, the editor is very grateful for the advice and assistance from the Senior Editor and the Assistant Editor of Springer, India.

Contents

Figures

Tables

Abbreviations

ACT	Action for Conflict Transformation
ADB	Asian Development Bank
AGMs	Annual General Meetings
APEC	Asia-Pacific Economic Cooperation
ARMM	Autonomous Region of Muslim Mindanao
ASEAN	The Association of Southeast Asian Nations
BCCSAP	Bangladesh Climate Change Strategy and Action Plan
BJP	Bhartiya Janata Party
BWDB	Bangladesh Water Development Board
CAB	Comprehensive Agreement on the Bangsamoro
CAFGU	Citizen Armed Force Geographical Unit
CC	Climate Change
CCDMC	City Corporation Disaster Management Committee
CDC	Communicable Disease Centre
CDM	Clean Development Mechanism
CED	Community Economic Development
CEDAW	Convention on the Elimination of Discrimination Against Women
CEGIS	Centre for Environmental and Geographic Information System
CEP	Coastal Embankment Project
CFF	Community Forestry Federations
CFUGs	Community Forest User Groups
CIFOR	Centre for International Forestry Research
CIPRB	Centre for Injury Prevention and Research Bangladesh
CoE	Council of Elders
CPI	Corruption Perceptions Index
CPP	Cyclone Preparedness Program
CPPIB	Cyclone Preparedness Program Implementation Board

CPP-NPA-NDF	Communist Party of the Philippines-New People's Army-New Democratic Front
CPRs	Common Property Resources
CRA	Community Risk Assessment
CRC	Convention on the Rights of Child
CRPD	United Nations Convention on the Rights of Persons with Disabilities
CRZ	Coastal Regulation Zone
CSDDWS	Committee for Speedy Dissemination of Disaster Related Warning/Signals
CSDRM	Climate Smart Approach to Disaster Risk Management
CSOs	Civil Society Organisations
CSPP	Conflict Sensitive and Peace-Promotive
CSR	Corporate Social Responsibility
CZMP	Coastal Zone Management Plan
DCH	Dhaka Community Hospital
DDMC	District Disaster Management Committee
DFIF	Department of International Development, UK
DFOs	District Forest Offices
DIPECHO	Disaster Preparedness ECHO
DM&RD	Disaster Management and Relief Division
DoF	Department of Forest
DRM	Disaster Risk Management
DRR	Disaster Risk Reduction
DWCRA	Development of Women and Children in Rural Areas
EIA	Environmental Impact Assessment
EPWAPDA	East Pakistan Water and Power Development Authority
FAO	Food and Agriculture Organisation of the United Nations
FDI	Foreign Direct Investment
FECOFN	Federation of Community Forestry Users, Nepal
FGD	Focused Group Discussion
FPOCG	Focal Point Operation Coordination Group of Disaster Management
GDP	Gross Domestic Product
GOI	Government of India
GOP UN-MDP	Government of the Philippines United Nations Multi-Donor Programme
GPS	Global Positioning System
HEB	High Energy Biscuits
HPB	Health Promotion Board
HTL	High Tide Line
ICESCR	International Convention on Economic, Social and Cultural Rights
ICIMOD	International Centre for Integrated Mountain Development
IDPs	Internally Displaced Peoples

ILG	Islamic Leadership in Governance
IMD	Indian Meteorological Department
IMDMCC	Inter-Ministerial Disaster Management Co-ordination Committee; National Disaster Management Advisory Committee
INGOs	International Non-Government Organisation
IOs	International Organisations
IPCC	Inter-Governmental Panel on Climate Change
IRFC	International Federation of Red Cross and Red Crescent Societies
ISRO	Indian Space Research Organisation
IUCN	International Union for Conservation of Nature
IWM	Institute of Water Modelling
JNA	Joint Needs Assessment
KCERP	Khulna Coastal Embankment Rehabilitation Project
KII	Key Informant Interview
KJDRP	Khulna Jessore Drainage Rehabilitation Project
LDCs	Least Developed Countries
LGUs	Local Government Units
LoC	Line of Control
MCC	Mindanawan Cocosugar Corp
MDBs	Multilateral Development Banks
MDGs	Millennium Development Goals
MGNREGA	Mahatma Gandhi National Rural Employment Guarantee Act
MILF	Moro Islamic Liberation Front
MNLF	Moro National Liberation Front
MODSAC	Meteorological and Oceanographic Satellite Data Archival Centre
MOE	Ministry of Education
MoEF	Ministry of Environment and Forests
MoFDM	Ministry of Food and Disaster Management
MoFSC	Ministry of Forests and Soil Conservation
MOH	Ministry of Health
MRD	Ministry of Rural Development
NAPSIPAG	Network of Asia Pacific Schools and Institutes of Public Administration and Governance
NCEUS	National Commission for Enterprises in the Unorganised Sector
NDMA	National Disaster Management Authority
NDMC	National Disaster Management Council
NEFEJ	New Era, Nepal Forum for Environmental Journalists
NER	North Eastern Region
NFA	Nepal Foresters' Association
NFP	National Forest Plan
NGO	Non-Government Organisation

NGOCC	NGO Coordination Committee on Disaster Management
NGOs	Non-Governmental Organisations
NPDRR	National Platform for Disaster Risk Reduction
NRSP	National Rural Support Program
NSCB	National Statistical Coordination Board Philippines
NTNC	National Trust for Nature Conservation
OECD	Organisation for Economic Co-operation and Development
OED	Operations Evaluation Division
OPAPP	Office of the Presidential Adviser on the Peace Process
PC	Paani Committee
PDALs	Peace and Development Advocates Leagues
PDAs	Peace and Development Advocates
PDCs	Peace and Development Communities
PDMC	Pourashava Disaster Management Committee
PDP	Philippine Development Plan
PESA	Panchayats (Extension to Scheduled Areas) Act 1996
PF	Panchayat Forest
PMOs	Project Management Offices
PPER	Project Performance Evaluation Report
PPF	Panchayat Protected Forest
PPP	Public–Private Partnership
PSD	Private Sector Developments
PWD	Persons With Disability
R&D	Research and Development
RCP	Representative Concentration Pathways (modelled emission scenario)
RECOFTC	Regional Community Forestry Training Centre
SAARC	South Asian Association for Regional Cooperation
SARS	Severe Acute Respiratory Syndrome
SCDF	Singapore Civil Defence Force
SDC	Swiss Agency for Development and Cooperation
SDGs	Sustainable Development Goals
SEZs	Special Economic Zones
SHGs	Self Help Groups
SIA	Social Impact Assessment
SSD	Social Service Department
TACs	Tribal Autonomous Councils
TCPD	Town and Country Planning Department
TRM	Tidal River Management
UDHR	Universal Declaration of Human Rights
UDMC	Union Disaster Management Committee
UN	United Nations
UNCAC	United Nations Convention Against Corruption
UNDP/GEF	United Nations Development Program and Global Ecological Fund

UNDP	United Nations Development Programme
UNESCO	United Nations Educational, Scientific and Cultural Organisation
UNFCCC	United Nations Framework Convention on Climate Change
UN-HABITAT	United Nations Human Settlements Programme
UNISDR	United Nations International Strategy for Risk Reduction
UPA	United Progressive Alliance
USA	United States of America
USAID	United States Agency for International Development
UT	Union Territory
UZDMC	Upazila Disaster Management Committee
VDC	Village Development Committee
VHF	Very High Frequency
VLCs	Village Land Councils
WB	World Bank
WFP	World Food Programme
WHO	World Health Organisation
WRI	World Resources Institute

Contributors

Dr. Rajat Agrawal is an Assistant Professor in Department of Management Studies, IIT Roorkee (India). His research involvement includes innovation and business incubation which is important to the CEO's role at the newly started Technology Incubation and Entrepreneurship Development Activity Centre at IIT Roorkee. He is a member of (i) the working committee of Intellectual Property Rights Cell, and (ii) the Board of Studies of IIT Roorkee. He has supervised many Ph.D. students. He has published more than 28 papers. His research areas are social entrepreneurship, sustainable supply chain management, green marketing, and MSME sector. He has contributed chapters in three different books published by IGNOU, Delhi and NIOS, Delhi. He has given more than 15 talks on various subjects from All India Radio. He has presented papers, chaired technical sessions in more than 50 national/international conferences. Dr. Agrawal is involved in a major research project from UGC, New Delhi to develop management concepts from ancient Sanskrit literature. He was an active member in a large project (Rs. 49.5 million) on Knowledge Economy sponsored by the Government of India at IIT Rorkee. He is also handling a project on Cycle Rickshaw to develop it as a tool for social entrepreneurship. He is also a part of a project to develop Ganga River Basin Environmental Management Plan. He is also involved in some international programmes such as ACARA challenge. He was a mentor for ACARA Challenge for IIT Roorkee where he helped IITR students to participate in many Acara Challenges. IITR team won this challenge twice in 2010 and 2011.

Prof. Dr. Akbaruddin Ahmad is the Chairman, Policy Research Centre.bd (PRC.bd) and Vice-Chancellor of Darul Ihsan University, Bangladesh. He is a scholar, management specialist with exposure in management of educational institutions of higher learning, financial institutions/banks and civil society organisations. He is a prolific writer and regular contributor to national newspapers and peer reviewed journals of international standards. He delivered lectures on BBA and MBA program and is the Editor-in-Chief of *Finance World*, a leading newspaper. He was also an IT advisor for Janata Bank, Rupali Bank and Bangladesh Computer Council. A commerce graduate from the Dhaka University and an MBA with major in Banking from the Edward University, USA, he obtained his Ph.D. in Banking and Finance from Bircham International University, Spain. He joined the Rahman Rahman and Huq, and S. F. Ahmed Chartered Accountants as Senior Consultant and worked there as a National Consultant with internationally reputed consulting organisations viz Booz, Allen, Hamilton and Nothon Associates, Management Consultants. He was a Senior Consultant for the Financial Sector Reform Project of the World Bank/USAID to restructure four nationalised commercial banks and the Bangladesh Bank. He was a Senior Consultant for Bangladesh Computer Council, Rupali Bank, Janata Bank and Director Operations of BCCI Bank. He attended several management courses at WANG Headquarters at Lowell, Mass, USA, several IT courses at Apple Computer Cupertino, California/USA, Athens, Cairo, London, Lowell, Tokyo and Cyprus. He also held earlier the position of Pro Vice Chancellor, Treasurer, Dean and Director, Institute of Business Studies. He has to his credit a large number of publications in current affairs, economics, banking, financial sector reforms, IT and tourism, and has written numerous scholarly articles and book contributions.

Dr. Charlita Andales-Escano is the Director III in Mindanao Development Authority (formerly Mindanao Economic Development Council), a primary sub-national agency in the Philippines mandated to promote peace and accelerate development in Mindanao, where she has served for 19 years now. A Doctor of Philosophy in Development Administration in the University of South eastern Philippines in 2012, she earned her Master in Business Administration and academic units in Public Administration in Cor Jesu College in Digos City, Philippines in 2000 and 2001, respectively. As a peace and development advocate, she has been involved in planning, investment programming, resource mobilisation, implementing various programs and projects including policy advocacy with funding from international donor agencies. She has conducted various researches and studies in infrastructure, transportation, budget, and social sciences subject matters. She also taught subjects in Engineering, Mathematics, Geology, Sewerage, among others as a Civil Engineer. She provides advisory and consultancy services to key decision-makers and policymakers in Mindanao. Her expertise includes: infrastructure, development planning, project/program implementation, monitoring and evaluation, policy advocacy and networking, human resource management and development, financial management, statistical analysis.

Kumkum Bharti is a doctoral student in the Department of Management Studies, Indian Institute of Technology, Roorkee (India). She has earned her Master Degree from National Institute of Technology, Surathkal (India) in Marketing and Finance. She was awarded Junior Research Fellowship in December 2010 by University Grant Commission of India. Her research area includes bottom of the pyramid and co-creation. She has attended more than 15 international conferences and published five papers in national and international journals like Emerald and Springer. She assisted in the project on 'Co-creation in Social Entrepreneurship' partnering with Copenhagen Business School, Copenhagen, Denmark and FGV,

Sao Paulo, Brazil from 2012–2014, and was awarded Noel Deerr Gold Medal Award of the Year 2012–2013 at the 71st Annual Convention of Sugar Technologists' Association of India (STAI) at Hyderabad for the research paper entitled *Vertical Business Integration: A Logical Prospective Evolutionary Step for Sugar Industry in India* under the management section.

Faiz Ahmed Chowdhury is a Research Associate at BRAC Institute of Governance and Development (BIGD), BRAC University in Dhaka, Bangladesh. His areas of research interests are sustainable development, climate change, international political economy, globalisation and governance. He has M.Sc. in International Relations from Jahangirnagar University, Dhaka. Before BIGD, he worked in think-tank Centre for Policy Dialogue (CPD) and Unnayan Onneshan, an NGO. His recent publications include Haque, Kazi et al. (2013) Civil Society Organisations in Climate Change Adaptation: Paani Committee's Movement for Tidal River Management in South–West Bangladesh. IGS Working Paper 17/2013. Dhaka: IGS; Chowdhury, Faiz (2013) Geo-politics, Democratisation and External Influence: The Bangladesh Case. IGS Working Paper 5/2013. Dhaka: IGS; and, Chowdhury, Faiz (2012) Climate Change Finance and Governance. IGS Working Paper 2/2012. Dhaka: IGS.

Mr. Zahrul Akmal Damin has a Bachelor Degree of Social Science (Honours) and Master of Public Administration from University of Science Malaysia. He is a lecturer at University Tun Hussein Onn Malaysia. Expertise in development and public policy enabled him to give a talk in public agencies. He is currently pursuing a Ph.D. at the University Utara Malaysia in public administration with a specialisation in public policy analysis.

Dr. Tek Nath Dhakal is a Professor at the Central Department of Public Administration, Tribhuvan University (Nepal) and works as Campus Chief. He is also the coordinator of 'Policy and Governance Studies (PGS) in South Asia: Regional Master and Ph.D. Programs'—a joint research project of the Department of Administration and Organisation Theory, University of Bergen (Norway), Central Department of Public Administration, Tribhuvan University (Nepal), Department of Continuing and General Education, North South University (Bangladesh), and Department of Political Science University of Peradeniya (Sri Lanka) financed by NORAD, Norway. He has conducted a number of research projects, including 'Status of the Implementation of Citizen Charter in Municipalities in Nepal', 'People's Trust in Public and Political Institutions in Nepal' and 'NGOs' Role in Livelihood Improvement in Nepal', including Institutional Analysis of Markhu and Hekuli VDC (Nepal). He has published several books and edited volumes, and many research reports both from Nepal and abroad. In addition, he has contributed around 50 articles in local and international journals. He also served as an editor of Nepalese Journal of Public Policy and Governance. He often participates in different national and international forums and conferences, and presents papers. He is also affiliated with a number of academic and professional networks, such as ISPA, ISTR, NAPSIPAG, EROPA, the Finnish

Society of Administrative Sciences and the Public Administration Association of Nepal.

Dr. Huong Ha is the Academic Coordinator of MBA and undergraduate business programmes at UON Singapore (University of Newcastle, Singapore campus). She was the Dean of TMC Business School and Director of Research and Development, TMC Academy (Singapore). She holds a Ph.D. from Monash University (Australia) and a Master's degree in public policy from Lee Kuan Yew School of Public Policy, National University of Singapore. She was a recipient of a Ph.D. scholarship awarded by Monash University, a Temasek scholarship awarded by the National University of Singapore and a scholarship awarded by the UN University/International Leadership Academy, and many other professional and academic awards. She has been an invited member of (i) editorial boards of many international journals, (ii) scientific and/or technical committees of several international conferences, and (iii) international advisory board of many associations. She has also been a reviewer of many international journals and international conferences.

Kazi Nurmohammad Hossainul Haque is a Senior Research Associate at the BRAC Institute of Governance and Development (BIGD), BRAC University in Dhaka, Bangladesh. He is a political sociologist with research interests in civil society, citizen participation, political participation, environmental governance and urban governance. He has an MA in Development Studies from International Institute of Social Studies (ISS) of the Erasmus University Rotterdam, The Hague and MA in Public Policy from Central European University, Budapest. His work experience includes research, programme management, advocacy, policy analysis and teaching. Some of his recent publications are: Haque, Kazi and Biswas, Sanjay (2014) *Diaspora Engagement in Local Politics of Bangladesh* in Md Mizanur Rahman and Tan Tai Yong (eds) International Migration and Development in South Asia, (Singapore: Routledge) (In press); Haque, Kazi et al. (2013) *Civil Society Organisations in Climate Change Adaptation: Paani Committee's Movement for Tidal River Management in South–West Bangladesh*. IGS Working Paper 17/2013. (Dhaka: IGS); Haque, Kazi et al. (2012) *NGOs as Drivers of Policy Change: Between Policy Autonomy and External Influence* in State of Governance in Bangladesh 2010–2011: Policy, Influence, Ownership, (Dhaka: IGS); Basu, Ipshita and Haque, Kazi (2012) *Urban Governance in Dhaka: An Overview* in State of Cities: Urban Governance in Dhaka, (Dhaka: IGS); and, Haque, Kazi (2012) The Political Economy of Land and Real Estate in Dhaka City. IGS Working Paper 2/2012, (Dhaka: IGS).

Major General Md Shafiqul Islam is the Executive Director of CRP, Bangladesh. Prior to this appointment he served for 34 years in the army in the communication and aviation branch. He was the Vice Chancellor of Bangladesh University of Professionals, and also served as the Military Secretary of the Bangladesh Army. During that appointment he also worked as the Chief Coordinator of the bio-metric National ID and Voter Registration Project, for the successful completion of which he was awarded the prestigious International People's ID World award in 2008 at Milan, Italy. General Shafiq has obtained his Master degree in Defence Studies, M.Phil. in National Security Studies and a Post Graduate Diploma in Management. He attended various courses in USA, UK, Indonesia and Thailand. He has participated in a number of seminars/workshops and written number of articles in different journals.

Jak Jabes is a Senior Fellow at the School of International Development and Global Studies, University of Ottawa since July 2013. Before returning to the University of Ottawa where he started his academic career, he was Visiting Professor at the Lee Kuan Yew School of Public Policy, Singapore. He is an experienced specialist on governance and public administration reform in "economies in transition," whose interests cover corruption, professional networks and the teaching of public management and administrative reforms. He has held senior positions at the Organisation for Economic Co-Operation and Development (OECD), the Asian Development Bank (ADB) and the United Nations Development Programme (UNDP). In these organisations his work focused on assisting governments in building partnerships, promoting regional capacity development initiatives, establishing knowledge networks and providing policy advisory and capacity building support. Dr. Jabes' professional life has been divided between academe and international civil service. After finishing his Ph.D. studies in social psychology at the University of Kansas he joined the then Faculty of Administration of the University of Ottawa where his research focused on work motivation, rewards and senior civil service management. He has been editor or member of the editorial board of several journals and he is the author, co-author or editor of a dozen books and 60 book chapters, articles, papers and reports.

Suriani Jamil attained a Bachelor of Business in Marketing and Management, and Masters in Business Administration from University of Newcastle, Australia. She enjoys studying about business in various 'bizarre' perspectives whereby sometimes, she would implement 'holistic learning'—integrate business knowledge that she has learnt in school into her hobbies. Her interest in research areas are in business and management; managing organisational change, business strategy/strategic management, politics and society in the context of globalisation, corporate social responsibility and sustainable development.

Ishtiaq Jamil is an Associate Professor at the Department of Administration and Organisation Theory, University of Bergen, Norway. His research interests include administrative culture, public policy, multi-level governance, and trust in public and political institutions. His work has been published in journals such as *International Political Science Review, Asian Survey, International Journal of Public Administration, Public Organisation Review, Journal of Commonwealth and Comparative Politics, International Review of Sociology, Journal of Contemporary South Asia, Journal of Middle Eastern and South Asian Studies,* and *South Asian Journal of Policy and Governance.*

Dr. Anishia Jayadev is an Assistant Professor in the Institute of Management in Government, a premier training institute for Government of Kerala. As a researcher, her areas of interest are *Sociology of Health, Gender Studies, Rural Development, Service Delivery and Land Governance*. She has also been associated as a resource person in the Full Bright Educational Partnership Programme, Claremont Graduate University USA and Department of Political Sciences, Kerala University. She has (i) conducted nearly 10 Training Needs Analysis of Government Departments of Kerala, (ii) been associated with Government of Kerala in drafting its NGO policy, (iii) been associated with Cochin University of Science and Technology (CUSAT) as an examiner for their M.Phil. Programme in ecological informatics, (iv) been a faculty for Post Graduate Diploma Programme in e-governance conducted collaboratively by IMG Kerala and IIITM-K. She has approved research guide under the Social Science faculty, University of Kerala, partnered with The Capacity Building for Poverty Reduction Programme funded by DFID, been associating with training needs analysis, designing of programmes, and conducted training in the Intensive Training Programme funded by the Government of India. She secured her Ph.D. from University of Kerala in 2001 under the topic Medical care and Rehabilitation of Cardiac Patients: A Sociological Analysis.

Prof. N. C. Kar has more than 25 years of academic experience in teaching and research in the business management stream in Utkal University and Xavier Institute of Management. He has written extensively on business, economic and social development areas.

Rabeya Khatun Kazi is an independent researcher based in Trondheim, Norway. She also works with Save the Children Norway as an activity guide for refugee children. She has M.Sc. in Sustainable Development from Uppsala University, Sweden. Her research interests are sustainability, environmental activism, development displacement, environmental governance and political ecology. Khatun studied political structure's impact on anti-dam protest mobilisation through comparative case studies of Sardar Sarovar Dam, India and Three Gorges Dam, China in her master's research. The dissertation title was *Political Structure and Anti-dam Protest Movements: Comparing Cases of India and China*. She is now converting the dissertation into journal article. Before becoming an independent researcher, she worked in Greenpeace Norway's frontline team and Amnesty International Norway's fundraising team.

Prof. Dr. Sarfraz Khawaja is a free-lance Policy Evaluation and Result-Based Monitoring Specialist and now works as individual consultant for the Higher Education Commission, Government of Pakistan. He earned his Ph.D. degree from University of Missouri, Columbia, USA and then joined University of Wisconsin, USA as a faculty member. He also taught at Government College Lahore, (now a university). His other teaching assignment was that of Chief Instructor (Professor) at the Civil Services Academy, Lahore. He served as Deputy Educational Advisor in the Ministry of Education, Government of Pakistan and Director, Academy of

Educational Planning and Management, Islamabad. His work experience extends to UNICEF as Planning, Monitoring and Evaluation Officer. He also worked for UNFPA as Management Advisor. Another key assignment was Monitoring and Evaluation specialist in Planning Commission, Government of Pakistan for three years. He was appointed member of Punjab Chief Minister Special Task Force on Higher Education 2003–2004. He also served as member of the Board of Directors, Centre for Pakistan Studies, Punjab University Lahore Pakistan 2007–2009. His Book *Non-Formal Education: Myth and Panacea* won National Book Foundation Award in 1994. He has been a consultant to several international agencies some of which include Swiss Development Cooperation, Asian Development Bank, Plan International, The World Bank, UNICEF, USAID, Aga Khan Educational Foundation, and Higher Education Commission. He was principal research partner in the BRIDGES project carried out in collaboration with Harvard University, USA (1988–1990). He has been a visiting speaker at Allama Iqbal Open University, Quaid-i-Azam University, National Institute of Public Administration, Foreign Services Academy, and National Defence College, Islamabad (now a university).

Associate Prof. Dr. Ahmad Martadha Mohamed had just finished his term as the Director of Quality and Strategic Planning, College of Law, Government, and International Studies, Universiti Utara Malaysia (UUM). Prior to this appointment, he was the Dean of Student Development and Alumni, UUM from January 2005 until June 2011. Besides administrative duties, he is also teaching Public Management courses at the undergraduate and graduate level. He has published extensively in local and international proceedings. His articles were published in several peer-reviewed and scholarly journals. He has also acquired several research grants from a host of national institutions. He was also a keynote speaker as well as invited speaker at the International Conference in India, Indonesia, and Malaysia. Currently, he is also a committee member for Network of Asia Pacific Schools and Institutes of Public Administration and Governance (NAPSIPAG) and Asia Pacific Society of Public Affairs (APSPA). He is the chief editor for *Journal of Governance and Development* (JDG). His expertise in the area of public administration and politics earns him invitation from local institutions such that he is a regular political analyst for Radio Television Malaysia (RTM) and Private TV Network Astro Awani as well as several independent news organisations. He has also appeared in Channel News Asia (Singapore) commenting on Malaysian general elections. He has extensively given talks at various governmental organisations that include Prison Department, Kedah Civil Service (KCS), BTN, PDRM, JASA, schools, and UUM covering topics such as Administrative Reforms, Strategic Planning, Managing Change, Conflict Management, Managing Diversity, Customer Service, Representative Bureaucracy, Pre-Retirement Planning Programs, and Succession Planning. He received a degree in Political Science from Southern Illinois University, Carbondale, USA, Master of Public Administration from Indiana State University, USA and Ph.D. in Political Science from Southern Illinois University, Carbondale, USA.

Saroj Nayak is a doctoral scholar, Sambalpur University and a postgraduate in anthropology from Sambalpur University. Saroj has around 19 years of experience in research, consultancy, capacity building, program/project execution, M&E and allied project management functions. He worked with Xavier Institute of Management, Bhubaneswar and handled different projects sponsored by bi-lateral and multilateral international and national agencies like DFID, World Bank, UNDP, EU, NABARD, SIDBI, LWS, CAPART, CORDAID, MEMISA, MESIRIOR, CARITAS, etc. in the institute. His basic experience relates to rural infrastructure sectors (energy and environment), bio-diversity and adaptation research and rural decentralisation. He has hands on experience in working with State Government, Central Government and private organisations. He was the team leader for strengthening rural decentralised governance system, and has developed the SEMF for the OCTMP. He has a key leadership position in the social transformation group of CTRAN.

Prof. Valerian Rodrigues teaches at the Centre for Political Studies in Jawaharlal Nehru University, New Delhi. His research interests include political philosophy, political ideas in modern India, and agrarian relations and politics. His books include, *The Essential Writings of B. R. Ambedkar* (New Delhi, Oxford, 2002) and *The Indian Parliament: A Democracy at Work* (co-author, B. L. Shankar) (New Delhi, Oxford, 2011).

Dr. Vikas Saharan obtained his Ph.D. from Jawaharlal Nehru University. His areas of interest are environmental issues and disaster management, education, health, gender, dalits, tribal, minorities, social movements, urbanisation, climate change, sustainable development, science and technology. He has been actively working on issues related to disaster management from almost a decade and has participated in numerous conferences and workshop related to the issue. He has also published numerous articles in eminent national and international journals.

Barendra Sahoo has over 20 years of experience in survey research. He has been working in the areas of bio-diversity assessment, environment assessment, and social assessment areas with several clients such as the World Bank, DFID, UNDP, etc. Since the last 15 years he has been working on issues relating to livelihood, water and sanitation, rural energy issues, micro-finance, health, education, mining and socio-economic areas. He has managed various assignments in these areas, and is a key member of the Social Transformation Group of CTRAN.

Dr. Vinay Sharma is an Associate Professor with the Department of Management Studies, Indian Institute of Technology, Roorkee, India, and has around 20 years of experience, in the areas of marketing, rural marketing, international marketing, business opportunity development, market development, brand development, IT enabled services and teaching for past 9 years. Vinay Sharma teaches marketing and the allied subjects, at various institutions including IIM Lucknow. His areas of interests also include poverty alleviation, rural market development through business development, market development and technology wherein he has proposed a model by the name *Affordability for the Poor and Profitability for the Provider* which, has been acknowledged at various platforms and published as a book, he has contributed an appendix on rural marketing in the 13th edition of Philip Kotler's *Principles of Marketing* (Prentice Hall). He is an associate and a member of the founding group of the Network of Asia Pacific Schools and Institutes of Public Administration and Governance (NAPSIPAG), New Delhi. He has been Associate Dean of CMES, UPES and was Associate Professor at IMT Ghaziabad. He has published and presented around 110 papers, chaired sessions at national and international platforms. He has developed 11 case studies based on primary research and has conducted more than 100 workshops, seminars, FDPs and MDPs. He is a member of the editorial board of international journals and is also the member of academic and advisory councils and Board of Governors of prestigious institutions. Presently, his engagements include Executive Committee member of IPR Chair at IIT, Roorkee and working group member of Ganga River Basin Management Environment Plan a pan IIT project. He also has done a project with CBS Denmark.

Ashok Singha is a doctoral scholar with Utkal University and a post graduate in Management from XIMB, Bhubaneswar. He specialises in strategic change management, governance and institutional development engagements. He has led initiatives in strategic climate change assessment, disaster risk management in several states including North East and J&K. He has worked closely with ADB, CIDA, DFID, GiZ, World Bank, UNDP and several other bi and multilateral agencies. Ashok has worked extensively on several policy formulation, programme development and impact assessment. He was involved in the formulation of two successful livelihood programmes (Western Odisha Livelihood Program, Odisha Tribal Empowerment and Livelihood Programme), Tejaswini Programme in MP and Maharashtra, review of Jiveeka Programme in Gujarat. He was also involved in the formulation of Odisha Modernizing Government Initiative, R&R Policy, and PPP Policy. He was part of the task force Bhumi. He has worked in missions in Japan, UK, Lao PDR and Kenya. His interest includes climate change, governance, clean-tech solutions at the bottom of the pyramid. He has advised several bilateral and multilateral agencies in the areas of governance, institutional development, sustainability, climate change, livelihood and enterprise value chains. He advises private sector companies on risk assessment, training and partnership structuring. He has also worked with several private sector companies in the areas of entry strategy, risk assessment, strategic HR processes, and collaborator assessment. Ashok is a small scale methodology expert involved in carbon trading and CER sale and project syndication. He is the co-author of a book titled "*The Forgotten Sector*" (New Delhi: Oxford & IBH Pub. Co), and contributor to a book on "*Governance of Rural Electricity System in India*" (ed. Prof. H. Panda, New Delhi: Academic Foundation), contributor to a book on "*Social Inclusion and India Infrastructure Report 2009, 2010*" (New Delhi: Oxford University Press); contributing to "*Social Inclusion: Pathways*" (Bangalore: Books for Change), and "*Governance Approaches to Mitigation of Adaptation to*

Climate Change in Asia" (London: Palgrave Macmillan). Ashok is the founder and Managing Director of CTRAN and leads social and environment practices.

Swapna Tripathy is a postgraduate and MBA holder with over 12 years of experience in working with the government; bilateral and multilateral aid agencies on various programmes/projects. Over the years she has gained experience in project identification, project preparation, project appraisal and project supervision. She has experience in issues relating to capacity building, project preparation, project correspondence and project assessment. She has hands on experience in working with state government, central government and private organisations, etc. She is a key member of social transformation group of CTRAN.

Vertika is an M.Phil. research scholar at the Centre for Political Studies (CPS) in Jawaharlal Nehru University (JNU), New Delhi. She is currently pursuing her M.Phil. thesis tentatively titled *The Concept of Agency in Feminism: Interrogating the Subject Constituted as Woman in Marriage and Family*. Her research interests include feminist theory, violence against women, issues of body politics and sexuality and agrarian studies. She presented her first paper titled *Critique of the Report of the Committee on State Agrarian Relations and the Unfinished Task of Land Reforms* at the 10th International NAPSIPAG Conference held in JNU, New Delhi (December 2013). She is a member of International Political Science Association (IPSA) and has been selected to present her paper titled *The Problematic Agency of the Disempowered Gendered Subject* at the 23rd World Congress of IPSA to be held at Montreal (July 2014). She has also worked as a Freelance Researcher in a Supreme Court Project on Ragging titled, *Psychosocial Study of Ragging in Selected Educational Institutions in India*. In addition she has pursued internships with the National Mission for Empowerment of Women (NMEW) and International Campaign for Justice in Bhopal (ICJB).

Dr. Vinita Yadav is an Associate Professor in Department of Regional Planning, School of Planning and Architecture, New Delhi where she is teaching subjects, such as Poverty and Development, Institutional Analysis and Governance and handling studio projects in the areas of Village, Block and District Planning. She has a total work experience of over 13 years in diverse consultancy organisations and in domains ranging from institutional strengthening, urban development, governance, infrastructure development, financial planning and capacity building. She has completed her Master's in Planning (with specialisation in Urban Planning) from School of Planning and Architecture, New Delhi. She obtained her Ph.D. from Jawaharlal Nehru University, New Delhi. She has co-ordinated projects in the field of urban and rural planning while working with the consultancy organisation J. P. S Associates (P) Ltd. Prior to this, she was a project manager in a non-government organisation (NGO) SEEDS India, New Delhi. She managed consultancy portfolio in an international NGO Praxis- Institute for Participatory Practices. She has used mixed approach in carrying out community centric planning and has managed projects funded by WHO, International Land Coalition, Bill and Milinda Gates Foundation, The Leprosy Mission, Goal India, Development

Authority and Municipal governments. Her book titled *Metropolitan Governance: Cases of Ahmedabad and Hyderabad* has been published by Copal Publishing Group, New Delhi.

Irina A. Zhernosenko is the Head of the "Culture and Communicative Technologies" Department of the I. I. Polzunov Altay State Technical University, Barnaul (Russia); Ph.D. in Cultural Studies, and associate professor of Ethics, Aesthetics, Theory and History of Culture; The Honourable Worker of General Education of Russian Federation. Her field of scientific and research interests are cultural history of Eurasia, Siberia; methodical development for teaching of world arts culture, history of regional culture. She has been teaching cultural science, history of world culture, culture of Siberia and Altai. She is the Scientific Editor of the electronic academic international journal *Evraziystvo: Philosophy. Culture. Ecology*. She is the author of more than 100 papers on the history of culture in Altai and Siberia, as well as study of sacred sites of Altai. The most significant works include: "Lectures on the History of Culture of Altai"/The educational materials for teachers. AKIP-KRO, Barnaul (2003, 120 p); "The sacred trailways of Altai"/The Training Manual for Instructors-guides./Ed. I. A. Zhernosenko—Gorno-Altaisk. ARTИКА, Barnaul (2008, 469 p); The collective monograph "The Multicultural Russian history and modernity: Siberia". Seventh book—St. Petersburg: Petropolis (2009, 452 p); "Altai reserved"/The Set of textbooks for schools in the Republic of Altai, vulture Russian Academy of Education/Ed. I. A. Zhernosenko—Gorno-Altaisk. ARTИКА, Barnaul (2009); The Set of textbooks for schools in the Republic of Altai: Textbook "World of reserved culture of Altai"/ Ed. I. A. Zhernosenko Gorno-Altaisk. ARTИКА, Barnaul (2009, 214 p); Textbook "World of the reserved nature of Altai"/Ed. R. V. Oparin—Gorno-Altaisk. ARTИКА, Barnaul (2009, 156 p); Textbook "Ecology and culture"/Ed. I. A. Zhernosenko—Gorno-Altaisk. ARTИКА, Barnaul (2009, 156 p); "Altai sacred: the religious shrines and the archeo-astronomic meanings of hypaethral temples": A collection of articles/Editors A. A. Tishkin, I. A. Zhernosenko—Barnaul: S.S. Zhernosenko's publishing house (2010, 144 p); The book "Altai—Belovodye—Shambhala in The Teachings of Nicholas Roerich". Barnaul : Five plus (2010, 235 p).

Notes on Reviewers

Prof. Dr. Akbaruddin Ahmad (Darul Ihsan University, Bangladesh) is the Chairman, Policy Research Centre.bd (PRC.bd) and Vice-Chancellor of Darul Ihsan University. He is a scholar, management specialist with exposure in management of educational institutions of higher learning, financial institutions/banks and civil society organisations. He is a prolific writer and regular contributor to national newspapers and peer reviewed journals of international standards. He delivered lectures on BBA and MBA program and is the Editor-in-Chief of *Finance World*, a leading newspaper. He was also an IT advisor for Janata Bank, Rupali Bank and Bangladesh Computer Council. A commerce graduate from the Dhaka University and an MBA with major in Banking from the Edward University, USA, he obtained his Ph.D. in Banking and Finance from Bircham International University, Spain. He joined the Rahman Rahman & Huq, and S. F. Ahmed Chartered Accountants as Senior Consultant and worked there as a National Consultant with internationally reputed consulting organisations viz Booz, Allen, Hamilton & Nothon Associates, Management Consultants. He was a Senior Consultant for the Financial Sector Reform Project of the World Bank/USAID to restructure four nationalised commercial banks and the Bangladesh Bank. He was a Senior Consultant for Bangladesh Computer Council, Rupali Bank, Janata Bank and Director Operations of BCCI Bank. He attended several management courses at WANG Headquarters at Lowell, Mass, USA, several IT courses at Apple Computer Cupertino, California/USA, Athens, Cairo, London, Lowell, Tokyo and Cyprus. He also held earlier the position of Pro Vice Chancellor, Treasurer, Dean and Director, Institute of Business Studies. He has to his credit a large number of publications in current affairs, economics, banking, financial sector reforms, IT and tourism, and has written numerous scholarly articles and book contributions.

Dr. Charlatan Andales-Escano (Mindanao Development Authority, Philippines) is the Director III in Mindanao Development Authority (formerly Mindanao Economic Development Council), a primary subnational agency in the Philippines mandated to promote peace and accelerate development in Mindanao, where she has served for 19 years now. A graduate of Doctor of Philosophy in Development Administration in the University of South eastern Philippines in 2012, she earned her Master in Business Administration and academic units in Public Administration in Cor Jesu College in Digos City, Philippines in 2000 and 2001, respectively. She has also undertaken short courses on Infrastructure Finance, Environmental Management, Project Management, Knowledge Management, Human Resource Management, Policy Analysis and Advocacy from international universities and local service providers commissioned by international donor agencies. As a peace and development advocate, she has been involved in planning, investment programming, resource mobilisation, implementing various programs and projects including policy advocacy with funding from international donor agencies. She has conducted various researches and studies in infrastructure, transportation, budget, and social sciences subject matters. She also taught subjects in engineering, mathematics, geology, sewerage, among others as a civil engineer. she provides advisory and consultancy services to key decision-makers and policymakers in Mindanao. Her expertise includes: infrastructure, development planning, project/program implementation, monitoring and evaluation, policy advocacy and networking, human resource management and development, financial management, statistical analysis.

Ms. Misa Aoki (Nara Women's University, Japan) Assistant professor of Nara Women's University Faculty of Human Life and Environment, Nara, Japan. Email: aoki.misa@cc.nara-wu.ac.jp.
Misa Aoki is currently an assistant professor of Nara Women's University Faculty of Human Life and Environment, Nara, Japan. She works on the issues relevant to behavioural changes of small-household farmers due to national or local government policies in Japan and other Asian countries. Specifically, she identified causes for agriculture policies and sustainable agriculture policies from the viewpoint of micro economics and new institutional economics. Earlier as an assistant researcher at Institute for Global Environmental Strategies she conducted research on climate change and agricultural policy implementation in Japan. She obtained her Mater degree from Kyoto University Graduate School of Global Environmental Studies in 2010.

Prof. Dr. Nasim Banu (Islamic University, Kushtia, Bangladesh) is Professor of Public Administration in the Department of Politics and Public Administration at Islamic University, Kushtia, Bangladesh. She received B.S.S. Honours and Masters in Public Administration from the University of Dhaka, an additional Masters Urban and Regional Planning from the Bangladesh University of Engineering and Technology (BUET). She earned Ph.D. in Monitoring and Evaluation the Public Sector Development Projects. She has over 24 years' experience in teaching. Her teaching responsibilities are comparative public administration, development studies, development administration, and project management. She held the position of Chairperson of the Department of Politics and Public Administration and has served as a member of Journal Committee, member of Development Studies Program and founder House Tutor, female hall of Islamic University. Presently she is academic council and faculty member of Islamic University; member, Professor and Associate Professor's selection board for the Department of Politics and Public Administration and advisor departmental journal committee and; member, teacher's selection board for colleges under National University of Bangladesh. She is also the Convener of Eve Teasing Prevention and Protection Committee, Islamic University, and Joint Coordinator for Study Group of Development Management of Public Administration and Governance Research Network. Her research interests include monitoring and evaluation of public sector development project, public policy, labour management and disaster management. She has published over 20 articles in reputed journals of home and abroad and participated several international seminars as paper presenter. She worked as a contributor to the 'Banlapedia' published by Bangladesh Asiatic Society. She is an active member of Bangladesh Institute of Planners, Bangladesh political Science Association, Bangladesh Public Administration Association, Dhaka University Public Administration Alumni Association and Dhaka University Alumni Association.

Mr. Tarakeshwar Dhurjati (Grant Thornton, India) Tarakeshwar Dhurjati is a Consulting and management professional with close to 14 years' experience in conceptualization, design, development and management of large scale transformation projects in the government sector in the areas of agriculture, biotechnology, police and law enforcement, Municipal administration, Healthcare etc. He is currently working as an Associate Director at Grant Thornton India and leads the Government and Social sector teams for management and delivery of Consulting and advisory services. Previously, Tarak had worked with PricewaterhouseCoopers Consulting and served as the national program coordinator for the Mission mode project of Ministry of Home Affairs, the Government of India implemented in 35 States and Union Territories of the country involving more than 22,000 project locations. Tarak is a certified Project Management Professional, an MBA from Indian Institute of Management, Ahmedabad and Masters in Biotechnology from Indian Agricultural Research Institute, New Delhi. Tarak is also a reviewer of technical articles for the Project Management Institute, India and contributes to writing consulting publications.

Dr. Jak Jabes (University of Ottawa, Canada) Jak Jabes is a Senior Fellow at the School of International Development and Global Studies, University of Ottawa since July 2013. Before returning to the University of Ottawa where he started his academic career, he was Visiting Professor at the Lee Kuan Yew School of Public Policy, Singapore. He is an experienced specialist on governance and public administration reform in "economies in transition," whose interests cover corruption, professional networks and the teaching of public management and administrative reforms. He has held senior positions at the Organisation for Economic Co-Operation and Development (OECD), the Asian Development Bank (ADB) and the United Nations Development Programme (UNDP). In these organisations his work focused on assisting governments in building partnerships, promoting regional capacity development initiatives, establishing knowledge networks and providing policy advisory and capacity building support. Dr. Jabes' professional life has been divided between academe and international civil service. After finishing his Ph.D. studies in social psychology at the University of Kansas he joined the then Faculty of Administration of the University of Ottawa where his research focused on work motivation, rewards and senior civil service management. He has been editor or member of the editorial board of several journals and he is the author, co-author or editor of a dozen books and 60 book chapters, articles, papers and reports.

Prof. Dr. Neena Joseph (Institute of Management in Government, Regional Centre, Kochi, India) Dr. Neena Joseph is Professor at Institute of Management in Government (IMG), Regional Centre, Kochi. She has her Masters in Business Administration from School of Management Studies, Cochin University of Science and Technology, Kerala. She joined Textile Corporation as a Management Trainee and while working there as Assistant Manager (Marketing) joined the Institute. She has her doctorate in Economics from Mahatma Gandhi University, Kerala. Her areas of interests are Management, Gender and Governance. She has conducted innumerable training programmes, workshops and seminars on these areas. She is a founding member of NAPSIPAG and has presented papers in international conferences including that of NAPSIPAG. She has published one book "Gender Related Problems of Women, Women's Empowerment and Panchayati Raj". She has published many articles on the 3 areas. She is in the Governing Board of Federal Institute of Science and Technology. She is on the Executive Board of Women Power Connect, Delhi and a trustee of Sakhi Women's Resource Centre, Trivandrum and Cultural Academy for Peace, Ernakulam. She has delivered many guest lectures in many national level and state level organisations including Government and NGOs. Currently she is facilitating the training programmes of Collegiate Education Department and Social Justice Department of Government of Kerala and heading the IMG Core Committee of Welfare oriented Departments of Government of Kerala. She is a member of the Research Committee of the Institute. She is an Indian citizen born on 18-08-1956.

Prof. Dr. Sarfraz Khawaja (Civil Services Academy, Lahore, Pakistan) is a free-lance Policy Evaluation and Result Based Monitoring Specialist and now works as individual consultant for the Higher Education Commission, Government of Pakistan. He earned his Ph.D. degree from University of Missouri, Columbia, USA and then joined University of Wisconsin, USA as a faculty member. He also taught at Government College Lahore, (now a university). His other teaching assignment was that of Chief Instructor (Professor) at the Civil Services Academy, Lahore. He served as Deputy Educational Advisor in the Ministry of Education, Government of Pakistan and Director, Academy of Educational Planning and Management, Islamabad. His work experience extends to UNICEF as Planning, Monitoring and Evaluation Officer. He also worked for UNFPA as Management

Advisor. Another key assignment was Monitoring and Evaluation specialist in Planning Commission, Government of Pakistan for 3 years. He was appointed member of Punjab Chief Minister Special Task Force on Higher Education 2003–2004. He also served as member of the Board of Directors, Centre for Pakistan Studies, Punjab University Lahore Pakistan 2007–2009. His Book "Non-Formal Education: Myth and Panacea" won National Book Foundation Award in 1994. He has been a consultant to several international agencies some of which include Swiss Development Cooperation, Asian Development Bank, Plan International, The World Bank, UNICEF, USAID, Aga Khan Educational Foundation, and Higher Education Commission. He was principal research partner in the BRIDGES project carried out in collaboration with Harvard University, USA (1988–1990). He has been a visiting speaker at Allama Iqbal Open University, Quaid-i-Azam University, National Institute of Public Administration, Foreign Services Academy, and National Defence College, Islamabad (now a university).

Prof. Dr. Sanjeev Kumar Mahajan (Himachal Pradesh University, India) Sanjeev Kumar Mahajan is presently a Professor in the Department of Public administration, Himachal Pradesh University, Shimla (India). He was formerly the Chairman of the same department. He has a Doctorate in Public Administration. His areas of specialisation include public enterprises and transport management. He has four books to his credit including *Performance of Public Undertakings in India—A Case Study*, Delhi: Devika Publications (1998), co-edited *Public Administration in the New Millennium*, New Delhi: Anamika Publishers & Distributors (P) Ltd. (2003), *Challenges in Governance*, New Delhi: Anamika Publishers & Distributors (P) Ltd. (2011), and co-authored *Financial Administration in India*, New Delhi: PHI Learning Private Limited (2014). He has been a noted contributor of research articles to various national and international journals. He has also presented and tabled valuable articles in different seminars and Conferences with in as well as outside India. He has also been awarded fellowship under Indo-Hungary Exchange Programme (2010).

A/Prof. Dr. Ahmad Martadha Mohamed (Universiti Utara Malaysia, Malaysia) had just finished his term as the Director of Quality and Strategic Planning, College of Law, Government, and International Studies, Universiti Utara Malaysia. Prior to this appointment, he was the Dean of Student Development and Alumni, UUM from January 2005 until June 2011. Besides administrative duties, he is also teaching Public Management courses at the undergraduate and graduate level. He has published extensively in local and international proceedings. His articles were published in several peer-reviewed and scholarly journals. He has also acquired several research grants from a host of national institutions. He was also a keynote speaker as well as invited speaker at the International Conference in India, Indonesia, and Malaysia. Currently, he is also a committee member for Network of Asia Pacific Schools and Institutes of Public Administration and Governance (NAPSIPAG) and Asia Pacific Society of Public Affairs (APSPA). He is the chief editor for *Journal of Governance and Development*

(JDG). His expertise in the area of public administration and politics earns him invitation from local institutions such that he is a regular political analyst for Radio Television Malaysia (RTM) and Private TV Network Astro Awani as well as several independent news organisations. He has also appeared in Channel News Asia (Singapore) commenting on Malaysian general elections. He has extensively given talks at various governmental organisations that include Prison Department, Kedah Civil Service (KCS), BTN, PDRM, JASA, schools, and UUM covering topics such as Administrative Reforms, Strategic Planning, Managing Change, Conflict Management, Managing Diversity, Customer Service, Representative Bureaucracy, Pre-Retirement Planning Programs, and Succession Planning. He received a degree in Political Science from Southern Illinois University, Carbondale, USA, Master of Public Administration from Indiana State University, USA and Ph.D. in Political Science from Southern Illinois University, Carbondale, USA.

A/Prof. Dr. Mohd Zin Mohamed (International Islamic University Malaysia, Malaysia) Dr. Mohd Zin Mohamed is Associate Professor at Kulliyyah of Architecture and Environmental Design (KAED). He has published several journal articles, book chapters and conference papers. He has been the lead consultant or Principal Investigator of the following projects "Rural Development Master Plan for Malaysia", "A study on Council Housing in Selangor and Its Future Direction", and "Analysis of Factors and Implications Towards Telecommuting as an Alternate Work Arrangement".

Dr. Klinpratoom Panyaping (Rajamangala University of Technology Lanna, Chiang Mai, Thailand) Dr. Klinpratoom Panyaping was educated on environmental management both in Thailand (Chulalongkorn University and Mahidol University) and abroad (TU Dresden, Germany), particularly on solid waste and hazardous waste management. In addition, she got scholarships from UN-UNESCO for training on environmental education and sustainable development for TVE in Indonesia, and related topics by ITASS on modernization process of Japan in Japan. She had experienced on education strategy for stakeholder on sustainable environmental management in USA as a Thai Fulbright Visiting Scholar. She has been worked in water, and air and waste management more than 20 years in the communities. Her expertise concerns with environmental project evaluation and utilisation of waste by using environmental technologies as well as waste management practices to minimise GHG emissions and to response to climate change in upper northern part of Thailand. She has also been a referee for the Journal in Thailand (TIC database) and abroad. She is working as Asst. Prof. at Department of environmental engineering, Faculty of Engineering, Rajamangala University of Technology Lanna, Chiang Mai.

Dr. Vidhyandika D. Perkasa (Centre for Strategic and International Studies, Indonesia) Vidhyandika D. Perkasa is a Senior Researcher at the Department of Politics and International Relations CSIS, Jakarta Indonesia. He is also the chief editor of Analisis journal, CSIS. The research of Vidhyandika Perkasa concentrates on issues of conflict, human trafficking, poverty, ethnic relations, social policy and local governance. He has conducted research on social policy, budgeting and institutional framework in Indonesia's MDGs Achievement' and Mapping of Policy Development Framework on Child Labour and Youth Employment Issues in Indonesia. Vidhyandika has also been involved in a research titled 'Global Migration and Access to Justice for Indonesian Domestic Women Workers'. The research observed the life-style and problems faced by Indonesian domestic workers residing in the UAE. He has also been conducting intensive research in the areas of governance and social accountability in Papua where he worker intensively with elements of civil society and government to promote better governance. His thesis probed into traders' economic rationality which was written with a life history method. In 1995, he obtained his Master degree from the Department of Social Policy and Planning in Developing Countries at the London School of Economics and Political Science, United Kingdom. He obtained his Ph.D. at the Department of Anthropology, School of Political and Social Inquiry at the Faculty of Arts, Monash University Australia in 2005. His doctoral research concerns poverty issues amongst Chinese community in West Java Indonesia. His thesis explores the links between poverty, race, ethnicity and identity, and how the poor Chinese community there deals with the vicious circle of poverty. His dissertation has been published by Lambert Academic Publishing, Germany in 2010 titled *Poverty in a Chinese Community in West Java Indonesia: Survival, Identity and Social Face*'.

Mr. Bigmal Raj Remi (Flinder University, Australia) Bimal Raj Regmi is a doctoral student at Flinders University, Australia. He has more than 15 years of academic and professional experience in climate change and natural resource management. His work covers the diversity of environment, development and climate change work. Mr. Regmi also has good exposure and experiences in climate change adaptation and mitigation issues. His specific expertise on climate change includes understanding of the legal and policy provisions and practices on climate change adaptation; governance of climate change at national and international level; biodiversity and natural resources cutting edge research and practices. He has contributed as author in number of publications on climate change and natural resource management.

Prof. Dr. David Rosenbloom (American University, USA) Distinguished Professor of Public Administration, Rosenbloom specialises in constitutional-administrative law, administrative theory, history, reform, and personnel management.

A major contributor to the field and a National Academy of Public Administration Fellow, his numerous awards include the Whittington Award for excellent teaching, Gaus Award for exemplary scholarship in political science and public administration, Waldo Award for outstanding contributions to the literature and leadership of public administration, Levine Award for excellence in public administration, and Brownlow Award for his book, *Building a Legislative-Cantered Public Administration*. He edited *Public Administration Review*, co-edited the *Policy Studies Journal*, currently serves on the editorial boards of about 20 academic journals, and is editor of the CRC/Taylor & Francis series in Public Administration and Public Policy. Author or editor of over 300 scholarly publications, he frequently guest lectures at universities in the US and abroad. Faculty appointments include the following universities: Kansas, Tel Aviv, Vermont, Syracuse, City University of Hong Kong, Hebrew (Israel), and Renmin (China). Rosenbloom served on the Clinton-Gore Presidential Transition Team for the Office of Personnel Management in 1992. His text, *Public Administration: Understanding Management, Politics, and Law in the Public Sector*, was ranked the fifth most influential book published in public administration from 1990–2010.

Dr. Rer. Nat. Claus-Peter Rückemann (Leibniz Universität Hannover and the Westfälische Wilhelms-Universität Münster, Germany) Dr. Rer. Nat. Claus-Peter Rückemann is a lecturer, researcher, and scientific member of the Leibniz Universität Hannover and the Westfälische Wilhelms-Universität Münster, Germany. He holds a University-Diplom degree in geophysics and a Doctorate degree for natural sciences in geoinformatics, informatics, and geosciences from the Faculty of Mathematical and Natural Sciences, WWU. He teaches Information Science, Security, and Computing at the University of Hannover, Faculty of Law, in the European EULISP Programme. Dr. Rückemann is the head of research of the LX Foundation and director of the Geo-Exploration and Information Consortium (GEXI). He is the founder and general chair of the International Conference on Advanced Communications and Computation (INFOCOMP), advisory chair and organiser of the GEOProcessing and CYBERLAWS conferences and ICNAAM symposiums. He is leading scientific advisor for international projects, High Performance Computing (HPC), natural sciences, information systems, geosciences, archaeology, environment, climatology, and governance. He is a member of international research grants committees, boards, expert panels, the Science and High Performance Supercomputing Centre and serves as scientific auditor, assessment reviewer, chairman, peer reviewer and editor. His research and teaching are focussed on computer science, Integrated Information and Computing Systems (IICS), and long-term multi-disciplinary knowledge resources and discovery, targeting sustainability in research and collaboration. For two decades Dr. Rückemann is in professional practice in natural and information sciences, advanced scientific computing and supercomputing, in research, management, and supervising positions. He is also on duty for HLRN, managing the largest supercomputer resources in Northern Germany. Dr. Rückemann is the founder of many research projects and author of numerous peer reviewed scientific publications, internationally awarded with several Best Paper Awards. He has been distinguished with the Grade IARIA Fellow of the International Academy, Research, and Industry Association for his scientific research on the state of the art improvement in HPC, information systems, distributed computing, and international collaboration.

A/Prof. Dr. Isaias S. Sealza (Xavier University, Philippines) Dr. Isaias S. Sealza teaches evaluation research in the program for Public Administration, of the Graduate School of Xavier University. For more than 25 years now, he has been directing research undertakings, mostly policy-oriented studies, on population and development, poverty, child labour, environment, natural resource conservation and development, peace and conflict, and, health issues, commissioned by national and international agencies. He has written, as single author or with collaborators, about a hundred research reports, journal articles, books, monographs and conference papers including *Community Modernization, In-Migration, and Ethnic Diversification* (Philippine Sociological Review, 1982), The *Cassava Industry in Bukidnon Province* (Philippine Studies, 1986) and *Mindanao Il Conflitto dei Poveri i movimenti d'ispirazione islamica nelle Filippine: trend attuali e reali prospettive di pace con i cristiani* (Revista di Intelligence, 2006). He completed recently a performance evaluation of a dairy farming program in Mindanao for Land O'Lakes International Development Foundation and the US Department of Agriculture, and a study of the humanitarian assistance for survivors of Tropical Storm *Washi* for the UNICEF. He is currently doing research on failed pregnancies for a regional field office of the Philippine Department of Health.

A/Prof. Dr. Vinay Sharma (Indian Institute of Technology, Roorkee, Uttarakhand, India) has around 20 years of experience, in the areas of marketing, rural marketing, international marketing, business opportunity development, market development, brand development, IT enabled services and teaching for past 9 years. Vinay Sharma teaches marketing and the allied subjects, at various institutions including IIM Lucknow. His areas of interests also include poverty alleviation, rural market development through business development, market development and technology wherein he has proposed a model by the name "*Affordability for the Poor and Profitability for the Provider*" which, has been acknowledged at various platforms and published as a book, he has contributed an appendix on rural marketing in the 13th edition of Philip Kotler's *Principles of Marketing* (Prentice Hall). He is an associate and a member of the founding group of the Network of Asia Pacific Schools and Institutes of Public Administration and Governance (NAPSIPAG), New Delhi. He has been Associate Dean of CMES, UPES and was Associate Professor at IMT Ghaziabad. He has published and presented around 110 papers, chaired sessions at national and international platforms. He has developed 11 case studies based on primary research and has conducted more than 100 workshops, seminars, FDPs and MDPs. He is a member of the editorial board of international journals and is also the member of academic and advisory councils and Board of Governors of prestigious institutions. Presently, his engagements include Executive Committee member of IPR Chair at IIT, Roorkee and working group member of Ganga River Basin Management Environment Plan a pan IIT project. He also has done a project with CBS Denmark.

Mr. Ashok Kumar Singha (CTRAN, India) Ashok Singha is a doctoral scholar with Utkal University and a post graduate in Management from XIMB,

Bhubaneswar. He specialises in strategic change management, governance and institutional development engagements. He has led initiatives in strategic climate change assessment, disaster risk management in several states including North East and J&K. He has worked closely with ADB, CIDA, DFID, GiZ, World Bank, UNDP and several other bi and multilateral agencies. Ashok has worked extensively on several policy formulation, programme development and impact assessment. He was involved in the formulation of two successful livelihood programmes (Western Odisha Livelihood Program, Odisha Tribal Empowerment and Livelihood Programme), Tejaswini Programme in MP and Maharashtra, review of Jiveeka Programme in Gujarat. He was also involved in the formulation of Odisha Modernizing Government Initiative, R&R Policy, and PPP Policy. He was part of the task force Bhumi. He has worked in missions in Japan, UK, Lao PDR and Kenya. His interest includes climate change, governance, clean-tech solutions at the bottom of the pyramid. He has advised several bilateral and multilateral agencies in the areas of governance, institutional development, sustainability, climate change, livelihood and enterprise value chains. He advises private sector companies on risk assessment, training and partnership structuring. He has also worked with several private sector companies in the areas of entry strategy, risk assessment, strategic HR processes, and collaborator assessment. Ashok is a small scale methodology expert involved in carbon trading and CER sale and project syndication. He is the co-author of a book titled "*The Forgotten Sector*" (New Delhi: Oxford & IBH Pub. Co), and contributor to a book on "*Governance of Rural Electricity System in India*" (ed. Prof. H. Panda, New Delhi: Academic Foundation), contributor to a book on "*Social Inclusion and India Infrastructure Report 2009, 2010*" (New Delhi: Oxford University Press); contributing to "*Social Inclusion: Pathways*" (Bangalore: Books for Change), and "*Governance Approaches to Mitigation of Adaptation to Climate Change in Asia*" (London: Palgrave Macmillan). Ashok is the founder and Managing Director of CTRAN and leads social and environment practices.

Dr. Yunxin Tu (Wuhan University, China) Yunxin Tu is a young Chinese legal researcher. He obtained a Bachelor of Law degree from Wuhan University, China and an LLM degree in International Law from University of Oslo, Norway and a doctoral degree of Constitutional and Administrative Law at Wuhan University, China. From 2009 to 2014, he initially attended and then coached several international moot court competitions in The Netherlands, China and USA. He has worked as a teaching assistant and lecturer in Wuhan University School of Law. From September 2012 to June 2014, he served at Wuhan University Research Centre of Constitutionalism and Rule of Law as a research associate. In 2013 he also acted as a legal advisor for the Standing Committee of the Local People's Congress of Hanyang District in the city of Wuhan in China. He has published journal articles on human rights, constitutional law and judicial reforms, law and democracy. He frequently contributes to newspaper/mass media with notable legal comments in China. His current research focuses on some controversial constitution-making issues in selected Asian countries including constitutional design, governmental legitimacy, *coup d'état*, self-determination and the issues related to human rights.

NAPSIPAG is Ten Years Old: An Agenda for the Next Decade

Jak Jabes

Introduction

Some 10 years ago, when I was at the Asian Development Bank (ADB), the idea of launching an Asia Pacific Network of Schools of Public Administration and Governance (NAPSIPAG) germinated in our minds. While, I was dubious that we would have financial support from the institution, my worries turned to be unfounded. ADB provided a generous technical assistance, and I ploughed on, taking time to slowly develop the idea, talking to few and many in international meetings and small groups, until we were able to gather over a couple hundred interested individuals from all over Asia and the Pacific region to a meeting in 2004 in Kuala Lumpur.

At this 10th conference of NAPSIPAG dealing in Growth Patterns, Disasters and the post 2015 Developmental Agenda, I would like to make three points.

First, it is important to be reminded of why NAPSIPAG was launched.

As an international civil servant, I was involved in the launching of two networks of schools of public administration, policy and governance with the support of the two of three multilateral organisations I worked for. In both cases my motivation was the need to help and support a geographic area which had been politically impoverished (Central and Eastern Europe after transition to market economy in the 90s) or economically underdeveloped (Asia and the Pacific, at least many countries). The aim in putting in place such Networks was:

- to foster collaboration among students, professors and administrators and to ensure that they could mutually benefit from academic and educational products produced in the region instead of depending on products that all came from the West;
- to address the knowledge management and regional cooperation gap within the Asia-Pacific region. Underlying this initiative was my belief that schools and institutes of public administration in developing countries play an important role as potentially powerful advisory resources to strengthen governance and public

management. These institutes have the advantage of being locally sustainable and powerful institutional change agents based on their local knowledge, acceptability and influence;

- to provide effective capacity-building services to member institutes on a sustainable basis; encourage the sharing of expertise and good practices; assist the member institutes in the continuing development of public administration theory and practice through research and other initiatives; and foster cooperation and collaboration between and among the member institutes and individuals in the pursuit of common interests.

Second, I want to underline some of the accomplishments.

232 persons representing 92 institutions from 26 countries attended the first conference in 2004. Clearly, for a network getting off the ground, this was quite an accomplishment. These conferences have continued, and we have reached the tenth of such conferences. This is an important achievement when travel funds are short and donor support not always readily available. An electronic journal was put in place with mostly an Asian editorial board and continues to publish. The 10th conference has seen participation from all over Asia, as in previous years. The topic of the conference dealing with land, water and disasters and how they should be addressed in the post 2015 development agenda resulted in numerous papers and a difficult selection. So we should take pride in these accomplishments.

Third, I want to share my views on the next decade for NAPSIPAG.

Asia and the Pacific is a vast, differentiated territory with countries that exhibit varied approaches to governance and public administration. It is natural that institutions involved in training, education and research in public administration and governance will first and foremost put emphasis on the national system in which they are embedded and impart an understanding about that system to their students. As well, they would also stress the need for reforming their systems. Change is at the forefront if we want to improve. And so, I am sure, institutions put emphasis on influencing the dialogue for change in their countries; and, in doing so look at weaknesses in their national systems, compare and contrast their systems to those of others, scan their environment to be on top of what is taking place in the regional as well as the larger international agenda. A good example is the tenth conference of NAPSIPAG. With a number of disasters striking Asia since the onset of the network, it is somewhat obvious that NAPSIPAG is devoting its conference in 2013 to managing disasters. From policy planning, to training to education there are many efforts that have taken place these last 10 years in the region regarding disasters, and this is an excellent platform to share this knowledge.

The psychologist that I was trained to be many years ago says that all people are alike but as well all people are different. I think this type of maxim more and more can be applied at the level of national public administrations. They are similar and different. In moving ahead in this region, the institutions that form NAPSIPAG need to be alert to developments in the larger environment, identify them and build their curriculum, research programs and interventions around them. In that sense, they have to be similar while being different because they have to be attentive to their national systems.

NAPSIPAG needs to set a future oriented agenda that focuses on identifying a number of themes that are crucial. In my modest view, this agenda can be the beginning of a conjoint effort of NAPSIPAG members among themselves, between themselves and other development agencies and for the improvement of their own institutions.

In setting this agenda we need to be cognizant of the pressing challenges that the region faces but take a more detached view and be more encompassing. Trends and fashions ought not to dictate pedagogy in public administration and governance; rather, the need to educate the future leaders and bureaucratic elites to face change and challenge should guide us.

The region where NAPSIPAG established itself was hit by serious financial crisis at the end of the twentieth century, and then by various natural disasters in the beginning of this the twenty-first century. Much growth has been experienced since then, although economies remain somewhat fragile because issues of good governance, corporate governance, and anticorruption among others have not been resolved. Progress has meant that strides have been made to reduce poverty and successes obtained but still much of the world's poor live in this region. Political uncertainty and conflicts plague countries and this in turn impacts what governments and institutions do. Infrastructure investments and service delivery are viewed today as the way out of poverty in Asia. But, when conflicts strike and states become fragilised problems augment.

More specifically, East and Southeast Asia have shown remarkable resilience after the Asian crisis in spite of multiple shocks. Behind the current existing optimistic outlook, there is an evolving new environment created by major new dynamics at work suggested by the fact that intraregional trade is emerging as a major growth driver for Asia. The Asian consumer is becoming more and more demanding and the two large economies with huge populations are showing effects of this demand.

Discussion

Many academics in Asia are heralding the Asian century (Mahbubani 2009). But, consensus does not exist even on this issue and success can be maintained if we continue policy reforms and strive for a peaceful and prosperous Asia and Pacific.

Against this unsettled background, what should be the thematic preoccupations of NAPSIPAG members? Let me briefly dwell on seven key themes:

1. Human Development, which, has been a key concern in the last part of the twentieth century. Great strides have been made in life expectancy, education and health. Income poverty is also in the process of being reduced. But the same cannot be said of advances in democracy where the balance sheet is

mixed. Conflicts have not subsided.[1] The international community has pledged to meet the MDGs (*Millennium Development Goals)* by 2015 while attempting to understand the underlying reasons for the slow progress at times and suggest solutions to resolve the problems. These include harmonising aid, better macroeconomic policy and nationally owned poverty reduction strategies. In other words, much remains to be done overall in advancing human development, reduce poverty and move from the MDGs to Sustainable Development Goals (SDGs). For the less developed countries this remains the key challenge.

2. Governance and Anticorruption. The word governance permeates every activity undertaken by a government these days. There is hardly any consensus on its definition. I like to use it in an enlarged sense of public administration. One Asian institution, the ADB has construed governance as resting on four elements of transparency, accountability, predictability and participation.[2] The Part of the good governance agenda these days is in deciding the role of the state and determining which level of government is responsible for exercising certain powers. The emphasis in Asia, for example on decentralization is a manifestation of this concern. Is decentralization a panacea for improving how citizens are administered? This is a governance sub theme worth exploring further.

   In anticorruption efforts the panorama of Asia and the Pacific is mixed. On the one hand some countries have addressed the issue head on and seriously for many years and see themselves ahead in this endeavour. They are cited as models to emulate. We can cite Singapore, Hong Kong, Australia and New Zealand among these. Then there are some middling countries where efforts continue but will require time to pay off such as Bhutan, Korea and Malaysia. But, unfortunately most other Asian and Pacific countries do not fare well and the view is that the lack of transparency, the failure of the judiciary to enforce the laws, the widespread bribery curtails development and is hurting the poor. It is paramount that NAPSIPAG members jointly or singly work on curricula and interventions, which aim to take the corruption scourge head on.

   Governments of the region are weary of international comparisons when it comes to good governance and anticorruption. Transparency International publishes a yearly report, which tracks countries against the Corruption Perceptions Index (CPI). Researchers at the WB (the World Bank) have established governance indicators and rank countries around a number of dimensions such as government effectiveness, regulatory burden etc. (Kauffman et al. 2004). This body of work should provide us with optimism because it shows that improvements are possible, even if a magic formula does not exist.

3. Partnerships: Most citizens do not understand how national governance processes and institutions work. However, as part of the democratic process it is important for interested citizens to understand and be able to influence the

---

[1] *Human Development Report* 2005, UNDP, New York.

[2] Governance: Sound Development Management, ADB, Manila, 1999.

direction of governance. Partnerships between civil service organisations and government have sprung up in response to fill this void but also because it is more and more understood that a government cannot satisfy all needs. Government agencies are unable to address the complexities, which they and their citizens face. But partnerships between the public and private sectors and the general public can help to integrate public administration outputs by identifying issues facing communities and taking joint action. So, my third theme would be partnership.

In the Pacific region we have observed weaknesses in the machinery of government of small island states (Mellor and Jabes 2004). I would argue that such systemic weaknesses also exist in Asia especially in the so called economies in transition to a market economy and democracy. To practice good governance an appropriate machinery of government is necessary for policy to be developed and for services to be delivered appropriately and effectively. A system is needed that permits efficient planning and decision-making through consultation between all tiers of government and citizens. The government of the future, of our immediate future, will have to be highly consultative.

Such partnerships and civil society involvement can influence corruption and hold public administration accountable for their exercise of power. Many international conventions aiming to reduce corruption stress this point. For example the ADB-OECD (Organisation for Economic Co-operation and Development) Anticorruption Initiative and its main instrument the Anticorruption Action Plan for Asia and the Pacific focus on civil society involvement through one of the three pillars on which the plan's action agenda is based. The United Nations Convention Against Corruption (UNCAC) also emphasised the importance civil society plays in curbing corruption. Civil society is active in the NAPSIPAG geography influencing governments and parliamentarians by providing them with evidence, survey data and complaints regarding corrupt practices of government officials.

4. Regional Cooperation: Evidence shows that through regional cooperation, which deals with pursuing collective actions among two or more countries, each country can individually improve development performance and strengthen its capacity to reduce poverty. Multilateral Development Banks (MDBs) have been active in this front and have increased their activities. In investing in infrastructure across countries, for example MDBs are advancing the regional cooperation agenda. Countries are working together in cooperation in intra-regional and global trade and investment. Regional cooperation has been shown to be effective in obtaining economies of scale. Post tsunami reconstruction, as is the case with the Aceh tsunami underlines the importance of regional cooperation when disaster strikes more neighbouring nations.
5. Private Sector Development (PSD): While tradition separates institutions of higher learning into departments, schools and institutes where subject matter is also taught in silos, vertically with little interaction, one should try to avoid this when training the future public servants. Schools of business administration usually shun schools of public administration and vice versa. At best each gives

the other lip service by involving a course on business government relations. Often, these are taught by business schools to show how deleterious government is to private sector development. It is time to work together in academic institutions. The geography which NAPSIPAG represents includes on the one hand countries which follow quite liberal policies in support of PSD and on the other hand countries which are transforming into a market economy, albeit slowly. Understanding of PSD, and how the state can help it should be a significant aspect of scholarship in the Asia Pacific region. The importance of regulatory environment and the need to have predictability has been shown to influence both Foreign Direct Investment (FDI) and PSD. As part of their own development agendas, countries of the region are requesting and responding to PSD. The multilateral institutions have made advancing PSD one of their key goals (Jabes et al. 2003).

6. Values: Much has been said of the importance of values in public administration. Some would argue, for instance that much of public administration in the European Union is based on shared values. Certainly, in the case of Europe while shared law exists, called the *acquis communotaire*, when it comes to civil service and laws governing civil service they are all operational at the national level. However, the civil services of countries which make up the European Union share similar values and goals and arrive at outputs and services to their citizens that resemble each other's even if their civil services rest on different laws and administrative traditions. Openness, transparency, neutrality, merit, value for money, loyalty, equity, and probity are some of these values that seem to govern public administrations. It would be important to stress such key values during the time students are educated in public administration schools and institutes of the region so that they understand administrative culture and what drives public bureaucracies.
7. Leadership: Asia's future will be shaped by its leaders. To bring change in public administration, leaders need to help develop a vision and mobilise public servants and ultimately the citizens toward this new vision. This region is undergoing rapid change and political transitions are occurring fast. Transformational leaders who can shape the technical, political and cultural pressures will be needed. Such leaders will have to reduce resistance to change, create a vision and mobilise commitment (Tichy and Ulrich 1984). Such leaders will have to understand the process of change. While there is no magic formula to train such leaders, creating a pedagogic environment conducive to their flourishment is an important goal for public administration training institutions. Leadership, too, is important to manage disasters when they strike, the theme of this conference and this book.

Conclusion

Each educational institution certainly needs to stress the traditions, culture, experiences and history of its country because these significantly influence governance. However, they also need to open the debate for reform by stressing the developmental needs of the country, pointing out good practice and influencing the debate on the degree to which good governance ideas and principles can be adapted to fit national conditions and political agendas. In a fast globalising world where information is readily available, it is not the raw knowledge anymore that is at issue but the environment that stimulates debate and discussion, and NAPSIPAG institutions have the duty to drive this process.

I do not for a moment suggest that this is a blueprint for public administration education in Asia. There will be the traditionalists and the change seekers each with their own way forward. Some will stress the need for disciplinary grounding in policy, statistics, economics and political science. Others will want to look at contemporary issues through case studies. Whichever way is chosen, my point about this thematic agenda is that the issues which I have outlined and the themes which I feel impact so heavily in this region need to be reflected in the courses taught (Jabes 2012). It is not difficult to do that in discipline based education as long as efforts to bring up these themes as study subjects are made.

Program design for NAPSIPAG institutions should ensure thematic emphasis across courses so that these seven themes and others, which are important, become familiar to students because they are stressed and studied across all courses and disciplines.

References

Jabes J, Collette M, van Heeswick JPM, Wescott C (2003) Governance, transparency, and attracting foreign direct investment in Asia. Global Forum on International Investment, Johannesburg

Jabes J (2012). Teaching in a public policy school: Some personal views. Jindal J Public Policy 1: 64–78

Kaufmann D, Kraay A, Mastruzzi M (2004) Governance matters IV: Governance indicators for 1996–2004. The World Bank, Washington D.C.

Mahbubani K (2008) New Asian hemisphere: the irresistible shift of global power to the east. Public Affairs, New York

Mellor T, Jabes J (2004) Governance in the Pacific. ADB, Manila

Tichy N, Ulrich D (1984) Revitalising organisations: The leadership role. In Kimberly JR Quinn RE (eds) Managing organizational transitions. Richard D. Irwin, Homewood, pp 240–264

Chapter 1
Land Use and Disaster Governance in Asia: An Introduction

Huong Ha

1.1 Introduction

The world population was about 2.5 billion in 1950, and 20 % of them lived in the urban areas (World Health Organisation 2014). The world population reached nearly seven billion now, and there would be nine billion people in this world by 2025 in which 7 out of 10 people would live in a city (Population Institute 2011; World Health Organisation 2014). This shows that the size of population would increase in general, and particularly in the urban areas in the coming decades. The population pressure increases, but land is still as limited as other scarce resources. How to manage the world population size and the land use proportionately so that the use/misuse of land would not be the cause of disasters is a burning question to world leaders. It would also be expected that the growth of population, the increase in demand for food and other consumable products, in both developed and developing countries, would create challenges on land use and planning, climate change mitigation and adaptation, and disaster risk reduction (Ha and Dhakal 2013).

The Asian region has become increasingly vulnerable to natural disasters, including floods, cyclones, storms, earthquakes, drought, typhoons and tsunamis. Given unsustainable changes being made in patterns of land use, catchment and coastal zones, increasing density of population, migration patterns and consumer culture across countries, the impact of disasters has increased manifold. Natural disasters always cause huge economic loss and result in severe death toll of thousands or even millions of people. Besides causing environmental degradation and damage to the ecosystems, natural disasters also cause short-, medium- and long-term socio-economic problems, such as resident displacement, loss of

H. Ha (✉)
UON Singapore (University of Newcastle, Singapore Campus), Singapore, Singapore
e-mail: artp6075singapore@yahoo.com

H. Ha (ed.), *Land and Disaster Management Strategies in Asia*,
DOI: 10.1007/978-81-322-1976-7_1

livelihood and family disintegration (Khan 2008). In addition, the pattern of land use, and the ways risk is managed have impacted on food security.

Governments in the Asian region, especially in Southeast Asia and South Asia, are at the crossroads of searching for new approaches to improve disaster governance as (Ishiwatari 2003) asserted that "governance is widely regarded as the key to reducing disaster risks" (p. 3). Governance refers to the execution of political, socio-economic, legislative and administrative power in the management of national affairs via the allocation of resources, mechanisms, structures, formal and informal instructions, and the interaction among various sectors (Ishiwatari 2003; Turnbull et al. 2013).

In this setting, there has been ongoing and intensive deliberation and discussion on how all sectors can jointly address challenges associated with disaster risk reduction and disaster management, including pre-disaster preparedness and post-disaster recovery and reconstruction since disasters can be conquered with efforts of the local community and the

> "commitment of organisational, political, professional, and sociological sectors.... Representative of various sectors of public and private level should be involved in the design of system of disaster managements", including the use of information technology to improve the effectiveness of communication before, during and after disasters (Islam and Chik 2011, p. 522).

In other words, all sectors also have to tackle other issues which are linked to disaster governance, such as climate change mitigation, land use and planning, food security, environmental degradation, public education, information technology for communication and so on. Reduction of the occurrence of disaster or governing disaster effectively could be addressed by effective land policies, and the mechanisms adopted for before and after disaster risk management. In fact, Tierney (2012) explained that "Disaster governance is nested within and influenced by overarching societal governance systems. Although governance failures can occur in societies with stable governance systems, as the governmental response to Hurricane Katrina shows, poorly governed societies and weak states are almost certain to exhibit deficiencies in disaster governance" (p. 341). In such cases, state and non-state sectors and their relationships, business enterprises, economic structures and activities, and societal transitions have great implications for disaster governance. Thus, it is imperative to examine and analyse how state and non-state actors can cooperatively contribute to the common cause of improving disaster governance in Asia. The motivation for us to embark on this project is twofold. Firstly, we wish to raise the awareness of all stakeholders of the critical aspects of disaster management. Secondly, we would want to call for more attention from policy makers, researchers and those who are interested in the topic to search for innovative and novel approaches to reduce risk and manage disasters.

The book includes the following main themes:

1. integrated land and water management, including administrative best practices, innovative governance approaches, multisector participation and impact upon sustainable human development

2. land management and food security
3. disaster governance, including disaster mitigation, prevention and control applications of information and communication technology in land planning, early warning systems, training and community participation in disaster management

This book includes 17 chapters, excluding the introduction chapter, contributed by authors who are academics, practitioners and researchers from different countries in Asia and from various disciplines. These chapters are classified under three following parts: (i) land management and food security, (ii) disaster management and (iii) community education and other issues related to disaster management. Four chapters are included in the first theme, six chapters are included in the second section, and five are in the third part. The next section describes the focus of each chapter in this book.

1.2 An Overview of the Chapters

First of all, it is impossible for us not to mention the Network of Asia Pacific Schools and Institutes of Public Administration and Governance (NAPSIPAG) in this edited volume since the debates and dialogues in the NAPSIPAG conference titled "Locked in Growth Patterns: Rethinking Land, Water and Disasters for the Post-2015 Development Agenda for the Asia Pacific" in 2013 in New Delhi, India, has inspired the editors and the authors to embark on this project. In other words, without NAPSIPAG, this edited volume may not be materialised.

Thus, "we need to remember why the Network was launched, how much was accomplished and its future objectives" as mentioned by Dr. Jak Jabes, the founder of NAPSIPAG, in the previous section.

Therefore, Dr. Jak jabes' note in this book summarises the development of NAPSIPAG from the launch in 2003 until now. Initially, NAPSIPAG was financially supported by the Asian Development Bank. Yet, NAPSIPAG has survived without any donor support for the past several years which is a great accomplishment for any non-profit research network. NAPSIPAG has been an excellent platform for dialogues among practitioners, researchers and academics regarding governance, public policy and public administration in the Asia Pacific region, extending to worldwide. Different countries have chosen different pathways for their political and socio-economic development. This was explained by Dr. Jabes that "some countries are lifting themselves from poverty but require support in managing their resources, involving their citizens and ensuring equity. Others are making inroads to move from authoritarian to more open and democratic regimes and they need support in ensuring inclusiveness, improving their fiscal situation, advancing citizens' rights" (Dr. Jak jabes' note). However, disasters may hinder the development process of the affected countries since disasters do not spare anybody or any country, either the rich or the poor. Hence, searching for better ways to mitigate risks and manage disasters is a rich agenda for further debate and discussion in disaster governance.

Part 1 of this volume starts with an interesting chapter titled *Land Reform: A Case Study in Kerela* (Chap. 2). This chapter revisited the definitions of land governance, land reform and examined the relationship between land governance and sustainable development. It also discussed (i) how land reform initiatives have been conducted, using Kerala state, India, as a case study, and (ii) challenges faced by the government of Kerala when executing land reform. The main objectives of land reform include redistribution of land to ensure land equity among the population via (i) setting a ceiling on the "absolute size of holdings", (ii) abolishing the "rice land tenancy" system and (iii) abolishing "tenancy in house garden lands" (Franke 1992, p. 85). The findings suggest that there are huge gaps between the theories and practice. Therefore, it is important to further examine the impact of land reform on various aspects since land reform may accrue benefits for certain groups of people, such as the landless and the poor, but may also result in further separation of the poor from the land.

Prolonged rainfall from a storm or a typhoon which produces too much water in a particular area can cause floods. Misuse and mismanagement of land can also cause natural disturbances, for example, landslides, mudslides, deforestation, etc., since "a landslide occurs when soil, rocks, trees, parts of houses and other debris is swept downhill …or general instability of the land" (California Department of Public Health 2014, para. 5). These together with other natural disturbances, such as earthquakes, volcanic eruptions and meteorite impacts, may lead to tsunamis. Thus, land use planning and management, addressed in Chap. 3, is a very important task in any developmental activities and risk reduction as well as disaster prevention. However, land management has always been a critical issue in any countries due to many reasons, such as limited land, various interests of different groups of stakeholders, lack of policy and regulation, ineffective implementation and enforcement of law and regulation and power struggles among various groups. The authors of Chap. 3 explained that the task of land reform in India has been abandoned for a long time until recently when the government of India introduced a comprehensive policy intervention to restart the land reform process in 2009. This intervention has been described in the Report by the Committee on State Agrarian Relations and the Unfinished Task in Land Reforms (2009) released by the Ministry of Rural Development, Government of India, Department of Land Resources (2009). The authors have selected the report as a subject of their research study given its uniqueness and all-inclusive spectrum to land reforms which establish the foundation and framework for research and discussion on "politics of land" in India, including a wide range of issues such as acquisition of agricultural land for non-agricultural perspectives, land use rights, compensation of land use and others (Tiwari 2010). The authors analysed both positive and negative aspects of this report with regard to the structure, the coverage and the content of the report and its policy implications. This chapter identified two main issues in the report, namely using forest land, and land acquisition for special economic zones in the country. These issues are pertinent to technical aspects, and they can be addressed by the *Forest Rights Act* (2006) and the *Land Acquisition Act* (2013), respectively. The report did produce positive recommendations which can promote the democratic process of agrarian relations. The report also emphasised the importance of stakeholder participation which would improve the land reform process in India.

Similar to India and other countries in South Asia, land use, control and management is a pressing issue in Pakistan. Not only the national politics and legislation and historical events but also national culture and traditional practice of land controlled by the elite minority and powerful landlords have made it difficult for the government to address land governance in this country (Mazhar 2004). The author of Chap. 4 explained that the elite minority and landlords did not only assert their power over land control but also wield their social and economic authority over the daily life and families of small and poor farmers (Chap. 4). As a result of land manipulation and management by a small group of people, prices and quantity of agricultural products have also been controlled and manipulated. The marginalised groups of people in the society do not have the ability to pay for food and other necessities. Thus, food security due to mismanagement of land, i.e. land tenure system, and property rights has become an issue in Pakistan notwithstanding the systemic corruption, chronic poverty and rural discontent. The authors asserted that food security in Pakistan was not an agricultural or technical problem, but it was the complexity of the current land distribution system manipulated by invisible landlords, which significantly has contributed to food deprivation of the disadvantaged and the marginalised people who accounted for about 30 % of the total population in the country. This chapter also proposed some strategies to address the issues and make policy recommendations to minimise food shortage in the country.

Chapter 5 is another interesting one discussing issues of food security in the context of Malaysia. Given the importance of food, food security and disaster risk management should be treated as twin brothers in research studies on disaster risk management. In 1984, the then Prime Minister of Malaysia, Mr. Mahathir, wanted to materialise his vision, i.e. to transform the country from an agriculture-based economy to an industrialised economy, via the introduction of several policies (Mahani 2004; Hirschman 2013). As a result of the new policy, a contraction of the agricultural sector has been observed which affects the level of local food supply (Hill et al. 2012; Anderson et al. 2013). This, in turn, has affected the amount of food availability and, therefore, posed a serious threat to food security. Thus, this chapter examined the effectiveness of the existing agricultural polices with regard to agricultural land use and management. The questions are (i) whether the current agricultural policy framework have been able to address issues associated with food security for the population in the country and (ii) what action has been taken by the government of Malaysia to address this shortage of food. This chapter also discussed new challenges and their implications for food security policies in Malaysia. Such challenges are "(1) realignment of national policies from agricultural based to industrial based, (2) conversion of agricultural land and (3) increased imports of food" (Chap. 5). The authors argued that traditional patterns of food insecurity could not be improved without drastic reforms of the current agricultural policies. In other words, without transformational change in the current policy framework on agriculture, food insecurity would continue to pose a serious threat to sustainable food production in Malaysia. The authors concluded that the government needs to play a more active role in carrying out strategic intervention to improve the state of food security in the country.

Chapter 6 discussed key natural disasters and their associated issues in Bangladesh. Common natural disasters in Bangladesh are floods, cyclones and earthquakes. Such disasters have greatly impacted several political and socio-economic aspects, such as health and well-being of the population, land use and planning, and economic activities of various sectors and industries. The root causes of disasters are varied, including ecological, biological, political and social and causes (e.g. hazards, poor planning, lack of resources, unsafe conditions, etc.) (Asgary and Halim 2011). The authors also examined disaster governance in the country which included institutional and regulatory frameworks and mechanisms to mitigate disaster risk. The authors also evaluated the current initiatives to manage disasters based on the three-sector governance model which promoted the collaboration efforts among government, the private sector and civil society organisations (CSOs). Finally, the chapter ended with some policy and technical recommendations to enhance the state of disaster risk management in the country.

Chapter 7 addressed a critical issue associated with administrative planning and political response to natural hazards and disasters. We all agree that natural disasters do not only affect people, their livelihoods, and ecological and social systems, but they also disrupt community activities and functions by damaging physical, economic and social capital in the affected areas. Disasters threaten people's lives, health and economic well-being, and thus, this requires urgent interventions of state and non-state sectors to minimise the negative consequences of disasters. The author of this chapter argued that "wide array of leadership and management competencies is required to manage these routine emergencies and catastrophes, under these challenging conditions people see to their leaders for their vision and direction to mitigate disaster impact" (Chap. 7). This chapter identified a number of factors which systematically and significantly affected the administrative capacities of the government of India to deal with disasters. This chapter also discussed leadership vision and capability building towards the ability to respond to hazards, and the correlation between these factors and the administrative bodies. Besides, the authors examined whether better procedures and practice implemented by government agencies can lead to greater capacity building and improve the ability to handle natural disasters. The chapter asserted that community engagement and effective self-organising processes could make a significant difference with regard to capacity building. The chapter also stressed that policy reform would enhance the whole disaster management process.

It is a serious mistake and a violation of human rights if the disability issues are not taken into consideration in disaster risks management. There are about 15 % of the population in the countries around the world living with some forms of disability (World Health Organisation and World Bank 2011). The poor and the people with disability (PWD), especially those in developing countries, are more vulnerable than others during disasters. This has been observed from the studies of Hurricane Katrina (USA, 2005), Cyclone Sidr (Bangladesh, 2007), Haiti Earthquake (2010), Haiyan typhhon (Philippines, 2013) and many others. During natural disaster, PWD do not only suffer the same impacts of natural disasters as other groups of victims, but they also do not have the ability to cope with the deterioration of environment. Many of the PWD are not able to flee from the affected areas, and they are usually neglected or abandoned during evacuation. Usually, they are often excluded from several initiatives regarding disaster preparedness, mitigation and interventions (International Federation of Red Cross and Red Crescent Societies 2007). Thus, Chap. 8 explained how the disability issues were disregarded in disaster risk reductions strategy in Bangladesh which has led to lapse in human security and social cohesion. It also proposed governance approaches to address such deficiencies in disaster risk management with regard to PWD. The author explained that although Bangladesh has developed comprehensive disaster management strategies, the exclusion of disability in such strategies, especially in the implementation phase, has deprived the rights of the PWD and affected their welfare and well-beings. The author suggested that this situation could be rectified via the inclusion of PWD at all levels and all stages in the disaster risk management national plan. Otherwise, these people would be marginalised from the mainstream, and their lives would be in danger, which ultimately leads to more social problems and social incohesion.

Moving away from disability issues, Chap. 9 focused on geopolitics which refers to the study of how geography of a country influences the politics, foreign policy and current affairs of the country. It also refers to the interaction between geographical location, politics and power distribution of countries (Hafeznia 2006). Geopolitics does not only affect international trade and diplomatic relations between countries, but it also plays a key role in disaster risk reduction, reconstruction and resettlement. Chapter 9 discussed the issues associated with geopolitics and community vulnerability, using Diu Island, India, as a case study. Diu Island, surrounded by the Arabian Sea, has frequently faced floods, cyclones and earthquakes in the past decades. Climate change has significantly impacted the island via several forms, such as submerge of land under sea water and increased vulnerability of the coastal community, causing loss of human lives and livelihood, whereas geopolitics affects the way disaster risk has been managed as well. This chapter examined how disasters have significantly affected the lives of the residents of the island. Various factors, such as governance, technology and partnerships between state and non-state sectors, have influenced the coastal community's ability to respond to such disasters. The chapter also discussed strategies and governance approaches to lessen economic and human loss caused by natural disasters in this island.

Different from other chapters, Chap. 10 elaborated how integrity can facilitate or prevent the preset goals to be achieved. This can apply to the process of disaster risk management. Importantly, this chapter examined the role of spirituality in the success or failure of the policy implementation and proposes that spirituality can be adopted to nurture commitment of participants in a project through sustenance of integrity. The chapter also emphasised the value of cocreation in the delivery of public service which could help stakeholders to achieve the objectives in disaster risk management.

There is empirical evidence that climate change has taken place, causing global warming and playing a havoc to several ecosystems. Anderson and Bausch (2006) commented that "natural disasters result when extreme events strike vulnerable areas: reducing vulnerability can reduce the impact of extreme events" (p. 1). Thus, Chap. 11 developed, interpreted and applied a climate smart approach to disaster risk management (CSDRM) to help stakeholders enhance their disaster preparedness. The authors of this chapter compared various aspects of the two disasters, namely the Glacial and Cloud Burst leading to Flash flood in Uttarakhand and the Super Cyclone Phailin in Odisha in India, in order to validate the CSDRM model in disaster management (World Health Organisation 2013). They concluded that a CSDRM approach could foster collaboration and coordination among government agencies, and between government and non-state sectors, such as the private sector and civil society. This approach provided the flexibility required during emergency and improves the adaptive capacity of users. One of the requirements of this approach was to engage all groups of stakeholders in the disaster management process. However, adopting this approach does not automatically ensure that all issues associated with climate change and disaster risk management can be addressed. Add-on activities, such as use of modern technology, improve human resource capability and so on, are required to enhance the effectiveness of this approach.

It is practically impossible for anybody or any country to stop or prevent natural catastrophes for several reasons. What we can do is to mitigate risk factors, prepare for disasters and reduce the impact of such disasters. One of the measures to address these issues is to enhance public awareness of disaster risk management in order to increase self-resilience before, during and after disasters. One of the measures to enhance public awareness of disaster risk management is to provide educational programs to the public (Ahmadi and Laei 2012). Thus, the first chapter in Part 3, Chap. 12, explained the importance of public education in disaster management. The authors of this chapter reviewed how different countries, such as Indonesia, Iran, Japan and Vietnam, have conducted public education programs with regard to risk reduction and disaster management. It also evaluated the effectiveness of public programs, using Singapore as a case study, based on the five elements of the *Wisconsin Model of Community Education*, namely (1) citizen involvement, (2) needs assessment and planning, (3) extended use of public education facilities, (4) interagency coordination and cooperation and (5) leadership and accountability (Wisconsin Department of Public Instruction 2014). The findings suggested that public education has been offered in Singapore, but more can be done to improve the public awareness of disaster preparedness and management (Chap. 12). Therefore, the authors proposed an integrated approach to public education, involving in both state (government) and non-state (private enterprises and civil society) sectors in the planning and delivery of public education.

Chapter 13 discussed climate change, but from a different dimension, i.e. civil society participation in environmental governance to address climate change issues and impact. The authors examined (i) how CSOs participate in local environmental governance through various forms of mobilisation and movement and (i) how CSOs could contribute to climate change adaptation. The authors addressed the research question via the analysis of a case study. This case study discussed Paani Committee's movement for mainstreaming of Tidal River Management (TRM) in environmental governance with regard to water resources and flood control in the south-west Bangladesh. The findings suggested that although relevant government agencies and the donors have continuously opposed the involvement of Paani Committee in environmental governance at the local level, this committee has successfully conducted several TRMs given the strong support and participation of local residents. The authors further explained that "successful cases of TRM complimented with people's mobilisation and policy advocacy drew attention of scientific and policy communities to the indigenous knowledge system. Thus, a handful of scientific studies were undertaken that produced scientific and engineering evidence vindicating the effectiveness of TRM as flood and water management approach suitable for south-west region's floodplain ecology" (Chap. 13).

The reason for inclusion of Chap. 14 in this edited volume is to allow the voice of the under-representative academics to be heard although the style and the contextual ideas may not be familiar with or highly regarded by many of readers. Most of us may not hear or know about the Altai Republic, whose neighbours are Mongolia, China and Kazakhstan, and her people. Therefore, this is a very good opportunity to learn about the experience of how indigenous people from remote places, such as Altai, have been educated with regard to ecological systems. Chapter 14 explained how educational projects, including the development of textbooks, to educate Altai people about ecological and environmental protection, have been carried out. The textbook developers have had to take into consideration various aspects, such as land use, natural science, humanity, spiritual culture and traditional customs, in order to educate the participants on how to preserve national culture and heritage given the rapid development of globalisation and modernisation.

Peace is another interesting element linked to risk reduction. In other words, there is a correlation between climate change and natural disasters, and peace building is also one of the issues associated with disaster management to some extent. Ferris (2010) asserted that "because the definition of a natural disaster is linked to the society's response capacity, state and social structures which are weakened by conflicts are less likely to be able to respond to the effects of a natural hazard, making it more likely that a natural disaster will result" (para. 7). For example, the weak Somali government did not have the capacity to respond to either the drought or flooding which has occurred in its country (Ferris 2010). Thus, Chap. 15 introduces a peace building model for the southern Philippines. The author employed qualitative research methods, namely focus group discussion, interviews with key informant, met synthesis and multiple case narratives, to address the research questions; that is, how conflicts could be prevented in the researched area. The author explained that "many conflict-affected and vulnerable communities have dysfunctional or weak local government units. These areas have persistent poor or lack of access to basic social services as evident on low human development index, high poverty incidence, and persistent volatile peace and security condition" (Chap. 15). The author proposed a peace building model for southern Philippines, adopted from the Action for Conflict Transformation (ACT) for Peace Program (Department of Foreign Affairs and Trade, Australia, n.d.). This model includes four strands, namely (i) social cohesion, (ii) basic social services, (iii) community economic development and (iv) conflict management. However, sustainable peace to prevent conflict may not be secured without good governance. In other words, this model must incorporate other elements of good governance.

The fourth chapter in Part 3, Chap. 16, on *Community Governance for climate change adaptation in Nepal* discussed on the role of community forest users' groups for governing community forest in Nepal. The authors concluded that the success of sustainable forest management and the climate change adaptation plan of action and securing socio-economic benefits for local communities depended on a number of factors. Such factors are land ownership, organisational capacity, capital, technical knowledge, legal management, clandestine loggers, market access, infrastructures development, managerial skills, economic returns and so on.

Given the strong linkage between climate change and disaster management, Chap. 16 revisited the concept of community forest users' groups for managing community forest in Nepal. The authors explained that this approach could help stakeholders achieve sustainable forest management, implement the national climate change adaptation plan and produce socio-economic improvement for local communities if only several issues are addressed. Such issues include land use, planning and ownership, the administrative and organisational capacity, human, social and physical capital, legal framework and compliance, and so on.

The last chapter summarises what has been discussed in other chapters with regard to land use, risk reduction, food security, disaster governance, disability issues, climate change and peace. It also provides some policy recommendations to address issues associated with land use and disaster governance.

1.3 Conclusion

It is important for all stakeholders in all sectors (government, the private sector and civil society) and at all levels (international, national, and sub-national) to acknowledge the fact that "prevention and mitigation measures may reduce the vulnerability and hazard while occurrence of natural disasters remain unavoidable" (Khan 2008, p. 663). Thus, efforts to prevent and mitigate disasters should go hand in hand with effects to address post-disaster issues, such as disaster recovery and reconstruction.

In conclusion, land and disaster governance is a challenging and dynamic process. Given the multidimensional challenges of disaster risk reduction and management, the governance process should be modified, adapted and enhanced to ever-changing conditions in both the external and internal environments of a country. Although international, regional and national contingencies have exercised their influence on the governance process, several problems associated with land use and planning, food security, and risk deduction have been observed. Therefore, the quest for innovative and new approaches for disaster governance should be an ongoing task.

References

Ahmadi A, Laei S (2012) Public education: compulsory and free? A paradox. Procedia—Soc Behav Sci 47:868–873

Anderson J, Bausch C (2006) Climate change and natural disasters: scientific evidence of a possible relation between recent natural disasters and climate change (IP/A/ENVI/ FWC/ 2005-35). In: Brussel: European parliament's environment, public health and food safety committee

Anderson K, Jha S, Nelgen S, Strutt A (2013) Re-examining policies for food securities. Food Sec 5(2):195–215

Asgary A, Halim MA (2011) Measuring people's preferences for cyclone vulnerability reduction measures in Bangladesh. Disaster Prev Manage 20(2):186–198

California Department of Public Health (2014) Know and understand natural disasters. Sacramento, CA, State of California

Department of Foreign Affairs and Trade, Australia n.d. ACT for Peace Program. Canberra: Department of Foreign Affairs and Trade, Australia

Ferris E (2010) Natural disasters, conflict, and human rights: tracing the connections. The Brookings Institution, Washington, DC

Franke RW (1992) Land reform versus inequality in Nadur village Kerala. J Anthropol Res 48(2):81–116

Ha H, Dhakal TN (2013) Governance approaches to mitigation of and adaptation to climate change in Asia. PalGrave Macmillan, London

Hafeznia MR (2006) Principles and concepts of geopolitics. Popoli Publications, Iran, pp 37–39

Hill H, Yean TS, Ragayah MZ (2012) Malaysia: a success story stuck in the middle? World Econ 35(12):1687–1711

Hirschman C (2013) Malaysia's development challenges: graduating from the middle. Asian-Pacific Econ Lit 27(1):163–165

International Federation of Red Cross and Red Crescent Societies (2007) World disaster report 2007: focus on discrimination, International Federation of Red Cross and Red Crescent Societies. International Federation of Red Cross and Red Crescent Societies, Bern, Switzerland

Ishiwatari M (2003) Disaster risk management at the national level. Asian Development Bank Institute, Tokyo

Islam SMT, Chik Z (2011) Disaster in Bangladesh and management with advanced information system. Disaster Prev Manage 20(5):521–530

Khan MSA (2008) Disaster preparedness for sustainable development in Bangladesh. Disaster Prev Manage 17(5):662–671

Mahani ZA (2004) Pemikiran ekonomi Tun Dr. Mahathir Mohamad dalam pembangunan negara dari perspektif korporat. In: Mustapa K, Abdul Rahman AZ, Hasliza S, Muhamed Nor AN (eds). Siri simposium berkala III: Ekonomi [Periodic symposia series III: Economic], Sintok: Institut Pemikiran Tun Dr. Mahathir Mohamad

Mazhar A (2004) Land, peasants and poverty: equitable land reforms in Pakistan. The Network Publications, Islamabad

Ministry of Rural Development (MRD), Government of India (GOI), Department of Land Resources (2009) Report of the committee on State agrarian relations and the unfinished task in land reforms. Ministry of Rural Development, Government of India, Department of Land Resources, New Delhi

Population Institute (2011) From 6 Billion to 7 Billion: how population growth is changing and challenging our world. Population Institute, Washington DC

Tierney KJ (2012) Disaster governance: social, political, and economic dimensions. Annu Rev Environ Resour 37:341–363

Tiwari R (2010) Politics of land. The Indian Express, Aug 30 2010

Turnbull M, Sterrett CL, Hilleboe A (2013) Toward resilience a guide to disaster risk reduction and climate change adaptation. Practical Action Publishing Ltd and The Schumacher Centre, Warwickshire, UK

Wisconsin Department of Public Instruction (2014) Wisconsin's Components of Public education. Madison, Wisconsin Department of Public Instruction, USA.

World Health Organisation (2013) Flash floods and land slides in Uttarakhand State, India. World Health Organisation, Geneva

World Health Organisation (2014) Urban population growth. Geneva: World Health Organisation. Available at http://Www.Who.Int/Gho/Urban_Health/Situation_Trends/Urban_Population_Growth/En/. Accessed 16 May 2014

World Health Organization and World Bank (2011) World Report on Disability 2011, World Health Organization, Geneva, Switzerland, ISBN 978-92-4-068521-5

Part I
Land Management and Food Security

Chapter 2
Land Reforms in Kerala: An Aid to Ensure Sustainable Development

Anishia Jayadev and Huong Ha

2.1 Introduction

2.1.1 The Issue of Land Governance and Land Reform

In any agrarian society, such as India, there is a strong linkage between land and societal status (Srinivasulu 2002). Land is the most valuable, imperishable possession from which people derive their economic independence, social status and a modest and permanent means of livelihood. In addition, land also assures land owners an identity and dignity and creates conditions and opportunities for them to realise social equality as stated in the Draft Land Reforms Policy by the Government of India (Department of Land Resources, Ministry of Rural Development, Government of India 2013). As a problem, insecurity of tenure has different meanings, i.e. the likelihood of forced eviction, inequitable land distribution between races and lack of secure jobs for farm workers. This may also refer to discrimination against the disadvantages, such as women and the minority, in favour of the rich and the powerful people.

Land governance is about policies, processes and institutions by which land, property and natural resources are managed (Enemark, McLaren and van der Mole 2010). Land reform requires the redistribution of wealth and power. Government plays the key role in this process because "it is the only actor with authority [and power] to command compliance" [International Fund for Agricultural

A. Jayadev (✉)
Institute of Management in Government, Vikas Bhavan Post, Thiruvananthapuram, Kerala, India
e-mail: facultyimgdrajdev@gmail.com

H. Ha
UON Singapore (University of Newcastle, Singapore Campus), Singapore, Singapore
e-mail: artp6075singapore@yahoo.com

H. Ha (ed.), *Land and Disaster Management Strategies in Asia*,
DOI: 10.1007/978-81-322-1976-7_2

Development (IFAD) 2006, p. 2]. International agencies, such as the IFAD, the United Nations Development Programme (UNDP) and the World Bank, explained that rural poverty cannot be eradicated without land reform and believes that poor people in the rural areas must have access to and control over land and other natural resources to eradicate extreme poverty (Enemark et al. 2010; Saturnino et al. n.d.).

Governments in different states in India have redistributed land by setting ceilings on how much land could be owned and leased to individuals. Land-holdings that were in excess of the ceilings were redistributed to peasants (UNDP 2008).

2.1.2 Research Objectives

The central theme of this chapter is to revisit the concepts of land governance, land reform, and the relationship between land governance and sustainable development. This chapter also investigates (i) land reform movement in Kerala, India, and (ii) challenges of land reform in this state. The chapter ends with some recommendation and a conclusion.

2.2 Theoretical Foundation and Research Methodology

The theoretical foundation of this chapter has been built on Marx's Theory of Alienation (Kaplan 1972). The most obvious argument in favour of land reform is equity. In a land-scarce country with a significant proportion of the rural population living below the poverty line, the case for ensuring that everyone has access to some minimum amount of land seems compelling from this point of view. In his works, Marx highlights that within the class system, there are two main classes, namely the bourgeoisie (capitalists) who are the upper class and have ownership of private property, and the proletariat (workers) who are the lower class and do not have ownership of private property (Rummel 1977). In his theory, Marx states that the working class invest their labour into the production of things to which they have no relationship. This theory holds good when we discuss about the land reforms and issues of the poor. Since they do not own the fruits of their labour, the products, they are therefore contributing to a world outside of their own and are thus alienated from the products. Land reform has been back in the international and national agendas. The findings of the study by Deininger, Songqing and Nagarajan (2007) suggested that there were "robust positive impact of land reform on income, consumption and asset accumulation" (p. 5). However, the impact of reform initiatives may decrease over time due to many political and socio-economic constraints (Deininger et al. 2007).

Thus, this chapter adopts a case study approach to explain why land reform is essential based on the above theoretical foundation and also examines the enactment of land reform legislation in the light of the sustainable development ideas as proposed by the Food and Agriculture Organization of the United Nations. Secondary data have been collected via scholarly literature and non-scholarly publications from reports, acts, documents published by governments, international and national non-governmental organisations. The state of Kerala, a tiny state in India, has been selected for this study because of its fabulous governance outcome. Kerala has received worldwide attention due to its remarkable achievement in significant reduction in poverty in the last several years. Land reform measures to address issues of social and human developments, and redistribution systems with reasonable social security have contributed to its success (Kannan 2005). Hence, it is worth to investigate the case of Kerala.

2.3 Literature Review

2.3.1 Land Reforms and Land Governance

Land, forest and other natural resources provide a platform for livelihoods and a basis of socio-cultural and religious practices (UN Economic Commission for Europe and Food and Agriculture Organization 2013). Secure tenure rights and equitable access to all relevant recipients are one of the means to eradicate hunger and poverty which, in turn, can support sustainable development and enhance the state of environmental protection. This helps people produce food for their own consumption and for earning their living. In this context, land reform refers to "the redistribution of land from those who have excess of land to those who have none, with the objective of increasing the income and bargaining power of the poor" in the rural areas (Jain and James 2011, p. 3). Land reform also embraces the way in which ownership, leasing conditions, sales and inheritance of land is regulated.

A working definition for land governance commonly used in discourses depicts it as the process by which decisions are made regarding the access to and use of land, the manner in which those decisions are implemented or not implemented and the way that conflicting interests in land are reconciled. Palmer et al. (2009) defined land governance as

> the rules, processes and structures through which decisions are made about access to land and its use, the manner in which the decisions are implemented and enforced, the way that competing interests in land are managed. (p. 9).

Land policies and processes, land-use planning, land reform and land administration reform, all are governance challenges. As proposed by the United Nations Human Settlements programmer (UN-HABITAT 2008), the nine principles of good land governance are summarised in Table 2.1.

Table 2.1 Principles of good land governance

• Security—Security of tenure with no forced eviction, land and property rights to all and an effective governance mechanism would contribute to addressing post-disaster and post-conflict vulnerabilities sensitively
• Sustainability—A balanced use of land would consider the socio-economic and environmental constraints and stakeholders' needs, and a land administration system which make land affordable to even the poor, and accessible to all relevant stakeholders. Specific capacity building to ensuring sustainability to all concerned is required. Proper land use and management can address the challenges of climate change and related consequences of natural disasters, food shortage, etc.
• Equity—Equity refers to supporting the poor, considering the gender differences, alienated and marginalised groups, such as tribes, the minority, the scheduled caste and the disabled and ensuring continuity of land rights
• Effectiveness and efficiency—This principle prompts service delivery by virtue of having a strong land administration, simplified rules and procedures by reducing the amount of paperwork and red tape
• Subsidiarity—It is important to decentralise decision-making and management through engaging the grass-root governance machinery, such as panchayat, local bodies and oorukkoottoms (tribal) in the process of land governance and administration
• Rule of law—Laws should consider all the different classes of needy or take due consideration of legal pluralism
• Transparency—Stakeholders should have free, fair and timely access to any information pertaining to rules, procedures, costs and decision-making processes. Digitisation of information and records would ensure this to a large extent
• Accountability—Government should prevent any forms of corruption regarding land-use planning and management. A number of common factors as a result of poor public land management have been identified. Some of the examples are "ambiguity in authoritative roles and responsibilities, a lack of accountability or methodology in the systems of allocation, appropriation, disposal or use of public land, and a lack of information on state assets" (Burns and Dalrymple 2008, p. 13). Burn and Dalrymple (2008) further explained that weak land governance has directly and indirectly affected citizens, hindered the socio-economic development process as well as the peace and national security
• Civic engagement or participation—It is important that government engages the public in land governance by frequent interaction with relevant groups of stakeholders and getting consensus in decision-making (Palmer et al. 2009)

2.3.2 Objectives of Land Reform: National and State Levels

At the national level, some key objectives of land reform in India are to (1) improve the productive capacity of land by improving the working and socio-economic conditions of farmers, who either are land owners or land leasers, so that they are more interested in investing in and improving agriculture, (2) ensure distributive justice regarding land use and ownership and create an classless society by eliminating all forms of discrimination and exploitation, (3) create a system of land ownership for all respective stakeholders and (4) redistribute income from the minority "high-income" earners to the majority "low-income" earners in order to stimulate the demand for consumer goods (Subramanian 2013).

The situation of land reform is different in different states in India. West Bengal and Kerala have introduced the notable land reform programs which have produced "a positive impact on agricultural production, poverty alleviation and economic growth" (Quizon 2005, p. 6).

The Kerala Land Reforms Act (1963) provides a legal foundation for imposition of the ceiling on land holdings. Actually, it "was inserted as Item 39 in the 9th Schedule to the Constitution" of India (Khanna 2008, p. 208). Land reform in Kerala aims to (i) weaken the control and power of landlords and ensure security of land tenure to the landless and poor farmers (Griffin, Khan and Ickowitz 2001), (ii) stimulate the growth of the agricultural sector via increasing productivity and output by eliminating feudal and semi-feudal systems of land control, (iii) develop rural markets via redistributing factors of production, such as land, and increase in public investment in rural farming, (iv) improve human development through greater investment in education and healthcare and (v) empower the minority, such as Dalits, women and tribal people, in order to address caste and gender oppression (Enemark et al. 2010; Subramanian 2013).

2.3.3 Land Governance and Sustainable Development

There is a link between land governance and sustainable development. According to the World Bank's declaration on land governance in support of the Millennium Development Goals (MDGs) (Enemark et al. 2010), sustainability in land governance can be achieved if government and relevant groups of stakeholders adopt the strategies stated in Table 2.2. Apparently, effective land governance can help countries achieve three out of eight goals of the MDGs, namely (i) reduction and elimination of "extreme poverty and hunger" by productive use of land, (ii) promotion of "gender equality" and women empowerment and (iii) ensuring "environmental sustainability" (United Nations 2013, p. 9).

2.4 Case Study of Land Reform in Kerala

2.4.1 Introduction About the State of Kerala

Kerala, also known as Keralam, is located at the extreme southern tip of the Indian subcontinent. The state, with an area of 38,863 km^2 and 1.18 % of India's landmass, "is situated between the Arabian Sea to the west and the Western Ghats to the east. Kerala's coast runs 580 km in length, while the state itself varies between 35–120 km in width" (Office of the Commissioner of Civil Supplies, Government of Kerala n.d., p. 9). Similar to other states and countries, land is one of the most invaluable resources of Kerala.

Table 2.2 Strategies for sustainable land governance

• It is important for government, working with relevant stakeholders, to "***provide*** transparent and easy access to land for all and thereby reduce poverty"
• Economic growth can be facilitated via "*securing* investments in land and property development"
• Government, working with non-state sectors, to search for strategies and mechanisms which can "*avoid* land grabbing and the attached social and economic consequences"
• In order to achieve the MDGs, it is important to "*safeguard* the environment, cultural heritage and the use natural resources"
• Land governance should "*guarantee* good, transparent, affordable and gender" responsiveness which can generate benefit for all recipients, including the most vulnerable groups, such as the disabled, the minority and the disadvantaged.
• Government and other groups of stakeholders should consider various factors when enacting and *applying* "a land policy that is integrated into social and economic development policy frameworks"
• Government, working with other sectors, should adequately and timely *address* the challenges associated with "climate change and impacts of natural disasters, food shortage, etc."
• Finally, state and non-state sectors should *recognise* the rapid increase in urbanisation which is one of the key challenges to "sustain future living and livelihoods"

Source Adapted from Enemark et al. (2010, p. 8)

Kerala is the state with the lowest population growth rate in India (3.44 %) and has a density of 819 people per km^2. The Human Development Index (HDI) of Kerala is 0.92, the highest score in the country, according to the Human Development Report 2011 (Office of the Commissioner of Civil Supplies, Government of Kerala n.d.). The state has the highest female literacy rate (92 %), and the lowest rate of maternal mortality in India (SA 2013). Interestingly, the state has more female population than male.

2.4.2 Land Use and Land Reforms in Kerala

Land reform is not only a national but also a state affair and is subject to the political will of the state government. In the agrarian structure in Kerala in 1951, the per capita net sown area was lowest among all states in India, i.e. 0.038 acres, compared to the average net sown area (1.09 acres) in other states in India (Besley 2000). Kerala also had "a much lower percentage of peasant proprietorship, a higher percentage of tenants and non-cultivating renters (absentee landlords) and a significantly high percentage of agriculture labourers than the all-India average" (Eashvaraiah 1993, p. 120). Thus, the agricultural structure of Kerala exhibited an altogether different pattern from the rest of the country. The agrarian system just prior to initiation of land reforms in the state had the following characteristics:

- There was a highly skewed distribution of land in support of the rich and powerful landowners (e.g. the Brahmin and the Nair) who cultivated their land with the help of Dalit tenants;
- There were a large number of landless agrarian households; and
- There were small, poor and powerless farmers who were allotted only small pieces of land for making their living (Dasgupta 1981).

Land is the essential and often only productive asset of poor people. Franke (1992) explained that the main initiatives had been carried out in Kerala during the land reform period are:

- Setting a cap (ceiling) on the absolute size of holdings and the excess would be distributed to those who did not have land previously;
- Abolishing the system of "rice land tenancy" which would lead to "the abolition of rental payments from actual operators to non-cultivating landlords" (Franke 1992, p. 85). The reform aims at preventing landlords from throwing tenants off the land.
- "The abolition of tenancy in house garden lands and thus the abolition of rents to the landlords who held title to them. As with provision 2, eviction stays kept tenants from being thrown off the land" (Franke 1992, p. 85).

The Kerala's land reform movement enables the Kudikidappukaran. This term refers to a person "who does not have an estate or any land exceeding five cents in value, and who has been allowed to use and occupy a small portion of land as his shelter by the lawful owner of the land, as owner or as tenant, with or without an obligation to pay rent" (Kerala Real Estate Legal Service n.d. para. 1) to retain the shelter and land and also purchase land surrounding their shelter to a limit. The land reform allowed the land users to compensate the land owners by paying only 25 % of the market value. The land occupants needed to pay only 50 % of the total value in 12 annual instalments, and the government took care of the balance 50 % (Franke and Chasin 1989).

Here, the aftermath of land reform was not only for the landless and the poor to have a new piece of land to cultivate or an increase in their family income but also to acquire the ownership right of the farmstead where they have been living on for a long time. This, to a great extent, provided some relief to the poor peasants. It should be reckoned that the reform measures had dismantled the fortress between different economic classes and minimised the caste inequality in Kerala society. To a great extent, landlordism and the *janmi* system had been removed, and ownership of land holdings had been extended to the majority (Krishnakumar 2004). It also provided protection to those who leased land from eviction and provided more land for building of housing for thousands of families (Krishnakumar 2004). Land reform also included social security schemes for workers in the agricultural sector. The Kudikidapppukars could now possess land ownership which was a unique feature of land reform in the Kerala state. The Kudikidapppukars and their families were free from fear and anxiety. It also abolished bonded labour system in land (Mishra 2002). The proportion of land owners who own fewer than 5 acres increased from 91 % in 1966–1967 to 96.7 % in 1976–1977, whereas the percentage of owners of bigger land areas (more than 5 acres) reduced. As a result, the areas of smaller land plot, i.e. below 5 acres, increased nearly double (Isaac 2008).

2.4.3 Hurdles in Land Reform

Even though there were factual data available regarding the emancipation brought about by enactment of land reforms in Kerala, there were a lot of stumbles towards ensuring a sustainable socio-economic, agricultural and environmental development in this state. Even though there were more owners of smaller land plots as discussed in the above section, huge land inequalities remain (Franke 1992). Land which has been appropriated could not be distributed to the landless. The distributed land remains fallow and is not adding the agricultural productivity of the state since agricultural practices have become disincentive to the farmers. Non-availability of disaggregated data in terms of gender, caste and class adds to the difficulty. In Kerala, during enactment of the reform, it was expected that there were four lakh Kudikidappukar, but the state barely received requests from one-fourth of the estimated number. Data state that in 1966–67, hundred thousands of acres of land were with the feudal land lords and religious institutions (Raj and Tharakan 1983). This would result in the inference that 40 percentage of the agricultural land was held by only 0.7 percentage of the population. It is to be doubted whether a reverse tenancy is happening in the state. For both agricultural and non-agricultural purposes, national and international companies have come to Kerala and accessed land on lease for long term. Some of them even purchased land for unimaginable price because the farmers and/or governance mechanisms have not ensured proper usage of land in a productive and sustainable manner. Even the community farming initiatives which could be successfully run by cooperative/self-help groups are grabbed by contract farming agencies where the small owners will be reconverted to mere agricultural labourers (Singh 2011; Singh and Asokan 2005). Even though it is proposed that contract farming agreements world provide the framers with quality inputs. *The Kerala Land Reforms (Amendment) Bill* (2002) explained that the *Kerala Land Reforms Act, 1963,* imposed a limit or a ceiling on holdings. Nevertheless, certain types of lands were exempted from the ceiling limit. They include plantations and private forests. Tea, coffee, cocoa, rubber, cardamom or cinnamon plants are categorised under plantation (Krishnakumar 2004; Narayan 2003). Since the supply of the above agricultural products, such as tea, coffee and cocoa, has increased whereas the prices of such products have decreased, landholders have little incentives to cultivate such products.

Lands cultivated with plantation will not get the exemption from ceiling limits, if they are used for non-plantation purposes (Krishnakumar 2004). Cashew estates having an adjoining extent of ten acres or more were originally not subject to the ceiling limits. However, the *Kerala Land Reforms (Amendment) Act, 1969* (Kerala Act 35 of 1969), removed this exemption, and this took effect from 1 January 1970. This has negatively affected the traditional cashew industry in Kerala. Besides, pharmaceutical plants have been in high demand all over the world. The owners of these plants have sought exemption of their lands from ceiling provisions. The relevant authorities have granted exemption to cashew and medicinal plants from the ceiling limits stated in the *Kerala Land Reforms Act, 1963* (Krishnakumar 2004). This means landholders can request for converting their lands from normal use to other plantation purposes in order to be exempted from the ceiling provisions.

This argument is categorically negated by experts like Khanna (2008) as the crops do not depend on the size of holdings. Khanna (2008) also fears that such exemptions would only lead to concentration of land in the hands of a few. This amendment was implemented in the year 2012 with minor modifications. Discussion and debates among officials in charge of the implementation of land reforms in the state have identified the following major gap depicted in Table 2.3.

2.4.4 Land Reform and Sustainable Development in Kerala

As discussed in Sect. 3.3.1, land reform, a major component of land governance in Kerala, can contribute to the achievement of some development objectives, including the achievement of the MDGs. Landlessness in the rural areas is often considered "the best predictor of hunger and poverty" since the poorest are usually

Table 2.3 Issues of land reform in Kerala

• The landless and the marginalised did not really benefit from land reform significantly. The real beneficiaries were still upper class and land owners since they could find ways to be exempted from being a subject of the ceiling provisions
• Plantation owners became a new class of landlords
• Some communities were totally left out, such as the fishing community
• Land reform activities could not improve agricultural productivity as there was no linkage between the distribution of land and development of infrastructure facilitating agriculture
• There was drastic declaim of paddy cultivation due to several difficulties faced by small farmers
• There were problems with boundaries of surplus land because of procedural irregularity and loss of survey record and other administrative procedures
• Due to time and resource constraints, information about resurveying and renumbering has not been updated, and thus, the status of land was not reflected in revenue records
• There was difficulty in locating holdings exceeding the ceiling limit due to manual maintenance of land records
• Land challenges are constantly evolving due to lack of clear solutions and social and institutional complexity. Also, multiple actors with various interests which did not converge with one another added into these challenges

landless or land-poor (Wickeri and Kalhan 2010, p. 1). Thus, provision of equal opportunities to access to land and other natural resources is a direct factor contributing to alleviation of hunger and poverty. This has addressed the first MDG which is "eradicating extreme poverty and hunger" (United Nations 2013, p. 9). The introduction of new land tenure has promoted equality among all classes and income groups, including gender equity. Such initiatives could promote women empowerment and make them more powerful in agricultural production (Allendort 2007). To a certain extent, land reform can achieve the third MDG, i.e. "promote gender equality and empower women" (United Nations 2013, p. 9). Rights to land are also linked to other access and resource rights, such as water, pasture, timber and non-timber forest. The outcome of this activity may partially meet the requirement of the seventh MDG, i.e. ensure environmental sustainability (United Nations Development Group n.d.). This emphasises that adequate supply of affordable land is critical to the prevention of the growth of new slums.

2.5 Further Steps for Effective Land Reforms–Policy Recommendations

Apart from challenges of land reform discussed in the above section, lands illegally amassed by encroachers were reclaimed by government. False documents have been detected and seized, and a lot of administrative and political actions have been initiated since 2010. Yet land grabbing and destruction of environment to a major extent have been observed during land reform in Kerala. Land reform initiatives had actually faced a lot of socio-political hardships. Given the challenges faced by the government of Kerala and the loophole of land reform legislation, some recommendations have been discussed below.

Firstly, land reform must be effective which should essentially lead to agricultural and sustainable development initiatives. The government of Kerala can help owners who have fragmented holdings to exchange holdings for productive land use. This can be applied to land owners who own wet/agricultural land but who are unwilling to do farming on their land. These land owners should be encouraged to sell or lease out land to small farmers who have no other means of earning a living without losing their ownership rights. Since this process requires a lot of political and administrative will and paper work, computerisation of land reform and administration is essential to achieve this objective (Meena et al. 2005). This process will not affect the interest of the rich land owners as cultivation on their land is not the main source of their income. New mechanisms to incentivise agriculture, such as cooperative farming and group farming, should be tested with. The state should aim for an agrarian structure characterised by small- and medium-size holdings and enable them to make agriculture remunerative.

Secondly, land use by the population, both in the rural and urban areas, vary due to the heavy constrain and pressure over the available land. There should be specific provisions for women, fishermen, slum dwellers and other marginalised groups in order to ensure equal opportunity for land access and land use. Also, legal reform must be accompanied by legal-literacy campaigns to ensure that all groups of recipients are aware of clauses/articles changes in land legislation (Bruce et al. 2007). Proper utilisation of the *Mahatma Gandhi National Rural Employment Guarantee Act* (MGNREGA), enacted in 2005, aiming at improving the income sources of the recipients, is recommended since the act aims to "provide for the enhancement of livelihood security of the households in rural areas of the country by providing at least one hundred days of guaranteed wage employment in every financial year to every household" (Sharma 2010, p. 7).

Thirdly, the state should revisit the reform act and its amendments, incorporating the public opinion, feedback and consensus. There have been a lot of amendments made under the consideration of various factors. Comprehensive revisit of acts and amendments would help to decide whether such amendments dilute the basic objectives of law.

Fourthly, to ensure sustainability, both environmental and developmental, the state should explore the possibility of coupling land reform with agricultural reform (Herring 2000). From environmental perspective, soil and water should be protected and used judiciously. Other techniques include enrichment of the land by proper irrigation through maintenance of channels, small waterways, etc.

Finally, there have been insufficient mechanisms to evaluate the effectiveness of land reform initiatives. Hence, the Land Governance Assessment Frame Work proposed by the World Bank can be utilised to gauge the far-reaching socio-economic consequences of enactment of land reforms in Kerala (Deininger, Selod and Burns 2012).

2.6 Conclusion

This chapter has discussed land use, land reform and the links between sustainable development and land management in general and in Kerala, India. It has explained how land reform has been carried out in Kerala, and the results of the reform. Land reform aims to redistribution of land and ensure land equity to all respective recipients. However, the findings suggest that not all benefits of land reform have been materialised due to many constraints and hurdles during the implementation process. The poor, the landless and the powerless still did not have insufficient land to cultivate and increase their standard of living due to legal loopholes. Therefore, a number of recommendations have been proposed to improve the effectiveness of land reform, namely (i) exploration of new ways to redistribute land via exchange of fragmented holdings or non-cultivated land and provision of incentives for agricultural development, (ii) promotion of equality regarding land use and opportunities for income generation, (iii) revision and modification of relevant acts and amendments, (iv) provision of legal education and (v) carrying land reform together with agrarian reform. It is difficult to measure the political and socio-economic success or failure of land reform, and therefore, collective efforts among various groups of stakeholders are required to search for innovation and novel mechanisms to implement land reform and evaluate the outcome of such land reform.

References

Allendort K (2007) Do women's land rights promote empowerment and child health in Nepal? World Dev 35(11):1975–1988

Besley T (2000) Land reform, poverty reduction, and growth: evidence from India. Q J Econ 115(2):389–430

Bruce JW, Garcia-Bolivar O, Hanstad T, Roth M, Nielsen R, Knox A, Schmidt J (2007) Legal empowerment of the poor: from concepts to assessment. USA: United States Agency for International Development, Washington, DC

Burns T, Dalrymple K (2008) Conceptual framework for governance in land administration. Paper presented at presented at the FIG working week in Stockholm, Sweden, 14–19 June 2008

Dasgupta R (1981) A cross sectional analysis of poverty and under-nutrition in rural India. Working paper no. 126 centre for development studies, Kerala, India

Deininger K, Songqing J, Nagarajan HK (2007) Land reforms, poverty reduction, and economic growth: evidence from India. The World Bank, Washington, DC

Deininger K, Selod H, Burns A (2012) The land governance assessment framework: identifying and monitoring good practice in the land sector. The International Bank for Reconstruction and Development/The World Bank, Washington DC

Department of Land Resources, Ministry of Rural Development, Government of India (2013) Draft national land reforms policy 18th July 2013. Department of Land Resources, Ministry of Rural Development, Government of India, New Delhi, India

Eashvaraiah P (1993) The communist parties in power and agrarian reforms in india. Academic Foundation, Delhi

Enemark S, McLaren R, van der Molen P (2010) Land governance in support of the millennium development goals: a new agenda for land professionals. The World Bank and the International Federation of Surveyors, Washington, DC, USA, Copenhagen, Denmark

Franke RW (1992) Land reform versus inequality in Nadur village, Kerala. J Anthropol Res 48(2):81–116

Franke RW, Chasin BH (1989) Kerala: radical reform as development in an Indian state. Institute for Food and Development Policy, Oakland, CA

Griffin K, Khan AR, Ickowitz A (2001) Poverty and the distribution of land. Department of Economics, University of California, California

Herring RJ (2000) Political conditions for agrarian reform and poverty alleviation (IDS discussion paper 375). Institute of Development Studies, Sussex, UK

International Fund for Agricultural Development (IFAD) (2006) Empowering the rural poor through access to land. International Fund for Agricultural Development, Rome, Italy

Isaac TMT (2008) Bhooparishkaranam eni enthu? Chintha Publishers, Thiruvananthapuram

Jain R, James S (2011) Land reforms in India: unfinished task. Centre for Legislative Research and Advocacy (CLRA), New Delhi, India

Kannan KP (2005) Kerala's turnaround in growth: role of social development, remittances and reform. Econ Polit Wkly 40(6):548–554

Kaplan D (1972) General and theoretical: Marx's theory of alienation. IstvánMészáros. Am Anthropol 74:1371–1372

Kerala Real Estate Legal Service (n.d.) Who is kudikidappukaran as per land laws in Kerala? Kerala Real Estate Legal Service n.d., Kerala, India. Available at http://www.bizandlegis.com/in/realestate/faq/category/kerala-real-estate-law/page/2. Accessed 15 May 2014

Khanna HR (2008) Legal classic: making of India's constitution. Eastern Book Company, Lucknow, India

Krishnakumar R (2004) Reversing land reforms. Frontline 21(4): online. Available at http://www.frontline.in/static/html/fl2104/stories/20040227003203400.htm. Accessed 13 May 2014

Meena SD, Thakur V, Shukla DR, Sisodia OP, Narula SK (2005) Computerisation of land records national perspective. In: Habibullah W, Ahuja M (eds) Land reforms in India: computerisation of land records. SAGE Publications India Pvt Ltd, New Delhi

Mishra L (2002) Annotated bibliography on forced/bonded labour in India. International Labour Office, Geneva, Switzerland

Narayan NC (2003) For and against the grain: politics of rice in Kerala, India. Institute of Social Studies, The Hagues, The Netherlands

Office of the Commissioner of Civil Supplies (n.d.) Government of Kerala Pilot project for implementation of direct cash transfer for Apl (state subsidy) scheme: pre pilot of the end to end computerisation of public distribution network in Kerala. Office of the Commissioner of Civil Supplies, Government of Kerala, Kerala, India

Palmer D, Fricska S, Wehrmann B (2009) Towards improved land governance. Food and Agriculture Organization of the United Nations, United Nations Human Settlements Programme, Rome, Italy

Quizon AB (2005) Agrarian reform programs in selected Asian countries: an overview. ANGOC policy discussion paper. International Land Coalition, Rome, Italy

Raj KN, Tharakan, MPK (1983) Agrarian reforms in Kerala and its impact on the rural economy: a preliminary assessment. In Ghose A (ed) Agrarian reforms in contemporary developing countries. St. Martin's Press, New York

Rummel RJ (1977) Marxism, class conflict, and the conflict helix. In: Rummel RJ (ed) Understanding conflict and war, 3rd edn. Conflict In Perspective, Beverly Hills, Sage Publications, CA

SA (2013) Mixed fortunes for the daughters of Travancore. The Economist, 28 March 2013. http://www.economist.com/blogs/banyan/2013/03/women-kerala

Saturnino MB Jr, Cristóbal K, Lodhi AHA (n.d.) Agrarian reform and rural development: historical overview and current issues. Institute of Social Studies (ISS) and UNDP, The Hague, The Netherlands, New York, USA

Sharma A (2010) Rights-based legal guarantee as development policy: a discussion of the Mahatma Gandhi national rural employment guarantee act (MGNREGA). International Policy Centre for Inclusive Growth (IPC-IG) Poverty Practice, Bureau for Development Policy, UNDP, Brasilia-DF, Brazil

Singh S (2011) Contract farming for agricultural development in India: a small holders perspective. In: Proceedings of workshop on policy options and investment priorities for accelerating agricultural productivity and development in India, India International Centre, New Delhi, 10–11 Nov 2011

Singh G, Asokan SR (2005) Contract farming in India: text and cases. Oxford & IBH, New Delhi, India

Srinivasulu K (2002) Caste, class and social articulation in Andhra Pradesh: mapping differential regional trajectories. Working paper 179. Overseas Development Institute, London. UK

Subramanian KV (2013) Land reforms remain an unfinished business. Planning Commission, Government of India, New Delhi, India

UN Economic Commission for Europe and Food and Agriculture Organisation (2013) Voluntary guidelines on the responsible governance of tenure of land, fisheries and forestry. UN Economic Commission for Europe, and Food and Agriculture Organisation, Geneva, Switzerland and Rome, Italy

UNDP (2008) Status report: land rights and ownership in Orissa. UNDP, India

United Nations (2013) Millennium development goal 8: the global partnership for development: the challenge we face. MDG Gap Task Force Report 2013. New York, UA, United Nations

United Nations Development Group (n.d.) Thematic papers on the millennium development goals. United Nations Development Group, New York, USA

United Nations Human Settlements Programme (UN-HABITAT) (2008) Secure land rights for all. United Nations Human Settlements Programme, Nairobi, Kenya

Wickeri E, Kalhan A (2010) Land rights issues in international human rights law. Malays J Hum Rights 4(10). Drexel University Earle Mack School of law Research Paper; Fordham Law Legal Studies research Paper No. 1921447

Chapter 3
The Indian State and the Unfinished Task of Land Reforms: A Critical Analysis

Vertika and Valerian Rodrigues

3.1 Introduction

> Land provides not only economic sustenance but also plays a key role in enhancing the prospects of asserting citizenship in much of rural India. Thus, the issue of land rights and access to natural resources is one which must be envisioned not in narrow economic terms (e.g., a unit of production) but as a basis for larger well-being of rural people (Ministry of Rural Development (MRD) and Government of India (GOI) 2009, p. 7).

The Report of the Committee on State Agrarian Relations and the Unfinished Task of Land Reforms from the Ministry of Rural Development is reflective of an understanding of how crucial is the resource of land for the realization of democracy for the people of the country. But the trajectory of State action and inaction in terms of the access of this resource to the people hints to us that the democratic potential of the land resource has been overshadowed by its developmental potential. And the same trajectory has also shown that the developmental project for which land is sought to be employed comes into contradiction with the democratic project, which land was envisaged to fulfil right from the time of independence. This has come to light more than ever before when land as a resource has come under immense pressure and contestation for more non-agricultural purposes as the drive towards industrialization as well as urbanization is gearing up. At the same time, agrarian unrest is escalating against increasing land alienation as well as positively demanding right to land both in the form of violent and non-violent movements across the country. These movements are often seen as a bottleneck for economic development, but at the same time, they have

Vertika (✉) · V. Rodrigues
Centre for Political Studies, School of Social Sciences, Jawaharlal Nehru University, New Delhi, India
e-mail: vertika.lsr@gmail.com

V. Rodrigues
e-mail: valerianrodrigues@yahoo.com

H. Ha (ed.), *Land and Disaster Management Strategies in Asia*,
DOI: 10.1007/978-81-322-1976-7_3

exposed the contradictions of the path of the neo-liberal model of development adopted after the 1990s and the demands it places on land resources and the exigencies of electoral democracy (Levien 2013). *The puzzle for the state is really how to balance these demands against other powerful interests in land and cater to a concept of growth along with that of justice*. This is the broader research question which this paper is concerned with.

To understand the political nature of the issue better, one might briefly want to look at the issue of land reforms historically. Effectively, the land question came to an end by the end of 1970s after the consolidation of a State which had started withdrawing from structural problems of the poor. The classic legislations on land reforms were passed under Nehru and Indira Gandhi; however, the problem remained that the Congress party's rural base comprised of the landed elite. Therefore, the legislations never saw the desired implementation. In the early 1980s, structural reform policies were abandoned for specific anti-poverty programmes and the land reform issue was more or less abandoned (Maiorano 2014). Another reason for the State giving up on the issue of land reforms altogether can also be understood by the changing class structure and the nature of the economy post the liberalization programme of the 1990s. Lerche (2011) proposes a very interesting argument as explanation of failure of agrarian transition in the Indian economy:

> With economic growth in society progressing extremely well, driven by sectors that are quite independent of agriculture, no dominant class interests outside of agriculture are pressing seriously for a more productive or more broad-based agrarian development of the kind Byres argues would represent a *proper* agrarian transition (p. 116).

Thus, on the one hand, structural reform policies, which had been met with considerable resistance over the decades, were no longer the central concern of governments since the 1980s who were more interested in populist anti-poverty programmes, and on the other hand, agriculture as a sector in the Indian economy was declining since the booming of the service sector specially after the neo-liberal reforms in the 1990s, which further eroded interest in the task of land reforms. In this context, a document like the Report of the Committee on State Agrarian Relations and the Unfinished Task of Land Reforms is pioneering in the history of governmental interventions on the issue of land and to that effect is a comprehensive demand repository for people's movements struggling for land rights. It makes reference to the reopening of the land question in 1990s at the Revenue Ministers Conference 1992. However, as the Committee notes since then the issue has made little progress (MRD, GOI 2009, p. 61). The United Progressive Alliance I (UPA I) Government that came to power in 2004 was pressurized into raising this question again as it was dependent on the support of the Left. This happened in view of the democratic call for better social security legislation in the wake of increasing neo-liberalization of the Indian economy. This Committee constituted in 2008, under the pressures of a people's movement, the *Janadesh* Movement of 2007, made a thorough and honest investigation into the issue of agrarian relations with a special focus on the issue of land reforms. Interestingly, many of the members of this Committee include activists and academicians of Leftist orientation.

If we look at the whole Report as such and the issue of land reforms retrospectively, it can be noted that the problem with interventions so far has been a lack of a comprehensive policy which deals with the different issues of land reforms together. It would also not be an exaggeration to say that the State governments have taken undue advantage of the autonomy left to them in dealing with this subject. And that is why we have such large variations and inconsistencies in agrarian relations across the country. This Committee has taken up the massive challenge of attempting to look at all the major issues on land as such together and provided exhaustive recommendations, which could lead to a very much desirable "National Policy on Land Reforms".

The people's movement, as a result of which this Committee was formed, had also led to the formation of a "National Council on Land Reforms". This Council apart from involving members from civil society working on issues of land reforms, also included Central-level Ministers, nine Chief Ministers and was led by the Prime Minister. Since 2009, when the present Committee under review submitted its recommendations, this Council has more or less seen very sluggish progress. This again reflects the age old problem of land reforms which scholars like Bandyopadhyay (1986) and Radhakrishnan (1990) have attributed to a lack of political will. However, another wave of people's movement—the *Jansatyagraha* 2012 led by the *Ekta Parishad*—has finally pressurized the Government into action which today can be seen as materialized into the "Draft National Land Reforms Policy" released on 24th July, 2013. We will come back to this newly circulated policy at the end of this paper.

In the light of the above background discussion and the challenges land policy is facing in neo-liberal times, the *research objective* of this paper is to enquire whether this Committee's recommendations on the issue of land live up to the twin objectives of the Indian State and its development strategy—i.e. growth and justice. Keeping this broad question in mind, a set of sub-*research questions* which are of prime importance to the question of land in India today can be identified as (1) whether the democratization of land can be acceptable to the present political dispensation? (2) How far can the land distribution in reality transfer the land ownership to the landless? and (3) What can be the strategy to implement the land reforms?

These questions will be examined in this paper which is divided into the following sections: (1) Theoretical Framework and Research Methodology—this section will outline the theoretical basis of understanding the nature of the Indian State and thereby reading the Committee's recommendations. It will throw some interesting questions with regard to the Indian State and will brief the readers about the kind of research methodology employed; (2) Discussion of the Report in the Light of Existing Literature—this section will encompass the analysis, findings and discussion of the recommendations of the Committee Report with the support of existing secondary literature. The section will be divided into a couple of subsections based on how the Committee Report has structured the different clusters of land issues in India and lastly; (3) Conclusion—which will summarize the highlights of the discussion as well as test them against our initial research objective and research questions.

3.2 Theoretical Framework and Research Methodology

The theoretical framework adopted for analysis in this paper takes into account the special circumstances of Indian democracy which draws on a combination of Marxist analysis of the State as well liberal theory and a rights-based approach employed to understand representative democracies. If we think about this whole episode of the State responding to a social movement or rather attempting to respond to it, looking at what kind of recommendations the Committee has made, it will be difficult to sustain a Marxist analysis of the State having a class character, as here we find that the State is contemplating action which would very much go against one of the most powerful of the ruling classes—the rich peasantry and the landlord class. While the recommendations seem to be less antagonistic towards the industrial ruling class interests, still the State here seems to be making a conscious effort to actually understand the kinds of demands people are articulating and take due cognizance of them. Thus, though the recommendations may not be radical in the utopian sense, but they could be said to be in the direction of destabilizing the existing agrarian relations and extending social justice springing from a rights-based approach.

It is very important to engage with them thus as if actually realized they could mean profound transformation which would open more space and give a stronger voice to people to raise much more destabilizing questions. But as we noted in the introduction, looking at the actual inaction on these noble intents so far it is hard to credit the State with the above kind of analysis. It is then a State which commits to intentions for extending people's rights, but attempts to avoid acting on them tactically as long as it can. It is trying to respond to demands but does not know how to deal with the powerful interests which strongly oppose such demands. It is felt that the Marxist analysis of relative autonomy of the State where it mediates between different ruling classes is also not enough to explain this action and inaction of the State. Real politics of a democratic State demands more explanation. But whether the liberal framework of the State arbitrating as a neutral body between competing interests is enough to fill the gap is questionable as the State seems to be claiming to be on the side of people's interests, but never seems to live up to it. This kind of a State which makes more promises than it ever fulfils, it can be argued is something specific to the Indian State—a product of demands of Indian democracy coupled with the actual class structure of the State.

As far as the research methodology is concerned, this paper's discussion and findings are mainly rooted in analysis and critique of one primary Report with the supporting evidence derived from either other policy documents or secondary literature. The focus therefore is on trying to highlight what in the authors' opinion are important recommendations or points raised by the Committee which lead us to some of the vexing problems in contemporary agrarian relations in India, and adding critical comments wherever they are called for. Having said this, it must be stated that the Report as a piece is largely a pro-poor document which takes the agenda of making land rights realizable seriously. Thus, it is not the main concern in this paper to provide a comprehensive summary of the recommendations. The research objective of the paper is such that it does not necessitate primary empirical work as that is inbuilt into the Committee's Report being discussed. However, getting into data in a detailed manner in the paper from the Report is also not the approach followed as the Committee itself acknowledges that its primary field data—the questionnaires it sent to the states received very inadequate responses. The Committee itself used secondary sources to supplement its empirical data.

3.3 Discussion of the Report in the Light of Existing Literature

If one takes a broad synoptic view of the recommendations of the Committee as such, they do try to tackle the problem of not just better legislation on land reforms, but also policy formulation on better implementation of those reforms. They do this by advocating a democratic process of transformation in agrarian relations. The overall tenor of the Report is to argue for an increasing participation and role of people's bodies such as the *Gram Sabha*, *Gram Panchayats* and Village Land Councils in tribal areas for the process of implementation of land reforms. This says a lot about what is envisaged as pivotal to realizing the unfinished task of land reforms. It is indicative of the fact that the team of the Committee realizes that the magnitude of the task of bringing about a change in agrarian relations through land reforms is so vast that it cannot be performed by the State alone, and the people have been given a central place whether in terms of consent required for any activity or for carrying it out. It is a slight shift from a purely bureaucratic and a top–down approach to a democratic approach to change. This is one of the most commendable strengths of the Report. In what follows below, an assessment of the Report is attempted. The review is done broadly under the subheads of the sub-groupings of the Committee. (The related issues of three subgroups have been combined under the third heading.)

3.3.1 Land Ceiling and Distribution of Ceiling Surplus, Government and Bhoodan *Land*

The Committee argues that in spite of the initiation of land reforms in terms of ceiling limits, these were largely unsuccessful. In many cases, especially till the time right to property was a fundamental right, the large scale of litigation as well as *benami* and fictitious land transfers meant that actual land transfers did not take place. The Committee expresses the root of the problem as "under locked interest structure developed amongst large land owners, the elite including the village elite and members of the bureaucracy" (MRD, GOI 2009, p. 27). Therefore, the Committee notes there is ample potential for reopening the debate on ceiling limits. According to its assessment, large areas were not under consideration like the exemptions which were granted to religious, educational, charitable and industrial institutions. Moreover, the ceiling debate had totally ignored change in improvement of lands due to irrigation and increase in productivity (MRD, GOI 2009, p. 28).

On this basis, the Committee recommends a new ceiling limit of 5–10 acres for irrigated land and 10–15 acres for non-irrigated land, to be decided by the concerned State governments. And with respect to allotment, not more than two acres of wet land and five acres of dry land to be allotted (MRD, GOI 2009, p. 28). However, the Committee, considering it also has to take into account economic growth considerations, uses the argument of the higher productivity of smaller farms which has been highlighted in data presented by Food and Agriculture Organization of the United Nations (FAO) as Dogra (2002) argues. These data, however, sharing widespread consensus have been contradicted by scholars like Patnaik (1979). She argues that this argument comes from a neo-populist view which fails to adequately differentiate between family-labour-based farms and capitalist farms which run on hired labour. The first thing to be considered is that many of the smaller farms are so small that they are lower than the per unit area on the basis of which productivity is calculated. Large numbers of farms, which are used as data to substantiate this argument, are below one acre. Thus, she argues that there is an inflation of productivity involved in the conversion of their productivity into per unit terms. Moreover, even in cases where smaller farms per unit are showing higher productivity per unit than large farms which mostly hire in labour, it is because the small farmer uses family labour and has the option of cutting down the investment which the large farmer has to make in the form of market rate of wage. She argues that they also choose to often cultivate crops which require labour-intensive techniques rather than capital-intensive techniques as they would not have access to them. But their production is barely enough to meet their subsistence needs and the large farmer cultivates for profit. Thus, she argues there is a fundamental fallacy involved in arguing for an inverse relationship between farm size and productivity. This debate is also ignoring totally the vast differences in land quality and facilities like irrigation.

Her argument has also been echoed by other scholars like Bandyopadhyay (1996) who base it on the fact of economies of scale as a result of application of modern technologies of production which means that large farms are much more productive than small farms. M. V. Nadkarni argues something similar when he questions the rationality of an indefinitely continuing process of redistributing land as "under the new conditions of technology driven production, the small farmer risks losing the greater part of his working capital and yet if he cannot adopt to the new technology the viability of his farm is reduced further". He thus raises the question of real radical land reforms of Marxist variety wherein "private ownership and control of land is abolished all together and the people of village become members of the production unit on equal status" (Nadkarni 1976, p. A145). However, politically the Indian State cannot be expected to be favourably disposed to such a position. And that is why Bandyopadhyay (1996) makes the argument for land reforms on the desirability of socio-economic equity in spite of the above problems.

This argument, however, can be strengthened by the overall positive impact on economic growth land reforms have as some scholars have argued citing evidence through analysis of data across the nation (Deininger et al. 2009). Their analysis goes on to substantiate the argument of the negative relationship between initial levels of inequality and subsequent growth in developing countries. While indivisible investment leads to credit market imperfections on the one hand, widespread inequality generates social tension and strife which anyway undermines growth prospects on the other hand. They conclude that even though the impact of land reforms is difficult to be generalized across the nation due to inconsistencies one can argue that overall land reforms had a positive impact on economic growth as well as equity and accelerated accumulation of assets in the form of physical as well as human capital. However, the impact of land reforms decreases with time. One might look at some other scholars like Ladejinsky (1964) who explain this slack in growth due to land reforms often being unaccompanied by other development measures required to bring in consistent transformation such as availability of credit resources, better marketing facilities and technical assistance. Therefore, it can be argued that a focus on redistribution alone is not going to bring in increased income to the poor peasants. We will look at these factors in a little more depth in the next section, as there are other significant issues with redistribution itself which the Committee raises.

The Committee further notes, "It also found a number of ceiling land beneficiaries not in possession and that significant portion of the area declared surplus is either not fit for cultivation or not available for distribution due to miscellaneous reasons" (MRD, GOI 2009, p. 27). This is the most vexing problem which the recommendations of the Committee are unable to deal with adequately. The problem of fictitious land transfers under *benami* which prevents land from being available for distribution is dealt with by recommending amendments to the *Benami* Transactions Act of 1989 and introduction of a Card Indexing System to prevent land going under fake names. However, what does one do about the issue of land which is allotted but not under actual possession? Or worse still when the landless are formalized as benefiting from land distribution when all they got was uncultivable land?

The Committee tries to deal with this issue by introducing penal provisions for this evasion—"Penal provision for non-submission of returns for ceiling surplus holdings should be strict and rigorous. A penal clause inserted within existing ceiling laws should make officers accountable and responsible for intentional lapses" (MRD, GOI 2009, p. 73). Such penal provisions though can be helpful, but they would also be controlled by some authorities, who are as susceptible to collusion as the ones who are already doing so. If they are, then adequate reason should have been provided as to how this will help the situation. Such is not the case however.

With respect to allotment of Government land also a similar problem is noted where the Committee recommends that, "the list of beneficiaries in fresh assignment should be selected by the *Gram Sabha*" (MRD, GOI 2009, p. 30). While participation of the *Gram Sabha* can make the process more transparent, it has not been clearly laid out on the one hand, and on the other hand, the core issue here really is that of political opposition to the idea of land reforms by classes which are in ownership of private lands or which are eyeing Government lands for their own purpose or have encroached upon them. The problem noted with the issue of *Bhoodan* Lands is also somewhat similar. People donate lands which are a liability on them and hence of no good to the beneficiaries. The owning classes do not want their rightful share of lands to be forcibly taken away. Why should they allow this? And on account of the power and the support structure they have, they ensure that they ultimately do not allow it.

Moreover, post the Green Revolution, the landowners are now capitalists and do not see it as less than a crime if the State is expropriating their property which they have developed and taken care of and which gives them the rightful profits they deserve. It is akin to asking the industrial bourgeoisie to part away with some ownership rights of their industry so that those who are unemployed or perhaps the working classes could also have a share. Thus, in a system where private property is not in principle abolished as such, it is difficult to deal with this argument even in terms of rights. Nevertheless, if the State commits to it as a matter of policy which will involve a democratic coercion (Mansbridge 1996) in its implementation how will one deal with the grassroots violence of the powerful against the marginalized? Bandyopadhyay argues on the lines of, "providing a minimum self-defence mechanism to prevent illegal physical eviction from the assigned land through violence by the erstwhile owners" (Bandyopadhyay 1986, p. A53). How will such a process materialize is the question. He and other scholars like Rao (1992) and Joshi (1978) all argue in favour of organizing the rural poor and creating a situation of countervailing power against the already powerful.

The sad reality of Indian politics today is that the rural poor find it very difficult to organize themselves in the absence of support of traditional political parties and the gradual marginalization of the Left. Brass (1994) while writing on the new farmers' movements which emerged in the 1970s as an after effect of the Green Revolution notes that most of the scholars who have researched in different regions of the country on the new farmers' movements conclude that their class basis is the landowning rich farmer who far from being supportive, is rather antagonistic to the interests of the rural proletariat and agricultural labourers and in most cases, barring a few of these organizations, did not enjoy their support. In spite of this, the sheer mass of support and the power which the new farmers' movements came to exert politically, one can argue, are further evidence of the political marginalization of the rural poor. Lerche (2011, p. 117) argues that their struggles today are much rather linked to demanding social security policies from the State such as the "National Rural Employment Guarantee and the National Social Security Scheme [National Commission for Enterprises in the Unorganized Sector (NCEUS) 2006, 2009] and are not based on strategies to address the enduring agrarian question". It is quite evident, therefore, that the organization of rural poor is nowhere even near to provide a challenge to the powerful landed interests and the resistance towards redistributive land reforms. This is a very important dimension which the Report does not even touch upon, but perhaps it was not in its scope being a Government Committee with assigned terms of reference to address this issue. Nevertheless, it could have raised this issue, but even that is conspicuously missing.

3.3.2 Tenancy, Sub-tenancy and Homestead Rights

The central question on tenancy which the Committee takes up is the variations under different State laws on tenancy it being illegal in some States. As the Committee recognizes, "About 90 % of the leased area is informal and unrecorded. The landless and the marginal farmers constitute the bulk (91 %) of those leasing in land" (MRD, GOI 2009, p. 14). It argues that such evidence calls for relooking the restrictions imposed on tenancy and their productivity. It cites a study by Haque and Deininger which argues that restrictions on tenancy would mean lesser incentive to lease out land, in which case land could be even left fallow if the owner is unable to cultivate it, or tenancies could go underground which would mean a high degree of insecurity for tenants who would be illegal, would not be able to access much institutional support and would have low productivity (MRD, GOI 2009, p. 94). Koshy (1974) while critiquing the policy under the Second Five-year Plan which allowed for eviction of tenants for resumption of lands for "personal cultivation" demonstrates how it led to tenants being evicted on a large scale especially with the coming of new technology which made cultivation profitable. The Committee is also aware of the problem of the reverse concentration of land which the lease market often leads to as small farmers with very little land unable to cultivate often lease it out to take up other occupations. Rao (1992) highlights how this became a widespread phenomenon with the coming of capitalist cultivation under Green Revolution when new technology led to benefits of large farms generating higher produce and making it equally difficult for smaller farms to viably cultivate their farms. It is in this light that the Committee recommends that tenancy and leasing should be allowed only within ceiling limits (MRD, GOI 2009, p. 33).

While the Committee should be appreciated in making this quite progressive recommendation, it raises two issues—the Committee also recommends that the State fixing of rent should be done away with and tenancy should prevail at the market level of rent which it calls "fair rent". Moreover, it also argues for automatic resumption of land upon expiry of lease and does not raise the question of occupation rights of the tenants except in the case of *bargadars* of West Bengal (MRD, GOI 2009, p. 96).

The first problem with allowing market determined rate to prevail is that leaving it open to the market forces, it immediately will lead to an escalation in the value of land and hence automatically of rent. In the rural sector in today's times under conditions of land hunger, competition over land and increased demand for land, there is no reason to assume that a market of land leasing will have rents which will be fair as we must not forget it is not only the poor who are competing for leasing in land, it is also the middle and even the richer farmers. Presuming that ceiling limits are actually imposed on even operational holdings there is still large degree of inequity in the leasing in market. The second problem with this recommendation is a larger problem with the concept of rent as such. It is inherently stabilizing the existing system of private property through leasing out and resuming land on end of tenure. Under conditions of capitalist agriculture, where personal cultivation by hiring in labour is quite profitable, leasing out of land will be profitable only when the rate of rent is higher than the profit which the farmer is earning by personal cultivation.

So it is not clear with what intent the Committee made this proposal as against State fixing of rent—was the State rent fixed too high and the Committee assumed the market will give the peasants a better deal, or did the Committee feel that the State-fixed rent was low and was not giving proper incentive to owners to lease out land? Either way, market determined rent is not going to help the real landless labourers to become cultivators. And in the absence of the Committee raising the issue of a progressive realization of ownership rights to tenants, a tenant will always face various kinds of problems associated with non-ownership of land, the primary among them being availability of credit. Swamy (1980a, p. 4) argues that, "inadequate access to institutional credit is far more critical a deprivation now than before since it denies the benefits of technological change to the rural poor" Citing various examples he provides good evidence of how the central axis of availability of institutional credit is land, and non-owning cultivators therefore find it difficult to come out of poverty and have to resort to traditional exploitative institutions like the usurious moneylender and the landlord. He outlines why even cooperatives which rely on credit of those who have it to lend it, fail to provide access to the poor as the powerful dominate them. Even government taccavi loans rest on the condition of land ownership. (Swamy 1980b)

In this context, the question arises again and again why does the Committee not talk about occupancy rights of the tenants if it wants to ensure real dignity through ownership of land to the vast number of landless in the country? The cultivator's relationship to the land has been argued to be the most critical factor in determining the productivity of the land (Ladejinsky 1964) and that is why scholars like Ladejinsky (1964) and Hanstad (2005) have argued for security of tenure at the least and converting tenants to owners as far as possible to be the ultimate goal. Others like D. Bandyopadhyay have made more interesting suggestions in this regard. According to him, "All non-owner crop sharing tillers of land should be recorded, prescribing a fair sharing of crop at 75 % (for the tiller) and 25 %(for the owner) and they should have heritable rights of cultivation without title to the land. The moment such recorded sharecroppers get certificate of sharecropping, they would become bankable. It would infuse institutional credit to augment both production and productivity" (Bandyopadhyay 2008, p. 42).

The recommendations of the Committee on the issue of *Homestead Rights* are quite good which envisage funds from the Centre and State on 75:25 sharing basis for allotment of homestead cum garden plots preferably in the name of women (MRD, GOI 2009, p. 98). The Committee's recommendations can be productively supplemented by Hanstad's research where he is arguing that such homestead cum garden plots should be at least 3,000 square feet to be feasible and additionally the government should develop some infrastructure around the village such as road, drinking water and electricity to ensure a meaningful transformation in the lives of the people. According to him, the 17 million landless families in India would require less than 0.5 % of India's agricultural land to provide dignified homes to them (Hanstad 2005, pp. 40–41). It is noteworthy that the recommendations of the Committee on women's right to land are evidence of a high commitment to moving towards gender justice in land whether it is the issue of joint title to allotted property to be cultivated (MRD, GOI 2009, p. 23) or preference to leasing in of land for women's cooperatives (MRD, GOI 2009, p. 16). The Committee must be commended on bringing up the gendered aspect of land in a manner which can go a long way in changing the lives of rural women if taken seriously.

3.3.3 Policies Related to Land, Alienation of Tribal and Dalits Lands, CPRs and Conversion of Agricultural Land for Non-agricultural Use

The problem of land alienation is especially acute with respect to the tribal lands and dispossession of Dalits which is leading to increase in landlessness in these two marginalized sections. The issue is also of pressing policy concern as there is widespread conversion of agricultural land for non-agricultural use which raises the issue of food security apart from the issue of rights of the displaced and affected population. Large sections of Common Property Resources (CPRs) which traditionally sustain village economies and tribes are getting fast depleted. These are the broad issues under which the Committee tries to look at some of the State policies which have created these problems or have attempted to address them. Its recommendations argue for strict protection to the CPRs, forest lands and as far as possible agricultural land against diversion and acquisition especially for Special Economic Zones (SEZs).

The issue of tribal land alienation in this regard is one which the Committee has made a sincere attempt to address. It says that data shows that though tribals constitute only 9 % of the country's population they have contributed 40 % of the total land acquired so far (MRD, GOI 2009, p. 8). The Committee notes with regret that the *Panchayats* Extension to Scheduled Areas Act, 1996 (PESA) has not been seriously implemented in any of the States (MRD, GOI 2009, p. 9). This Act holds immense potential to grant people control over their resources as it grants the local people's bodies the power to control local economies and act as units of self-government. This Act, however, gains special relevance for the issue of land as it postulates that the *Gram Sabha* needs to be consulted before any land can be acquired (GOI 1996). The Committee points out, however, that this Act is applicable only to "scheduled areas but a large part of the tribal population lives outside scheduled areas. Therefore, the provisions of PESA should be applicable mutatis mutandis to villages/areas where there is a sizable tribal population or where majority of the population consists of Scheduled Tribes" (MRD, GOI 2009, p. 20).

While this is a very important recommendation and must be taken immediate cognizance of by all the States, there are some issues with the PESA which the Committee has not been able to bring out. Pal (2000) points out that there is a problem within the Act itself when at many places it authorizes the *Gram Sabha* and/or *Panchayats* for crucial purposes like making recommendations prior to grant of mining licenses or land acquisition. She argues that though it can be interpreted that the *Panchayats* are supposed to implement the decision of the *Gram Sabha*, yet the ambiguity in the clauses of the Act can lead to conflicting interpretations. Thus, even though the spirit behind PESA is based on participatory democracy which lies really in *Gram Sabhas,* the Act leaves room for this spirit to be violated. However, while the intent of PESA is surely a very empowering and radical vision, caution needs to be exercised while celebrating such consultations with *Gram Sabhas*. As some scholars note, such consultations can easily become social licenses for corporate companies who are interested in tribal lands for their projects as the *Gram Sabhas* do not have a right to refuse a project, and nobody is legally bound to either address the issues raised or accept their decision (Lahiri-Dutt et al. 2012).

Though the Committee recommends that the principle of land for land be made mandatory for all acquisitions of tribal lands and be strictly enforced (MRD, GOI 2009, p. 37) and amendments in the Land Acquisition Act be made to restrict acquisition of landed property in the tribal areas, (MRD, GOI 2009, p. 142) this does not take away from the fact that such acquisitions are taking place under the clause of "eminent domain" and that the tribals are strongly resisting such take-over of lands. However, it must be noted that after the enactment of the *Land Acquisition Act 2013*, the principle of land for land has been institutionalized in case of takeover of land for irrigation projects (MLJ, GOI 2013, p. 40) and special provisions have been made which safeguard the interest of tribals, for instance, the consent of the *Gram Sabha* has been made mandatory in scheduled areas (MRD, GOI 2013b, pp. 14–15). So even though till now private companies try their best to avoid the *Gram Sabha* consultations as the *Gram Sabha* may not be prepared to consent to the project at all, things are going to be more difficult for them now. Here, we can see a direct legislative impact of the recommendations of this Committee in the form of this revised law on Land Acquisition.

This brings us to the *larger question on land use* which the Committee repeatedly throws up. It is quite an interesting proposition which the Report (MRD, GOI 2009, pp. 45–46) makes—

> The Committee opines that the Land Reforms cannot be carried out appropriately unless there are *land use plans of village, states and the nation*. Such land use plan should capture the overarching concerns: ecological, food production, livelihood and allocating land for industry and development purposes. The land use plan can be developed and executed involving people, States and Central governments, and dedicated non-governmental organizations. Thus, absence of a long term perspective is the cause of land related contentions observed throughout the country. Furthermore, absence of long term land perspective on land and land use plan have led to improper recognition of common property resources in the country. This has also contributed to rampant conversion of agriculture land for non-agricultural purposes having detrimental effects.

This argument is a running thread throughout the Committee Report. It is interesting to note that the Committee takes up such an ambitious vision to a long-term solution to the current problem in development debates. It is also interesting that the Committee does not see this as a top-down bureaucratic exercise and in its larger tone attempts to make this also participatory. It is also aware that due to large variations in States and acute unevenness in land, a uniform policy may not be feasible but certain guidelines can be provided on the basis of which the States should undertake this exercise (MRD, GOI 2009, p. 175). For this purpose, it recommends revitalization of Land Reforms Council at the Centre and Land Policy Boards in the States as urgent (MRD, GOI 2009, p. 9). However, the question can be raised could such an approach again use the language of participation to legitimize and formalize a certain type of developmental agenda, as one does not need to cite evidence to say that the State's track record in terms of decisions of development has not lived up to participatory ideals even if it has committed itself to them in its policy.

Interestingly, it will be pertinent to point out here that in the context of the 2014 General Elections in India the mention of a National Land Use Policy has been made by the Bhartiya Janata Party (BJP) in its Election Manifesto. It announced that such a policy will be driven to balance the food production goals with economic goals of the country and its management would be by the National Land Use Authority at the Centre and its State branches at the State levels (BJP 2014, p. 28). Given the fact that this party has won a majority on its own in the Lok Sabha and given the support it has elicited from corporate houses during the elections, such a Land Use Policy might have dangers of foreclosing the land question and demands framed from alternate imaginations of development. A credible policy on the issue should cater to developmental purposes of both agriculture and industry, while at the same time playing a major role in ameliorating rural poverty by protecting indiscriminate transfer of agricultural land to private developers and for projects like SEZs. Is it not better for a democracy to have messy relationships which leave room for struggle and alternate visions to be raised rather than any kind of totalizing framework which would make thinking outside it a far too restricted possibility?

3.3.4 Modernization of Land Management

The existing land records are so antiquated and unreflective of ground reality that they have turned into instruments of exploitation of the weak rather than guaranteeing them enabling rights. The Committee places the *Gram Sabha's* role as central in the surveys of land records, and it places most of the rights to enforce those records in the *Gram Panchayat*. It argues for half-yearly or annual updating of land records. The records are to be approved by the *Gram Sabha* before publication and will be with the *Panchayat* for public access to them. In addition, it recommends social auditing of survey operations to check rent-seeking behaviour. To modernize land records, the Report makes exhaustive recommendations with regard to computerization of land records, digitization of maps and various other kinds of technology which can reduce the scope of manipulation and human error with respect to land records. A concept of Land Banks comprising of Self-Help Groups (SHGs) of landless workers is introduced which can make use of cultivable wastelands, and other unclaimed lands with rights to them on a priority basis (MRD, GOI 2009, pp. 144–202).

The Committee is also positive about the idea of *Nyaya Panchayats* (MRD, GOI 2009, p. 44) which Bandyopadhyay (2005) has been arguing for making the system of dispensing justice with respect to disputes in land records people-centric and participatory. It is evidence of the understanding of the Committee of the limitations of the legal system which moreover the people find difficult to access and which is not in a position to evaluate ground situations as well as the *Panchayats*. In the light of the various functions it entrusts the *Panchayats* to be responsible for, it also makes recommendations for the capacity building of their personnel and elected representatives in different skills required for maintenance of land records including skills in computation, data management and maintenance of records. It also outlines detailed recommendations for establishment of such training institutions (MRD, GOI 2009, pp. 200–201). All this would be possible as the Committee makes a case for bringing in revenue administration under the Planning Commission (MRD, GOI 2009, p. 24). It is due to this recommendation that the Committee feels there will be funds for implementation of the rest of its exhaustive recommendations on building up of a modern infrastructure for land management. However, having entrusted such great faith in village *Panchayats*, one should keep in mind that they can be dominated by locally dominant interests especially considering the caste hierarchy in our country. It is then that the powers vested in the *Gram Sabhas* which would actually make for moving towards what Joshi (1978) calls a restructuring of rural administration with shifting the power base into the hands of people and possibly empowering slowly the sections on the margins.

3.3.5 Land Management in North-Eastern States

The situation of the North-Eastern Region (NER) as the Committee notes is peculiarly different from the rest of the country. Traditionally, this region is characterized by more of a community ownership in land and governed by customary laws. The pattern of these customary laws moreover is very uneven across different States and within the States themselves. So the Indian State laws have a differentiated sphere of application, and the customary laws of the community have proper rights to co-exist. In fact, bodies like the Tribal Autonomous Councils (TACs) are formulated with the aim of creating a harmony between State administration and traditional tribal institutions (MRD, GOI 2009, p. 49). However, the Committee Report notes that these bodies have largely failed to perform the role they were supposed to and have begun to be dominated by political parties and State funds in their activities with respect to carrying out the process of development in tribal areas. With increasing effects of urbanization, there are also tendencies of increasing individual cultivation and ownership of land but these are not granted formal titles (*pattas*) to land. This creates a problem in their being able to derive the benefits of the formalized institutions (MRD, GOI 2009, pp. 49–55). The problem is especially acute with respect to the Land Acquisition case when rights to compensation are severely affected in the absence of formal rights to land (MRD, GOI 2009, p. 244).

Thus, the puzzle for the State has been, "whether to expedite privatization of landholding or to encourage community holding for a better management of the greatest resource base of the land and forest, given the contradictory forces operating within the community" (MRD, GOI 2009, p. 51). This Committee comes to the conclusion that protecting and further strengthening the community basis of the rights is very important as that is the form of organization of the NER and its weakening has led to erosion of self-governance and creation of an inegalitarian relationship (MRD, GOI 2009, p. 52). It can be said that the Committee took the most judicious position on the north-eastern States under its capacity and scope by making recommendations for more power to Village Land Councils (VLCs) in self-governing the land management system. The Committee also pointed out that the issues of the north-east are deeply complex and would require in depth research with formulation of problems to make more specific and concrete suggestions. Such an agenda was not under the terms of reference of this Committee, and thus, this issue is wanting for further comments.

3.4 Conclusion

Any conclusive comment on a Report which deals with most of the pressing issues concerning land in today's times is difficult concerning the vast canvas the Committee tries to cover. Having said this it must be noted that there are issues which the Committee deals with in some detail which have not been discussed in this paper. Two such broad issues are Forest Lands and Land Acquisition for SEZs. However, the reason behind this occlusion is that these issues involve more technicalities than the scope of the Committee itself could handle and have been much better addressed in the Forest Rights Act, 2006 and the Land Acquisition Act, 2013, respectively. Forest issues mainly concern the tribes, and the issue of land alienation of tribes has been discussed in some detail. Similarly, some relevant aspects of land acquisition have also been touched upon but a detailed discussion of these laws is out of the scope of this paper and was not really relevant to the overall issue which the paper is trying to deal with mainly—i.e. Land Reforms.

Therefore to sum up, one can argue that large parts of the recommendations of the Report are positive and can be said to have aspirations of democratization of agrarian relations. The extent of advocacy of participation by people's bodies like *Gram Sabhas* and *Gram Panchayats* in the process of implementing land reforms would impart great deal of agency to the people if the words on paper were to be taken seriously. It would also make the task of implementing land reforms much more decentralised and consequently practical. Thus by extension, it can be argued that this Committee tries to look at the unfinished task of land reforms as an opportunity to answer to the demands for democratization and succeeds in the same to a large extent. It does so while keeping in mind the demands of economic growth and efficient use of land as far as possible with issues like productive size of land holdings, market friendly legislation on tenancy, proper utilization of CPRs, modernization of land management and minimizing land acquisition of agricultural land.

On the other hand, the ultimate question which must be raised is does the discussion of this Committee having covered most of the issues concerning the politics of land today put a closure on the land question as such? Some comments were made with regard to this question while discussing the Committee's proposals on a National Land Use Policy. What could be said in addition to the fact that ascribing any kind of finality to the debate as it is now would be undemocratic in itself is that, this discussion on land rights is framed with the larger background assumption of a system of private property firmly in place. The question they are then dealing with is how to make it fairer. To be fair to the members of the Committee, it must be reiterated again that their analysis is quite sensitive to the poor and is quite radical in the sense of the strong case it makes for engendering land relations to increase access and hence agency.

But if we look at the overall Report with the kind of recommendations it has made it raises the question of endless legislation and monitoring of land under a private property system. Legislation is never divested of an inherent violence, and it is precisely that violence which will never allow the conflict on land to rest. The political situation across the country is nowhere near to accepting the radicalism of land reforms suggested within this Committee. Thus, the opposition to this attempt at changing relations will beget a great deal of more violence. The conundrum is that once again those who rest their hopes in the coercive power of the State to overcome this opposition are liable to get far less than even what the law proposes, due to their weak and marginal, almost helpless condition of political organization.

So should we instead talk about raising the issue of much more radicalized land relations with total abolition of private property to salvage the masses from this constant cycle of violence? If the political situation is not able to realize less radical land reforms, where will we find the possibilities of realizing a further radical agenda?

For the time being our best hopes lie in making a case of implementation of the lesser radical reforms as only that can empower the masses and provide them with capital to raise further questions. But as Lenin posed the pressing challenge of redistribution of land—it has the threat of making radical classes conservative and aligned with the system of individual private property as it encourages more private ambitions. That is what happened to the new class of farmers which benefited from the initial land reforms. Will the same happen again to the next class of dispossessed? And will it further differentiate the marginalized while benefiting some and leaving out the others? One can only speculate. But having made the criticisms with respect to policy, we will argue for immediate political commitment for realization of these recommendations with some modifications in actual policy. It will then provide us fresh grounds of analysing new problems and raising further questions. What is necessary, however, is the requisite political will by the State to overcome narrow class interests in order to implement such an actual policy of land redistribution in favour of the larger majority of the marginalized farmers and agricultural workers.

As argued before as well, recent policy initiatives taken by the Government of India are reflective of steps needed to take seriously the recommendations of this Committee. The recently passed amended Bill on Land Acquisition in 2013, specifically tries to implement the envisioned democratic decision-making with respect to any projects on Land Acquisition undertaken by the State whether on its own or in partnership with the private sector. It has introduced provisions for a Social Impact Assessment Study to be conducted by the *Gram Sabha* of the village where Land Acquisition is proposed to be implemented. The Report of the *Gram Sabha* has to be made available publicly at the end of which a Social Impact Management Plan has to be formulated by the Government which includes ameliorative measures with respect to the issues highlighted in the Report (MLJ, GOI 2013). However, the law off course grants some exemptions in cases of urgency specified in particular sections. Nonetheless, this gives us immense scope for empowering the people in decision-making and not just decision sharing and is a sincere attempt by the Indian State in balancing the twin objectives of growth and justice, in order to carve out *growth with justice*.

Another outcome of the Report which we have just analysed as mentioned in the beginning of this paper is the Draft National Land Reforms Policy released on 24th July 2013 by the Department of Land Resources, Ministry of Rural Development for purposes of discussion. A preliminary glance at the policy seems to suggest that most of the recommendations of the Report of the Committee on State Agrarian Relations and the Unfinished Task on Land Reforms have been accepted (MRD, GOI 2013a). But it will be an interesting and a very important exercise to undertake a detailed examination of this policy and the new Land Acquisition Act, on another occasion as the Government has finally been pressurised to begin the process of translating the long pending demand of just Land Reforms and a prospective Land Use Policy into action.

References

Bandyopadhyay D (1986) Land reforms in India: an analysis. Econ Polit Wkly 21(25/26):A50–A56

Bandyopadhyay R (1996) Global review of land reform: a critical perspective. Econ Polit Wkly 31(11):679–691

Bandyopadhyay D (2005) Nyaya Panchayats: the unfinished task. Econ Polit Wkly 40(51):5372–5375

Bandyopadhyay D (2008) Does land still matter? Econ Polit Wkly 43(10):37–42

Bhartiya Janata Party (BJP) (2014) Election Manifesto. Ek Bharath Shreshtha Bharat

Brass Tom (1994) Introduction: the new farmers' movements in India. J Peasant Stud 21(3):3–26

Deininger K, Jin S, Nagarajan HK (2009) Land reforms, poverty reduction, and economic growth: evidence from India. J Dev Stud 45(4):496–521

Dogra B (2002) Land reforms, productivity and farm size. Econ Polit Wkly 37(6):532–533

Government of India (GOI) (1996). The provisions of Panchayats (extension to scheduled areas) Act 1996, No 40 of 1996. http://hppanchayat.nic.in/pdf%20files/Pesa.pdf. Accessed on 15/04/2014, 24 Dec 1996

Hanstad Tim (2005) Improving Land Access for India's Rural Poor: a Review of Past Efforts and Recommendations for the Road Ahead. Rural Development Institute, New Delhi

Joshi PC (1978) Land reforms implementation and role of administrator. Econ Polit Wkly 13(39):A78–A83

Koshy VC (1974) Land reforms in India under the plans. Soc Sci 2(12):43–61

Ladejinsky W (1964) Agrarian reform in Asia. Foreign Aff 42(3):445–460

Lahiri-Dutt K, Krishnan R, Ahmed N (2012) Land acquisition and dispossession: private coal companies in Jharkhand. Econ Polit Wkly XLVII(6):39–45

Lerche J (2011) Review essay: agrarian crisis and agrarian questions in India. J Agrarian Change 11(1):104–118

Levien M (2013) The politics of dispossession: theorising India's "land wars". Polit Soc 41(3):351–394

Maiorano D (2014) Continuity amid change in India's political economy from 1980 to 2004. Econ Polit Wkly XLIX 44(9): 44–54

Mansbridge J (1996) Using power/fighting power: the polity. In: Benhabib S (ed) Democracy and Difference: Contesting the Boundaries of the Political. Princeton University Press, Princeton, pp 46–66

Ministry of Law and Justice, The Gazette of India (MLJ, GOI) (2013) The right to fair compensation and transparency in land acquisition, rehabilitation and resettlement Act, 2013, No 30 of 2013. New Delhi, 27th Sep (2013)

Ministry of Rural Development (MRD), Government of India (GOI) (2009) Department of Land Resources. Report of the Committee on State Agrarian Relations and the Unfinished Task in Land Reforms, New Delhi

Ministry of Rural Development (MRD), Government of India (GOI) (2013a) Department of land resources. Draft national Land Reforms Policy, New Delhi

Ministry of Rural Development (MRD), Government of India (GOI) (2013b). The right to fair compensation and transparency in land acquisition, rehabilitation and resettlement Act, 2013, An Overview. New Delhi, 8th Sep (2013b)

Nadkarni MV (1976) Tenants from the dominant class: a developing contradiction in land reforms. Econ Polit Wkly 11(52):A137–A146
Pal M (2000) Panchayats in fifth scheduled areas. *Econ Polit Wkly* 35(9)19:1602–1606
Parishad E (2012) http://www.ektaparishad.com/. Accessed on 15/04/2014
Patnaik U (1979) Neo-populism and marxism: the chayanovian view of agrarian question and its fundamental fallacy. J Peasant Stud 6(4):375–420
Radhakrishnan P (1990) Land reforms: rhetoric and reality. Econ Polit Wkly 25(47):2617–2621
Rao VM (1992) Land reform experiences: perspective for strategy and programmes. Econ Polit Wkly 27(26):A50–A64
Swamy D (1980a) Land and credit reforms in India: Part One. Soc Sci 8(11):3–13
Swamy D (1980b) Land and credit reforms in India: Part One. Soc Sci 8(12):46–64

Chapter 4
Cacophony to Euphony: Land Tenure System and Food Security in Pakistan

Sarfraz Khawaja

> *Hunger is exclusion—exclusion from the land, from income, jobs, wages, life and citizenship. When a person gets to the point of not having anything to eat, it is because all the rest has been denied. This is a modern form of exile. It is death in life…*
>
> Josue de Castro

4.1 Introduction

The focus of this paper is to analyse the land tenure system in Pakistan and how it impinges on the food security in the country. The unjust and massive land holdings by absentee landlords directly and negatively affect the yield of major crops, wheat and rice. It is not an agriculture problem alone, rather it involves the complexity of land distribution system which contributes to food deprivation of marginalised population which constitutes about 30 % of the population (total population 180 million) living below the poverty line. The changing climate and weather patrons also pose challenges to an already mal-governed system. However, this paper would mostly address the issue of land tenure and food security in the context of governance in peasant and landlord equation. The performance of agriculture sector has been impressive

> Despite the fairly robust performance of agriculture over the 1990s, where its growth rates stayed above the population growth rates and were fairly reasonable, rural poverty increased significantly over the same time period. In 1990-91, rural poverty, according to government Head Count figures, stood at 25.2 percent. By 2000-01, the figure had gone up to 38.99 percent. This is a massive increase of almost 15 percent (Bari 2003, p. 13).

Such a situation has been the reflection of poor governance devoid of equitable distribution system for better access of food to the unreached. Pakistan has been considered as an agriculture economy though it still imports food. The total

S. Khawaja (✉)
Civil Services Academy, Lahore, Pakistan
e-mail: sarfrazshk@yahoo.com

H. Ha (ed.), *Land and Disaster Management Strategies in Asia*,
DOI: 10.1007/978-81-322-1976-7_4

geographical area in the country is 196.7 million acre (2.47 acre = one hectare). The area suitable for agriculture is about 72.7 million acres, and the cultivated area is 54.2 million acre (inclusive of irrigated + Barani areas). Out of this total area, about 48.2 million acre is under irrigation system and Barani area is 4.0 million acres. There is potential to bring 20.3 million acre of land under irrigated agriculture (Agriculture Statistics of Pakistan 2009, p. 2). It could happen if irrigation system is to be extended and made efficient.

The hypothesis of this research is that the efficient governance of land would contribute to high production of crops and strengthen the financial position of small farmers, and food availability would be facilitated to the unreached. What happened in Russia when the land reforms were undertaken with the purpose to empower the land labourers and peasant 'But all these things combined to bring about sweeping rearrangements in the relationship between the landlords and the peasants who rented their land. The final outcome was serfdom for the mass of Russia's rural population. The origins of this transformation of a free peasantry into serfs bound to the will of their masters' (Blum 1971, p. 219).

Land tenure system and property rights are critical aspects that undermine the food security in Pakistan notwithstanding the systemic corruption, chronic poverty and rural discontent. The landed aristocracy consists of about 15 % of land owners' class but has vast holdings of 85 % of land. The population of tenant farmers and labourers is 85 %, but their share of earnings is 15 % of agriculture income. As a result, the tenants and labourers live in persistent and dire poverty. This imbalance between the landlords and tenants in terms of land holdings and earnings leads to discontent and deprivation. Food insecurity in terms of physical availability, access and distribution is the direct and indirect result of land tenure system. This has to be addressed through the amendment in the land tenure system and good governance. The government has no priority either at policy or plan levels to address the issues. The quick fix solution of food subsidy is neither sustainable nor manageable.

4.2 Research Questions

The objective of this chapter is to address the following research questions.

1. How the existing land tenure system directly or indirectly impinges upon the food security?
2. What specific issues to be addressed and measures to be taken to mitigate the problem of food insecurity?

4.3 Right to Food

The central theme of the paper is that the provision of food security is not a charity by the government or civil society. It is an issue of human rights, and right to food is mandatory for every living human being. It is obligatory for the state to fulfil this commitment. The Universal Declaration of Human Rights (UDHR—Article 25(1), 1948) clearly mentions the right to food as

1. States must refrain from violating the right to food, e.g. discrimination against women and forced displacement from lands. This can also be used for the international perspective—governments of developed countries should not knowingly violate the right to food of citizens in other countries through, e.g. trade rules.
2. Third parties must refrain from violations, and government has a duty to regulate third parties—this relate the corporate violations, regulatory frame work and to armed opposition groups.
3. States must take positive action to progressively realise the right to food for all—relates to food security policies, trade issues, etc.

The right of food is set out in the International Convention on Economic, Social and Cultural rights (ICESCR—Article 11.1, 1966), the convention on the Elimination of Discrimination against women (CEDAW Article 14g/h, 1979) and the convention on the rights of child (CRC Article 24C, 1990). The commitments made in UDHR, ICESCR and CRC were reiterated by UN General Assembly in February 2002 that right to adequate food and the fundamental right to everyone to be free from hunger so as to be able fully to develop and maintain their physical and mental capacities (UN General Assembly Resolution No. 56/155, 2002).

In the 1973 Constitution of Islamic Republic of Pakistan (Pakistan Constitution Article 13(d), 1973), several provisions of different articles re-emphasise the implementation of human rights and ensure provision of basic necessities of life including food for the citizens of Pakistan.

4.4 Methodology

The methodology involves the review and analysis of relevant literature. These include the reports of specialised international organisation such as Food and Agriculture Organisation (FAO) of the United Nations, World Food Programme (WFP), World Bank in addition to the local documents include reports and records of Poverty Alleviation Fund (This is a Pakistan Organisation), National Rural Support Program (NRSP), Pakistan Economic Survey, Agriculture Policy and Five-Year Development Plans.

Methodology also includes some interviews with key stakeholders including big landowners, small farmers, civil servants dealing with Agriculture and Food Ministry and visit to an urban slum where very poor population lives.

4.5 Literature Review

4.5.1 Land Tenure System

Historically, land tenure system in Pakistan carries the baggage which rulers of India left for subcontinent. They won the military and political support of the recalcitrant principalities. The local chiefs or *sardars* of far-flung Indian states became very powerful. The land given to them to administer and collect revenue became their property, and they obtained the ownership rights of the land. The distribution or bestowing of land by the Mughal rulers made them economically prosperous and politically arrogant. The *sardars* also acquired land and manipulated emperors under the artificial threat of rebellion at local or state level. India had been such a vast empire that it was not possible to control its far-flung principalities without the support of local chiefs. The Mughals not only lost their glory but also their will and vigour to control India as its rulers because of their internal discords and lack of proper policy for people at large and finally tried to create the images of grandeur by constructing buildings of unparalleled and unique architecture. The land tenure system was the outcome of this backdrop. The irrigation system for crop fertilisation, the communication system for transportation and trade and the useful contacts with the foreign powers, mostly European, were attracted and visited India to see its grandeur. The Taj Mahal is a testimony to this grandeur. There are hundreds of high-quality buildings designed and constructed all over India in the form of forts, mosques, tombs, etc.

The British were no different than Mughals as far as their strategy to control far-flung areas of India was concerned. They extended and gave huge areas of land to local chiefs with the key objective to win their support and minimise the chances of rebellion. The relationship between and among the different rulers of Indian states also played an important role in devising the land tenure system. Although the British rulers in the beginning did not forfeit the land ownership rights, soon they realised that it was not possible for them to collect revenue from Indian states; thus, they modified their policy and became more flexible in granting land rights to *sardars*. To win the political support of different chiefs/sardars, the British distributed the land directly to those influential who proved their loyalty to the British rulers. This practice of land distribution continued during the British rule in India till 1947 when the partition of the subcontinent was mutually agreed among the Hindus, Muslims and British notwithstanding the horrible bloodshed, looting and hatred on both side of border between India and Pakistan.

4.5.2 Types of System

In more specific terms, there have been three kinds of land tenure system prevalent in Pakistan.

1. *Ryotwari* System
2. Single-Ownership Cultivation System
3. Joint Ownership Cultivation System

1. Ryotwari System
   In this system, government distributes the land among farmers with fixed-term time duration. The ownership rights are also transferred to the farmers with tacit approval of the state to transfer this land to their heir if need be. The system was devised because the land was in abundance and cultivators were not available in sufficient number to till the land. The government encouraged the cultivator to grow crops on the land which was not owned by the cultivators. A nominal rent was paid to the state establishment. This system was devised to bring maximum land under cultivation mostly in Sindh and Punjab where the land was in abundance. Although the Ryotwari system was legally abolished in early 1940, many cultivators occupied the land and became landlords. Such land was cultivated by tenants known as *haris* with no legal protection of their tenancy rights.
2. Single-Ownership Cultivation System
   This is the most prevalent system in Pakistan. One private individual or one family owns huge amount of agriculture land. This ownership could run into thousands of acres consisting of several villages, settlements and *taulkas*. The bread and butter of the residents of these areas directly or indirectly depended on their loyalty and subjugation of the landlord. As a common practice, the landlord/family of the landlord retains some land for cultivation by themselves. But major portion of land, about 80 %, has been distributed into small parts for cultivation by the tenants. The ownership of the land remains the exclusive right and privilege of the landlord. This has been done for two main reasons: first, to keep the exclusive control and dominance on the land and secondly to keep the tenants subjugated not only in financial terms but also control them in making social and family decisions about education, employment, marriage, etc., of their children. There are two kinds of tenants. One consists of these, whose forefathers were also tenants and provided greater support to the landlord for cultivation of land. They are called 'occupancy tenants' because they continue to occupy land and maintained steady and confident relations with landlords. The other category is the 'tenants at will' which means that landlord has the exclusive and absolute authority to eject the tenant at will without any reason and loss of income. The number of such tenants is quite large.
3. Joint Ownership Cultivation System
   In this practice, the real owner is the state. The land is the property of the state but is distributed among the influential and big lords. This middle tier of landlord plays an important role in the society. On the one hand, government wins the loyalty of this small, but very powerful group and simultaneously they gave nominal revenue income to the government. Most of the members of this group play dominant role in the politics of Pakistan, and most of them have become members of legislative assembly at national and provincial levels.

Secondly, these landlords/politicians give land to the tenants/haris on contract on year-to-year basis or in some cases for longer period where the tenant enjoyed a greater confidence of the landlord. These tenants remain at the mercy of landlord and could be ejected at any time. This state of flux never motivated the tenants to use modern agriculture techniques, make more investment on land, and use trained labour, better equipment, methods of market economy, and improved fertilizer to enhance yield. Hence, the power is concentrated in the hands of those small elite groups of landlords and they accrue benefits from the government by using the state land and exploiting tenants to get more crops from them.

Analysis

One Land tenure system in Pakistan is dominated by absentee landlords. The land is concentrated in the hands of feudal and landlords. The tenants have been exploited and deprived of their due share of income. The fragmentation of land has become dominant as tenants are given small portion of land for cultivation. About three-fourth of the land is cultivated by tenant farmers and one-fourth by the owners.

Two Exploitation of tenants is dominant because of the absence of any land tenure rights for them. They remained unskilled labourers; control and decision-making in their lives mostly rests with landlord because of the fear of ejection. The landlords charge higher rents, make no investment in land in the shape of better equipment and training of tenants, and maltreat the tenants. This is being done on the basis of the collaboration between landlord and state authority. Such power was either bestowed or obtained by landlord from the government. The agriculture labourers do not have other economic opportunities as the industrial sector is unable to provide them employment because they neither possess necessary skills nor basic education.

Three The uneconomic farming units in Pakistan is also responsible for lower agriculture productivity; that the lands are subdivided and fragmented, the law of inheritance mentions that land of deceased owner is divided amongst his heirs; small farmers also are busy in sale and purchase of small pieces of land which means division of land into smaller segments which are uneconomical as agriculture units.

4.5.3 Food Insecurity and Land Tenure System

In Pakistan, agriculture contributed about 21 % of GDP during 2012 and provided employment opportunities to 45 % people and sustained 60 % of rural population (Finance Division 2012, p. 17). Of the 15 million jobs created between 2000 and 2009, more than a third were in agriculture and the remainder in services and manufacturing. Only 21 % of women compared with 82 % of men participated in

Table 4.1 Production and yield of wheat

Year	Area		Production	
	('000 ha)	% change	(000 tons)	% change
2007–2008	8,550	−0.3	20,959	−10.0
2008–2009	9,046	5.8	24,033	14.7
2009–2010	9,132	1.0	23,311	−3.0
2010–2011	8,901	−2.5	25,214	8.2
2011–2012	8,666	−2.6	23,517	−6.7

Source Economic Adviser's Wing (2012), p. 22

the labour force (Hussain 2012). The proper use of land is to utilise it to banish hunger, malnutrition and make it efficient and productive sector of economic viability. But 25–30 % of the population in Pakistan is living below the poverty line despite increase in GNP which is US$1400 (Hussain 2012). About 30 % of the population is malnourished. The key question would be as to why the food security is declining in Pakistan and how it is linked with the existing land tenure system?

Pakistan has population rate of 2.2 % which means a net addition of 3 million persons every year. During the period from 1948 to 2010, population has increased fourfold, but the production of wheat and other major food crops has increased to 2.9-fold. The country's consumption requirement of wheat is approximately 26 million tons per year, and the production is 23 million tons (Hussain 2012). The production of wheat decreased in 2011–2012 (Table 4.1). The smuggling of wheat to neighbouring countries and poor management of storage and distribution make the situation complex. Good governance system needs to be implemented.

4.5.4 Capacity to Purchase Food

There are three variables of food insecurity which has direct correlation with land tenure and agriculture output. First is the physical availability of food. It is thought that enough food is available in the country for entire population. Second is the access to food which is assumed that marginalised population has enough buying power or resources to purchase food for consumption. Third is the distribution aspect which means that faulty management has not deprived the rural poor and urban slums to obtain food stuff which is otherwise available but does not reach to the poor. The issue of availability is less serious, notwithstanding the poor policies of government to export wheat in 2007 and import wheat of about 3 million tons in 2008. The real problem is the purchasing power of the poor population and distribution aspect. The declining purchasing power of poor is reflected below.

Table 4.2 demonstrates that 31.5 % of the income is being spent on the consumption of wheat/floor alone. Other necessary food items for household cannot be ignored, i.e. vegetable, lentils, sugar, milk, etc. Then, there are other important

Table 4.2 Expenditure incurred on wheat consumption

Year	Consumer price of 1 kg of wheat/floor (Pak Rupees)	Inflation rate (%)	Monthly minimum average earning by unskilled worker (Pak Rupees)	Average family size	Monthly minimum average consumption/cost of floor per family
2007–2008	25	7	5,000	5	
2008–2009	27	8	7,000	5	
2009–2010	30	9	8,000	5	
2010–2011	30	8	9,000	5	
2011–2012	35	8	10,000	5	90 kg/Rs. 3,150

Note This information has been extrapolated/compiled from different sources which include government reports, small-scale research studies done in universities and some media reports

items, i.e. utility bills, house rent, education, medicine, transport, etc. The 'utility bills', house rent and medicine are such items on which the expenditure is unavoidable. The crunch would be on food, education and transport. About 30 % of the population is living at this subsistence level which made them food insecure.

The production of rice in the country is declining because of land fragmentation ignoring the need of agriculture farming. There is sufficient data available which determines that Pakistan is food insecure. The structure and efficiency of land tenure system determines the level of food security. In the matrix of land tenure, the first variable is the ownership of land which means who owns it and what is the nature of ownership? Secondly, what are the conditions of cultivation of land, whether cultivation is done by landlord or the tenant? Third, what is the position of state regarding the ownership of land and the formula or condition of crop distribution between the owners and cultivators? In the prevailing land tenure system in Pakistan, the landlord possess unlimited powers and keep the tenants at their mercy not only in economic terms but also socially in making family decision regarding the education, marriage and employment of their children. An efficient irrigation system is a prerequisite for higher production of crops and enhances the capacity of agriculture sector.

4.6 Findings and Discussion

With the globalisation of the market economy and curtailing of subsistence agriculture, food security depends on the capacity of purchasing power of consumer. It is those without purchasing power who will go hungry amidst a world of plenty.

As a result of the land tenure system

> the majority of Pakistan's rural poor are neither tenant farmers nor farm owners. Farmers (including both owners and tenants) comprised only 43 percent of households in the

bottom 40 percent of the rural per capita expenditure distribution in 2004-05. Non-farm households (excluding agricultural labourers households) accounted for slightly more than half (52 percent) of the poor. Overall, agriculture (including both crop and livestock production) accounts for only about 40 percent of rural household incomes; the poorest 40 percent of rural households derive only about 30 percent of their total income from agriculture (Arif 2004, p. 13).

4.6.1 Poverty

Poverty is the mother of all deprivations. It impacts the purchasing power of poor people and is a major hindrance to access enough resources to produce or buy quality food. Poor farmers use conventional farming techniques, do not afford fertiliser and labour-saving equipment and above all possess very small farms. All these factors limit food production; hence, it has direct implication on poor farmers. It is not only the agriculture production which is affected negatively but also the cash strapped suffered more of incidence of disease, access to education opportunities and other basic needs.

4.6.2 Non-equity of Food

During 1987–2007, food poverty incidence in the country demonstrated that about one-third of the households were living below the food poverty line and they were not meeting their nutritional requirements. The incidence of food poverty is higher in rural areas (36 %), than in urban areas (26 %). Urban and rural areas, however, did not differ much in terms of calorie intake per capita, and the differences across the four provinces were also not substantial. The problem lies in the non-equity of food distribution within each of these categories and even within the members of the household.

4.6.3 Water Needs

The availability of sufficient water is directly related to the output of agriculture. The production of 1 kg of wheat needs an average of 100 L of water and similarly 500 L of water for producing 1 kg of rice. The small farmers and tenants do not get their due share of water for varying reasons—i.e. inefficient irrigation system and highhandedness of big landlords (Food Security 2004).

4.6.4 Calamities

Disasters in the shape of frequent floods or droughts because of changing environmental conditions and cyclones destroy large amount of food. The conflicts in the form of war, ethnic violence, border skirmishes and terrorism also destroy food production and storage. The conflict between opposing groups or soldiers also has negative effect to the growth and access of food. The destruction of food production and storage is also a consequence of earthquakes and other calamities.

4.6.5 Gender

Agriculture and food production is generally considered as a male domain, but the role of women as food producers, processors, traders and income earners plays an important role for the health of the family. But the lower social and economic status of women deprives them to access education, training, land ownership, decision-making and credit. All these factors limit the role of women to improve their access to and use of food.

4.6.6 Health

Health factor impinges directly on the performance of agricultural production. An unhealthy labour is less able to make use of the food that is available. A sick or hungry mother gives birth to underweight children and subsequently frequent illnesses. Poor health has direct correlation with learning disabilities, reduced resistance to extreme weather and above all loss of working hours every day.

4.6.7 Food Security and Agriculture

Agriculture growth is necessary but not sufficient condition to mitigate food insecurity. Such findings have been suggested in the World Bank report (2007). However, agricultural growth remains an important factor to raise the income of small farmers and minimise food insecurity. Agriculture growth has played a major role in the country's development and continues to be crucial for overall growth and poverty reduction. The report on rural poverty suggests the following options to reduce poverty:

- Progressive public policy and effective implementation would promote efficient and sustainable agriculture growth. The incomes of small farmers would increase and improve the purchasing power. It would generate growth linkages in the rural non-farm economy.

- Develop and support an enabling environment for the rural non-farm sector to enhance employment and incomes, and improve rural public service delivery infrastructure. The importance of health and education sectors has to be developed to serve as a foundation for growth and to increase household welfare and food security.
- The governance of rural institutions has to be improved through decentralisation and strengthen local demand for enhanced accountability. For the establishment of good governance, the key indicators are transparency, accountability, participation and predictability.
- Most vulnerable sections of the society are marginalised and poorest of the poor. They must be empowered and protected through social mobilisation, safety nets and by facilitating access to productive assets for income-generating activities for poverty reduction and food security.

4.7 Conclusion

The reform in the land tenure system or the redistribution of land is not the Panacea for food security. It is neither a philosophical debate between neoliberalism and landed aristocracy. The argument is that the prerequisite for food security is the availability of sufficient food, and secondly, and more important is the distribution of food items on equitable bases. This means that the poor and marginalised population would have either sufficient cash to purchase food or there must be some system that ensures that they get the food as per the requirement of family. The requirement is usually determined by the daily calories intake of each member of the family. It means that growth and development have to be marinated in such way that everyone gets the sufficient share of pudding for daily intake. But this can only happen if enough food is available, and the first step in this direction is the reform and modernisation of land tenure system.

In this context, a hotly debated issue during the Indian election of April/May 2014 is the 'The Gujarat Model' of development. It has been analysed that the Gujarat Model is viewed in the light of the wave of neoliberalism 'What does it (neo-liberalism) mean? The only interest of most capitalists is to maximise their profits regardless of the damage they do to the economy' (Hensman 2014). This argument has been further elaborated that 'Hundreds of thousands of farmers, agricultural workers, fishermen, and Dalits were displaced and land was allocated to corporations. According to some estimates, 16,000 workers, farmers and farm labourers committed suicide by 2011' (Nasir 2014).

The eradication of malnutrition and ensuring food security is an important element of a debate going on between the economists. One group focuses on growth, while the other group, led by Amrayta Sen, Noble Laureate in economics, stresses 'inclusive growth and redistribution' (Nasir 2014). Adequacy of food supply, access to food and equity of food distribution require special interventions and enabling environment for ensuring household food security.

The availability of enough food as per the needs of each member of a family is necessary but not sufficient for the improvement in the quality of life because 'Quality of life as an approach emphasised the capacity of individuals to develop their capabilities to overcome constraints to achieving systemic goals. The quality of life approach was seen to place the individual at the centre of development' (UN Economic and Social Commission for Asia and the Pacific 1990, p. 62).

The policy and strategic options to reduce food insecurity can thus be concluded:

- **Sustainable and efficient utilisation of the natural resources**
  - **Land**: Land-related problems like depleting soil fertility, soil erosion, water logging and salinity need immediate attention to improve and maximise the yield of crop.
  - **Water**: The efficient and effective use of irrigation water is most crucial to the future of Pakistan. It is also necessary that the government should plan for the future needs, taking into account agricultural, domestic and industrial demand of the future and impact on the environment (UN Statement on Food Security in Pakistan 2000).
- **Efficient use of inputs**
  - **Seed**: Improved seed is one of the important factors in crop productivity enhancement, but unfortunately, this category of seed coverage is within low range of 14–20 % for wheat and other major food crops.
  - **Fertiliser**: Proper use of organic and inorganic fertilisers is also critical for maintaining soil fertility to enhance agriculture productivity. Fertilisers should be used according to the site-specific requirements of crops.
  - **Credit**: Due to the financial limitations, the small farmers are largely dependent on credit to procure agriculture inputs. The existing credit procurement system is complicated and not in easy access to small farmers. There is an immediate need to simplify such procedures (UN Statement on Food Security in Pakistan 2000).
- **Production of major food crops**
  - **Wheat**: The yield range for the wheat is in the range of 0.8–5.5 tons/ha. This clearly indicates that through improved management, the gap can be reduced, and correspondingly substantial increase would be obtained in the production.
  - **Rice**: The yield of rice is about 2.0 tons/ha against expected yield of 2.56 tons/ha. This increase in yield could be easily achieved through good management practices at farm level. Rice productivity can be increased through increasing plantation intensity, mechanised transplanting and reducing post-harvesting losses.

- **Oil seed**: The local production of the oil seed is low, and the country imports a large quantity to fulfil the demand of the fast-growing population. There is a need to exploit the potential of the oil crops by creating more attractive and conducive environment (UN Statement on Food Security in Pakistan 2000).

- **Enhanced income generation**
  There is better potential to increase the yield of cash crops mostly vegetables by using better seeds, ensuring the water availability and preventing the extinction of agriculture land for making housing societies in the name of development. This approach would make the availability of essential food items specially vegetables more affordable to be sold in the nearby semi-urban areas and cities. This would also increase the purchasing power of women who can afford to work in the nearby fields close to their homes. The cash crops grown in excess of the needs of closer proximities would be sold in urban areas which could be helpful for more cash-strapped population.
- **Imbalance of resources management**
  There is a constant struggle on the part of small landowners and small farmers to get their fair share of resources which includes water, fertiliser, pesticides and credit, but the big landlords take the bigger share much more than the size of land they own. It means that financial and land resources have been skewed in favour of big landlords. In addition, there has been unjust resources distribution which reflects the condition of the poorest of the poor. An equitable distribution of resources would facilitate the overall growth of agriculture in the country, and the deprivation of the poor would be minimised.

4.8 Mitigation Strategies

One of the major challenges of the developing economy of Pakistan and prevailing system is to bring changes and modify necessary laws. This is a great challenge but quite possible and has been done by many countries, and the best example is that most of the countries in East Africa region (Malaysia, Indonesia, Singapore, Thailand) demonstrated this change efficiently and effectively.

4.8.1 Paradigm for Change

There are five key elements for bringing a meaningful and successful change.

First is POLITICS: The first major step is the enunciation of policy by the government. Policy formulation is a political activity which is undertaken with the help of professionals. But decision regarding the policy provisions is usually undertaken by politicians in most of the countries. Hence, it is pertinent that any change in the system or a sector has to be carefully crafted, taking the key stakeholders on board. The announcement of policy does not mean much unless it is owned by the political decision-makers at all levels of political hierarchy. Once this step is adequately and appropriately managed, then the time comes for its legislation.

> A favourable political environment is very essential to the speed and effectiveness of the economic reform process as rapid achievement of the goals of economic liberalisation depends upon the sequential implementation of a range of interrelated policy measures. Bangladesh had reached the second phase of political consensus on reforms in the 1990s and the first stage of convergence on reform policies (Palit 2007, p. 69).

Second is LAW: Policy needs legislation which requires the approval of legislative body to make it a law. The policy enactment is primarily a decision about its constitutionality that extents the authority for implementation by the government and other stakeholders. Sometimes, the decisions of the superior judiciary are also sought for legitimating purposes. Policy implementation requires government structure, legal sections, financial resources and executive authority. The most important of these elements is the finances. Where it comes from? It is provided by the tax payers. The taxation policy and tax collection mechanism remains the prerogative of government. Legislation is not a theoretical exercise or academic preposition rather a deliberate effort to obtain the legal force required for policy implementation.

The right of land labourers and peasants were not legislated and protected which further empowered the landlord. 'Like India, Pakistan inherited much of its labour legislation from the colonial period. Subsequently, even though the government signed the 1948 ILO convention on the right of workers to organise and the 1949 convention on collective bargaining, detailed legislation on workers' rights was not enacted until the 1969 Industrial Relations Act' (Banuri 1991, p. 214).

Third is MANAGEMENT: Wherever policy needs to be implemented it requires not only the force of law but also equally important is the mechanism through which it would be implemented. It needs a working and efficient management structure. The critical mass (trained manpower), finances, technical support, necessary equipment/machinery and above all capacity and skill to manage the policy implementation are necessary requirements. The bureaucratic set-up and bureaucracy are often targeted by politicians and public at large for policy and program failures, yet they remain the linchpin and backbone of bureaucratic structure. The criticism would not improve any policy implementation unless the system is made efficient by adopting timelines and financial control and effective through professional and regular training of different cadre to achieve laid down objectives.

> The government should capitalise on the self-help trend existing in the communities through training the Local Government and Rural Development Department staff to successfully mobilise the community and get them involved in the Rural Water Supply

and Sanitation schemes. The media must be used to demonstrate to the communities the high probability of success in combining the self-interest factor with the self-help factor (Community Management in Rural Water Supply and Sanitation Systems 1995, p. 61). This methodology and approach could be extended to negotiate the relationship between land lords and peasants.

Fourth is MONITORING and EVALUATION: Monitoring and evaluation are both important activities but usually either misunderstood or ignored. Monitoring is done of the activities, and evaluation is undertaken juxtaposing objectives, be it a policy, program or a project. If the monitoring and evaluation is not done, then you cannot tell success from failure. If you cannot see success, you cannot reward it. Hence, the success loses its meaning and in the process, failure may be rewarded. If you cannot recognise failure, you cannot correct it. If you cannot see success, you cannot learn from it.

Fifth is INSTITUTIONAL GOVERNANCE: Some efforts are being made at government level to minimise the gap between food insecurity and poverty. The government has established the National Food Security Council which consists of representatives from federal, provincial and local governments. It has yet to be determined how effective and efficient it would be in performing its function of ensuring the availability of food for deprived population and developing mechanism for equitable distribution of food and the coordination between and among the three levels of government, i.e. federal, provincial and local.

The second challenge is discussed as follows. The Ministry of National Food Security and Research (new name given to the Ministry of Food and Agriculture under 18th Amendment of the constitution of 1973) is launching the zero hunger programme in collaboration with the WFP worth US$1.6 billion to address the food security objectives. 'Under this Programme the Ministry shall donate up to 500,000 metric tons of wheat per year and World Food Programme intends to negotiate with local producers to exchange part of the donated wheat for High Energy Biscuits (HEB) and the similar products manufactured in Pakistan factories for distributions through WFP operations to primary school children, siblings of malnourished children and vulnerable populations especially children at risk of malnutrition. The fund will also be converted to prepare fortified wheat flour for distribution which aimed at combating food insecurity in Pakistan' (Finance Division 2012, p. 17).

4.9 Recommendations

Some recommendations are discussed below.

1. One of the most important interventions is to change the land tenure system. It has to be transparent based on the system of accountability and participation of stake holders. This also includes the system of food distribution for marginalised population.

2. It is necessary to revisit, redesign and implement the prevailing land tenure system to ensure the equitability of land distribution. The small farmers and tenants cannot be left at the mercy of absentee landlords with large land holdings.
3. There is need to amend the laws for civil land adjudication, mechanism for conflict resolution and mitigation procedure for both and between landlords, tenants and small farmers.
4. Landless (peasants or tenants) should have access to land which would be possible if the government-owned land is given to them. Similarly, women-headed household should be given preference to own land and use for cultivation.
5. The efficient management and distribution system of food among the poor and deprived population demands serious attention by the state authorities along with donors and NGOs.
6. Food security is closely linked by (a) reforming the land tenure system, (b) better pricing for agriculture outputs, (c) facilities for scientific research, (d) efficient marketing and transport services and (e) timely availability of seeds, fertiliser, pesticide, water and agriculture machinery.
7. The equitable distribution of food is necessary but not sufficient condition to minimise the number of poor who are food insecure. The option of good governance as mitigating strategy needs to be applied.

References

Arif M (2004) Land, peasants and poverty: equitable land reforms in Pakistan. The Network Publications, Islamabad

Article (2004) Food Security. Business Recorder, Karachi

Banuri T (1991) Economic liberalisation: no Panacea. Oxford University Press, New York

Bari F (2003) Growth and poverty paradox. The Journal. National Institute of Public Administration, Karachi

Blum J (1971) Lord and peasant in Russia from the ninth to the nineteenth century. Princeton University Press, New Jersey

Economic Advisory Wing (2012) Pakistan Economic Survey 2011–2012. Finance Division, Government of Pakistan, Islamabad

Hensman R (2014) Economic and political weekly. Bombay

Hussain (2012) Jobs and urbanisation. The NEWS International, Lahore

Local Government and Rural Development Department (LG&RDD) (1995) Community management in rural water supply and sanitation system. In: Proceedings of the national workshop on RWSS systems, Mirpur, AJK

Nasir J (2014) The Gujrat model. The News International, Lahore

Pakistan Bureau of Statistics (2009) Agriculture statistics of Pakistan 2008–2009. Statistics Division, Government of Pakistan, Islamabad

Palit SK (2007) The impact of socio-economic reforms of governance: The Bangladesh experience. Malaysia. J Admin Gov. National Institute of Public Administration INTAN

United Nations Convention on the Rights of the Child (1990) General Assembly Article No. 24C of CRC. United Nations, New York

United Nations Economic and Social Commission for Asia and the Pacific (1990) Guidelines on methodological approaches to the conduct of a regional survey of the quality of life as an aspect of human resources development. United Nations Economic and Social Commission for Asia and the Pacific, New York

United Nations (1979) Entity for gender equality and the empowerment of women. Article 14g/h CEDAW. United Nations, Geneva, Switzerland

United Nations Office of the High Commissioner for Human Rights (1966) International Convention on Economic Social and Cultural Rights, Article 11.1. General Assembly Resolution 2200A(XXI). United Nations, New York

United Nations Statement on Food Security in Pakistan (2000) United Nations System in Pakistan, Islamabad

World Bank (2007) Rural poverty report on Pakistan. World Bank, Washington DC

Chapter 5
Industrialization and Sustainable Food Security: New Challenges for Malaysia

Ahmad Martadha Mohamed and Zahrul Akmal Damin

5.1 Introduction

Government of Malaysia in 1984 decided to transform the country from an agricultural country to an industrial country. Introduction of industrial policy was in part a result of Mahathir's Vision 2020 for Malaysia to become a highly developed nation by the year 2020 (Hirschman 2013). According to Mahani (2004), the government aims to make Malaysia as an industrialized nation by focusing on high-value products. As a result, it causes a contraction in the agriculture sector and hence affects sustainable food security (Hill et al. 2012). Indeed, the agriculture sector is the main food supply to deal with the increasing demand in tandem with population growth (Anderson et al. 2012). This has led to the increase in the threat to food security, especially in terms of food availability. However, this does not mean that the government does not take action to prioritize this issue. Food security is emphasized in every agricultural policy. For example, staple foods such as paddy, fruits and vegetables are grown in most states (Ministry of Agriculture and Agro-based Industry 2012). Nevertheless, there is still a shortage of food supply compared with the increasing demand of the population (Mahmudul et al. 2013). Ismail (2011) even harshly criticizes the food security policy by lamenting its failure to ensure sustainable food production. In light of this shortage, government has to import critical commodities from various countries such as Australia, China and Thailand (Ministry of Finance Malaysia 2011). Furthermore, the rapid increase in population to 30.05 million populations in 2014 also puts

A.M. Mohamed (✉) · Z.A. Damin
School of Government, College of Law, Government and International Studies, Universiti Utara Malaysia, 06010 Sintok, Kedah, Malaysia
e-mail: martadha@uum.edu.my

Z.A. Damin
e-mail: zahruld@uthm.edu.my

H. Ha (ed.), *Land and Disaster Management Strategies in Asia*,
DOI: 10.1007/978-81-322-1976-7_5

pressure on Malaysia at the same time. Building upon this scenario, this study aims to examine whether the existing agricultural policies are able to restore the situation through more effective land management. This is important since the introduction of the policy by the government to enhance the agricultural production of food can minimize the threat in relation to food security.

5.2 Literature Review

5.2.1 Conceptual Framework of Food Security

The concept of food security has been in existence for the first time officially in the Food and Agriculture Organisation of the United Nations (FAO) in 1974 (Maxwell 1996). However, it was debated as early as in 1937 by the International Committee on Nutrition (Padilla 1997). This concept arose from the concern over the amount of food offerings and global stability. But in 1994, food security has seen more serious attention when Haq introduced the concept of human security in the United Nations Development Programme (UNDP) (Gasper 2005). Introduction of human security has been seen as a new dimension in creating a well-being society (Acharya 2001; UNDP 1994). Until now, the concept of food security as defined by FAO has undergone several evolutionary improvements to take into consideration the idea previously introduced by Haq and the World Bank. Recently, FAO (2002) defines food security as ‘a situation that exists when all people at all times have access to physical, social and economic to adequate food, safe and nutritious food to meet their needs and have food preferences for an active lifestyle and healthy life’ (Devereux et al. 2012; Erb et al. 2012). In light of this new development, Bizikova et al. (2013) have pointed out that success in addressing threats to food security is dependent on four key pillars for food. It is food availability, food access, food utilization and food stability (Cafiero 2013; Ericksen et al. 2009).

1. Food availability: This pillar emphasizes the individual’s ability to obtain sufficient food in the quantities required. In this case, the availability of food is dependent on the provision of food by the country in meeting the needs of its population (FAO 2006). Typically, there are three main methods applied by countries to ensure the availability of food security. First, 100 % of food production in the country; second, 100 % of food imports from foreign countries; and third, a mixture of both methods of food production in the country (FAO 2006).
2. Food access: This pillar emphasizes the ability of individuals to access and obtain food (FAO 2008). Self-sufficiency for their use can be achieved through one or a combination of methods such as the production of food by an individual and by the individual stockpiling food, buying food, feed or food aid loans (Gross et al. 2000). However, the ability of an individual to access and acquire food depends on income and food prices. Meanwhile, at the national

level, access to food refers to food production by country and supported by the importation of food if it is not enough (Inter-American Institute for Cooperation on Agriculture 2009).

3. Food utilization: This pillar emphasizes the usefulness of food taken by the beneficial food to meet the nutritional needs of the individual self (International Federation of Red Cross and Red Crescent Societies 2007). In this case, the usefulness of nutritional food for themselves is determined through adequate food intake. This is due to the fact that adequate nutrition is a concern for an individual in maintaining the efficiency of the body and at the same time avoiding any disease. Among other things, this pillar is also concerned about health care, especially in life. To meet this need, the ability to get clean water and food-processing knowledge were given priority (United States Agency for International Development 2007).
4. Food stability: This pillar emphasizes the individual's ability to obtain sufficient food at all times (Ericksen et al. 2011). Aimed at guaranteeing food stability, this pillar is critical in ensuring sustainable food production. This includes the risk of contingencies such as the unavailability or lack of food due to the economic crisis, political crisis or unexpected climate changes, resulting in flooding and drought. Among other things, the stability of the food also refers to enough food at each cycle of seasonal foods (Stamoulis and Zezza 2003).

However, food security in this paper will be explored within the provision of traditional food production. In doing so, the focus would be on government policies in agriculture particularly in regard to food production that involves paddy, fruits and vegetables.

5.2.2 Challenges Against Food Security in the Industrial Era

Government's ability to defend the Malaysian food security in the industrial is often being questioned despite the emphasis on agricultural-food policy. This is pertinent after taking into account the inability of the government to accommodate the growing demand for food in the country. Following these concerns, the paper will analyse the issue of food security based on three critical challenges: (1) realignment of national policies from agricultural based to industrial based, (2) conversion of agricultural land; and (3) increased imports of food.

5.2.2.1 Realignment of National Policies from Agricultural Based to Industrial Based

Focusing on industrial sectors highlighted by the government starting in 1984 had sunk agriculture output (Rajah Rasiah 2011). The impact of this action can be felt until now despite several agricultural policies being introduced to increase the food production. This policy change causes the direction of Malaysia's development plans to change from an agricultural based to an industrial based (Chamhuri et al. 2005). The focus on industrialization has seen an increased in export for manufacturing products such electrical goods, automotive, and textiles. However, after the 1997 economic crisis, the government has demonstrated a strong commitment in creating food security. Various policies have been developed such as the Third Agricultural Policy, National Biotechnology Policy and Agro-food Policy as a reflection of the importance of this issue. Various strategies adopted include the development of high-value agriculture, agriculture-driven research and development (R&D) as well as increase the productivity and competitiveness of the agriculture sector (Ministry of Agriculture and Agro-based Industry 2011; Ministry of Science, Technology and Innovation 2005; Ministry of Agriculture of Malaysia 1999). However, the continuing focus on the development of a more vibrant industrial sector has again caused the agricultural sector to take a back seat. Specifically, the development expenditure budget allocation for agricultural purposes has been declining steadily for the past few years. On the other hand, the industrial sector has seen an incremental increase in its budgetary allocation over the years (Martadha and Akmal 2012). It is unavoidable that industrial sector is perceived to be in a better positioned as the engine of economic growth compared with agricultural sector (Kamarudin et al. 2006). Because of that the country has steadily experienced a decline in food production and other agricultural activities.

5.2.2.2 Conversion of Agricultural Land

The rapid development of the industrial sector resulted in competition for agricultural land for food (Muhammad et al. 2011). For example, existing agricultural land had to compete for the use of food crops and commercial crops. Table 5.1 shows a comparison between an area of major food crops of paddy, vegetables and fruits with commercial crops of rubber and oil palm. Commercial crops were found to have a large area planted instead of food crops. In fact, there are many existing lands for food production have been taken and modified for commercial agricultural purposes (Fatimah and Mad Nasir 1997). The increased in commercial food production has, in turn, reduced the land for agricultural purposes (Ruhaidini 2013). The issue becomes worse as the agricultural lands are converted for residential areas and industrial sites. This is because from 1997 to 2012, a total of 106,000 ha of paddy fields or the equivalent of 50,000 football fields has been converted to residential areas (Aruna 2012; Dayang Hajyrayati 2012; Golam Hassan 1998). It is further heightened by the findings of the Department of Statistics (2012) that found that rural–urban migration is by 6.4 % between 2010 and 2011 and 21.9 % due to the boom in industries.

Table 5.1 Comparison of total area major food crops planted with commercial crops (000 ha)

Crop	1985	1990	1995	2000	2005	2010	2011
Paddy	649	662	666	400	611	677	683
Vegetables	14	31	36	42	77	52	53
Fruits	119	177	244	345	379	239	240
Rubber	1,948	1,836	1,679	1,344	1,259	999	998
Palm oil	1,447	2,029	2,539	3,376	4,051	4,853	5,000

Source Compilation from different available resources such as Malaysian Department of Statistics and Ministry of Agricultural

5.2.2.3 Increased Imports for Food

Although the importation of food from abroad can increase the availability of food in the country, it also gives the opposite effect. The risk of rising prices and the threat of reduction or termination may occur if the delivery of food exporting countries faces crisis such as flooding, drought and rising food prices worldwide (Ahmad and Sani 2012; Qureshi et al. 2013). Exporting countries will produce the food for the consumption of their own people before the commodity is exported. The magnitude of such crisis occurred in 2007 and 2008 (Braun 2008) when the prices of food jumped excessively. Early signs of this crisis occurred when grain production faced difficulties in meeting global demand, hence causing the prices to triple (Mittal 2009). Similarly, Vietnam as the world's second largest exporter, followed by China and India, has banned the export of paddy completely in early 2008 (Slayton 2009; Tenebaum 2008). This situation creates panic in food importing countries as they are vulnerable to food insecurity (Koizumi 2013). Thus, dependency on external food is a worrying scenario particularly that Malaysia is a net importer of food (Fatimah et al. 2010; Lembaga Pembangunan Pelaburan Malaysia 2013). This issue is even more serious when the food trade balance shows a deficit figure. From 2001 to 2005, the percentage of this deficit had increased by 42.12 % from 1995 to 2000, 46.12 % before the big swell to 72.00 % in the period from 2006 to 2010 (Kementerian Pertanian dan Industri Asas Tani Malaysia 2012).

5.3 Theoretical Framework

Arguments to the food security policy setting are based on the theory of elites. The theory expounded by Pareto, Mosca and Michels in the late nineteenth century is that the community is a pyramid and classified according to the function of society itself. According to Berberoglu (2005), those who are on the top layer are an elite group that serves as a temporary government to the lower layers of the society. This theory asserts that although an elite group is a small group but they have power in the government, especially in the agenda-setting and decision-making

(Kirby et al. 2000). Thus, the elites are responsible for the use of power to meet the needs and desires of the community of the governed. This is due to the power used by an elite group to pursue their interest to achieve a specific goal. The elites are using power as a tool to protect their interests and influence society (Brym and Lie 2010). According to Hinrichs (2003), a political system that is occupied by an elite group is also closely associated with the development of food security policy agenda. The study by Jayne et al. (2002) shows that this elite group has the autonomy to exploit public resources and become involved in decision-making (Arcand and Wagner 2012). Consequently, the reality of the agriculture sector especially food production is subject to the intervention and influence of this influential elite group (Sosya et al. 1999). Thus, the elite group involved in influencing and formulating policy should recognize the importance of social welfare on food security (Dover 1995). This can be accomplished by putting more resources to ensure sustainable food production (Jaffe Lopez 2010).

5.4 Research Methodology

The approach used in this study is a qualitative approach. In examining the issue of food security, there are some researchers such as Gallaher et al. (2013), Hoddinot et al. (2013), and Riley and Legwegoh (2013) who have used a qualitative approach to their study. In terms of research design, this study relies heavily on a content analysis as used in a similar study by Sneyd et al. (2013). However, in this study, the sources analysed include planning documents and development. Specifically, the analysis involves scrutinizing 5-years development plan of the Fifth Malaysia Plan (1986–1990) to the Tenth Malaysia Plan (2011–2015). Justification for the selection of these documents is that they contain the basic policy of Malaysia such as goals, strategies and achievement of national food security. The study will examine the documents, select, screen and set the text content of the words, figures and printed message based on the theme of agriculture particularly food crops. It is analysed with reference to step, approach, strategy, planning or implementation undertaken by the country to increase food production through better management of agricultural land for paddy, vegetables and fruits.

5.5 Findings and Discussions

Although the government shifts its focus to industrial based, the issue of food security is still paramount to creating stability in the country (Ahmad Zubir et al. 2010; Noorfazreen and Asmak 2011). The findings of the content analysis showed that there were 46 measures taken to increase the production of paddy, vegetables and fruits through land management as Table 5.2. Specifically, 20 policies (43.48 %) focused on the management of land for paddy cultivation, 15 policies

Table 5.2 Land management for the purpose of the food security policy for the production of paddy, vegetables and fruits (frequency)

	5 MP	6 MP	7 MP	8 MP	9 MP	10 MP	
	1986–1990	1991–1995	1996–2000	2001–2005	2006–2010	2011–2015	
Paddy							Total
Focused 8 paddy field	2	1	–	–	1	–	4
Plantation estate	1	–	2	–	–	–	3
Group farming	1	2	2	–	–	–	5
Change of land use	1	1	–	–	–	–	2
Agriculture integration	1	–	–	–	–	–	1
Efficient farm management	–	–	2	–	–	–	2
Contract farming	–	–	-1	–	–	–	1
Levelling of land	–	–	1	–	–	–	1
Land consolidation and rehabilitation	–	–	–	–	–1	–	1
Total	6	4	8	0	2	0	20
Fruits							
Intensive land use	–	2	–	–	–	–	2
The use of idle land	–	1	–	–	–	–	1
Zoning	–	–	1	–	–	–	1
Estate management	–	–	1	–	–	–	1
Production areas	–	–	–	1	–	–	1
Agricultural technology park	–	–	–	1	–	–	1
Horticultural garden city	–	–	–	1	–	–	1
Satellite farms	–	–	–	1	–	–	1
Agriculture integration	–	–	–	1	–	–	1
Opening up new land	–	–	–	1	–	–	1

(continued)

Table 5.2 (continued)

	5 MP	6 MP	7 MP	8 MP	9 MP	10 MP	
	1986–1990	1991–1995	1996–2000	2001–2005	2006–2010	2011–2015	
Fruit production zone	–	–	–	–	1	–	1
Cluster development	–	–	–	–	1	–	1
New production zones	–	–	–	–	1	–	1
Farm contracts and strategic alliances	–	–	–	–	–	1	1
Total	0	3	2	6	3	1	15
Vegetables							
Horticultural crops	–	2	–	–	–	–	2
Planting in situ	–	1	–	–	–	–	1
Cultivation in the highlands	–	1	–	–	–	–	1
Cultivation on ex-mining land	–	1	–	–	–	–	1
Farming without soil based	–	–	1	–	–	–	1
Hydroponic	–	–	–	1	–	–	1
Aeroponic	–	–	–	1	–	–	1
Opening up new land	–	–	–	1	–	–	1
New production zones	–	–	–	–	1	–	1
Farm contracts and strategic alliances	–	–	–	–	–	1	1
Total	0	5	1	3	1	1	11

Source A compilation of data over the years from various government documents

(32.61 %) focused on the management of land for the cultivation of fruits and 11 policies (23.91 %) focused on the management of land for the cultivation of vegetables.

5.5.1 Land Management for Paddy Cultivation

Under the management of land for paddy cultivation, there are nine items identified that serves as the basis for land management for food. The items in question are eight paddy field (4 frequency), plantation estate (3 frequency), group farming (5 frequencies), change of land use (2 frequency), integrated farming (1 frequency), efficient farm management (2 frequency), contract farming (1 frequency), levelling of land (1 frequency), as well as land consolidation and rehabilitation (1 frequency). All items are accounted for another 20 or equivalent to 43.48 % frequency. In the context of focusing on the cultivation of paddy, the government has attempted to optimize the use of land to paddy cultivation by carrying out twice a year. Similarly, an area of 9,000 ha of agricultural land paddy was provided with drainage and irrigation between 1986 and 1990 (Malaysia 1986). But before that the government has developed a similar facility for the land area of 69,000 ha between 1976 and 1985 (Malaysia 1976, 1981). At the same time, the government has also encouraged the exchange of small-scale paddy farming to the implementation of a mini estate, group farming and contract farming production techniques to create a more efficient and more effective land management. This effort continues to increase revenue through better land management, more efficient water management, use of high-yielding seeds, and increase the intensity of cultivation and post-harvest loss reduction (Malaysia 1991). Thus, a more systematic land management can indirectly increase food security for paddy production.

5.5.2 Land Management for Fruit Cultivation

Meanwhile, under the management of land for the cultivation of fruits, there are 14 items that serve as the basis for this theme. The items concerned are intensive land use (2 frequencies), the use of idle land (1 frequency), zoning (1 frequency), estate management (1 frequency), production areas (1 frequency), agricultural technology park (1 frequency), horticulture garden city (1 frequency), satellite farms (1 frequency), agricultural integration (1 frequency), opening up new land (1 frequency), fruit production zone (1 frequency), cluster development (1 frequency), new production zone (1 frequency), as well as farm contracts and strategic alliances (1 frequency). All of these items contributed to 15 frequency or equivalent to 32.61 %. Influenced by soil management practices for paddy cultivation, land management for fruits also exhibits the same trend of change in land-use patterns of small scale to large scale. Cultivation of fruits has also been introduced with intensive land management through zoning, agricultural technology parks, mini-estate crops, group farming and contract farming. These lands have been opened by the authorities of the regional development along the state government. As a result of these efforts, new lands were opened for the purpose of increasing the area planted with fruit. Between 1985 and 2010, covering an area of 120,000 ha of land has been devoted to the cultivation of fruits (Malaysian Department of Statistics 2001, 2011). This production was done on a large scale in food production zones and satellite farms for fruits and focused on the type of selected fruits that have potential for export (Malaysia 2001, 2006). Even so, the government at the same time also encourages the private sector to exploit idle lands for cultivation of fruits to encourage horticultural crops.

5.5.3 Land Management for Vegetable Cultivation

However, under the management of land for the cultivation of vegetables, there are ten items that serves as the basis for this theme. The items concerned are horticultural crops (2 frequency), planting in situ (1 frequency), cultivation in the highlands (1 frequency), cultivation on ex-mining land (1 frequency), farming without soil based (1 frequency), hydroponic (1 frequency), aeroponics (1 frequency), opening up new land (1 frequency), new production zones (1 frequency), as well as farm contracts and strategic alliances (1 frequency). All of these items contributed to 11 frequency or equivalent to 23.91 %. The results of the analysis also found that soil management techniques on a large scale have an effect when it is extended in planting vegetables. Between 1985 and 2010, 38,000 ha of land have been devoted to the cultivation of vegetables (Malaysian Department of Statistics 2001, 2011). Thus, the opening of new land, the introduction of new production zones and the development of contract farming will be able to give high returns in the context of the production. By means of large-scale agriculture, the government will ensure the provision of extension services provided by the government will improve vegetables production. Yet at the same time, the use of existing land for vegetables has also been enhanced. The use of modern farming technique incentives such as horticulture, hydroponics and aeroponics has been developed to achieve the goal without land-based agriculture. Among other things, the government also optimizes the mined land and high land for vegetable farming.

5.6 Conclusion

The emphasis on the production of paddy, fruits and vegetables through land management proves that food security is important. This factor is evident when the government seeks the optimal use of agricultural land. Improved infrastructure, such as irrigation systems as well as improved soil management, and change in focus from small-scale farming to large-scale farming, such as the implementation of a mini estate, group farming and contract farming, are among government initiatives to increase food production. This is done so that the existing agricultural land for paddy, fruits and vegetables can be managed better by applying modern farming techniques and technology. However, the government should create awareness and education among farmers so that they can instead focus on a larger scale for food production. This step is important because it can increase the growth of agricultural production of food. In addition, the government has also opened up new agricultural land as permanent food production zones and satellite farms. Preparation for funding agriculture, modern agriculture management techniques and systematic cultivation, and use of new technology are among recent strategies introduced by the government to increase food production. In sum, the modernization and commercialization of agricultural sector are taken to ensure a sustainable food security for the nation for years to come. This is done amidst the vision of the country to become a highly developed nation by the year 2020.

References

Acharya A (2001) Human security: east versus west. Int J 56(3):442–460

Ahmad Z, Sani AS (2012) Pertanian 'benteng pertahanan negara' (Agriculture as a 'national defence'). Berita Harian, 27 Sept

Ahmad Zubir I, Chamhuri S, Rospidah G (2010) Keselamatan makanan negara: Analisis perbelanjaan awam dalam sektor padi (National food security: analysis of public spending in the rice sector). Paper presented at International Seminar on Economic Regional Development, Law and Governance in Malaysia and Indonesia, Pekan Baru, 7–9 June 2010

Anderson K, Jha S, Nelgen S, Strutt A (2012) Re-examining policies for food securities. Paper presented at Asian Development Bank (ADB) Economics, Manila, Feb 2012

Arcand JL, Wagner N (2012) Elite capture revisited: does community driven development improve inclusiveness? Evidence from Senegal. Retrieved from http://www.researchgate.net/publication/257029589_Elite_Capture_Revisited_Does_Community_Driven_Development_Improve_Inclusiveness__Evidence_from_Senegal/file/3deec5243e0935fa83.pdf

Aruna P (2012) Price of progress. The Star, 30 March

Berberoglu B (2005) An introduction to classical and contemporary social theory, 3rd edn. Rowman & Littlefield Publishers Inc, Maryland

Bizikova L, Roy D, Swanson D, Venema HD, McCandless M (2013) The water- energy-food security Nexus: Towards a practical planning and decision-support framework for landscape investment and risk management. The International Institute for Sustainable Development

Braun JV (2008) Food and financial crisis: implications for agriculture and the poor. International Food Policy Research Institute, Washington

Brym RJ, Lie J (2010) Sociology: your compass for a new world, 2nd edn. Wadsworth Cengage Learning, Belmont

Cafiero C (2013) What do we really know about food security? Paper presented at The National Bureau of Economic Research

Chamhuri S, Surtahman KH, Norshamliza C (2005) Ekonomi Malaysia [Malaysian Economic], 6th edn. Pearson, Petaling Jaya

Dayang Hajyrayati AK (2012) Impak Pembangunan Tanah ke atas Strategi Penghidupan Kaum Seping di Ulu Belaga, Sarawak (Land Development Impact on Livelihood Strategies the Seping ethnic in Belaga, Sarawak). Paper presented at international conference on social sciences and humanities 2012, Universiti Kebangsaan Malaysia, Bangi

Department of Statistics Malaysia (2012) Migration survey report. Department of Statistics Malaysia, Putrajaya

Devereux S, Eide WB, Hoddinott J, Lustig N, Subbarao K (2012) Social protection for food security. High level panel of experts on food security and nutrition (HLPE). Retrieved from http://typo3.fao.org/fileadmin/user_upload/fsn/docs/HLPEII/Social_Protection_V0_report.pdf

Dover MA (1995) Social security and national security in a cold war world. Retrieved from https://www.google.com.my/#q=Social+security+and+national+security+in+a+cold+war+world

Erb KH, Mayer A, Kastner T, Kristine-Elena Sallet KE, Haberl H (2012) The impact of industrial grain fed livestock production on food security: an extended literature review. Paper work (136) social ecology, Institute of Social Ecology. Retrieved from http://www.uni-klu.ac.at/socec/downloads/WP136_Webversion.pdf

Ericksen PJ, Ingram JSI, Liverman DM (2009) Food security and global environmental change: emerging challenges. Environ Sci Policy 12(4):373–377

Ericksen P, Thornton P, Notenbaert A, Cramer L, Jones P, Herrero M (2011) Mapping hotspots of climate change and food insecurity in the global tropics. The CGIAR Research Program on Climate Change, Nairobi

Fatimah MA, Mad Nasir S (1997) Sekuriti makanan dan isu tanah pertanian (Food security and agriculture land issues). Paper presented at Consumer Association of Penang: land—emerging issues and challenges, Pulau Pinang

Fatimah MA, Fauzi MJ, Khanif MY (2010) Agenda polisi sekuriti makanan Malaysia (Malaysian food security policy agenda). Paper presented at Bengkel Mengarusperdana Pertanian dalam Model Baru Ekonomi Malaysia, Bangi, 9–10 Nov 2010

Food and Agriculture Organization of United Nations (2006) The state of food and agriculture. Food and Agriculture Organization of United Nations, Rome

Food and Agriculture Organisation of United Nations (2008) An introduction to the basic concepts of food security. Retrieved from https://www.fao.org/docrep/013/al936e/al936e00.pdf

Gallaher CM, Kerr JM, Njenga M, Karanja NK, WinklerPrins AMGA (2013) Urban agriculture, social capital, and food security in the Kibera Slums of Nairobi, Kenya. Agric Hum Values 30:389–404

Gasper D (2005) Securing humanity: situating human security concept and discourse. J Hum Dev capabil 6(2):221–245

Golam Hassan AA (1998) Ketidakseimbangan agihan industri dan migrasi penduduk di negeri Kedah (Imbalance in population distribution and migration industry in the state of Kedah). Analisis 5:99–118

Gross R, Schoeneberger H, Pfeifer H, Preuss HJA (2000) The four dimensions of food and nutrition security: Definitions and concepts. Retrieved from https://www.foodsec.org/DL/course/shortcourseFA/en/.../P-01_RG_Concept.pdf

Hill H, Yean TS, Ragayah MZ (2012) Malaysia: a success story stuck in the middle? The World Economy 1687–1711

Hinrichs CC (2003) The practice and politics of food system localisation. J Rural Stud 19:33–45

Hirschman C (2013) Malaysia's development challenges: graduating from the middle. Asian-Pacific Econ Lit 27(1):163–165

Hoddinot J, Wheeler RS, Berhane G, Handino M, Kumar N, Lind J, Taffesse AS, Tefera M (2013) Implementing large scale food security programs in rural Ethopia: Insights from the productive safety net program. In: Rahmato D, Pankhurst A, Uffelen JG (eds) Food security, safety nets and social protection in Ethiopia. Forum for Social Studies, Ethiopia

Inter-American Institute for Cooperation on Agriculture (2009) IICA's definition of food security. Retrieved from http://www.iica.int/Esp/Programas/SeguridadAlimentaria/Documents/SeguridadAlimentarias_Quees_Eng.pdf

International Federation of Red Cross and Red Crescent Societies (2007) Global food security assessment guidelines. International Federation of Red Cross and Red Crescent Societies, Geneva

Ismail K (2011) Dewan Rakyat (House of Parliament). Bil 38

Jaffe Lopez MI (2010) Institutional change in Cuba's agricultural sector. Doctoral thesis, Institute of Agricultural Economics and Social Sciences in the Tropics and Subtropics, University of Hohenheim

Jayne TS, Govereh J, Mwanaumo A, Nyoro JK (2002) False promise or false premise? The experience of food and input market reform in eastern and southern Africa. World Dev 30(11):1967–1985

Kamarudin O, Rosman M, Mohamad Shukri J (2006) Hubungan eksport terhadap peningkatan produktiviti sektor pembuatan di Malaysia: Satu kajian Granjer Causality (Export contacts to increase the productivity of the manufacturing sector in Malaysia: A study of causality Granjer). Shah Alam: Universiti Teknologi MARA

Kementerian Pertanian dan Industri Asas Tani Malaysia (2012) Perangkaan Agromakanan 2011 (Agrofoods statistics 2011). Putrajaya: Bahagian Perancangan Strategik dan Antarabangsa

Kirby M, Kidd W, Koubel F, Barter J, Hope T, Kirton A, Madry N, Manning P, Tringgs K (2000) Sociology in perspective, AQA edn. Heinemann Educational Publishers, Oxford

Koizumi T (2013) Biofuel and food security in China and Japan. Renew Sustain Energy Rev 21:102–109

Lembaga Pembangunan Pelaburan Malaysia (2013) Industri Barang Makanan (Food products industries). Retrieved from http://www.mida.gov.my/bm/index.php?page=makanan

Mahani ZA (2004) Pemikiran ekonomi Tun Dr. Mahathir Mohamad dalam pembangunan negara dari perspektif korporat. In: Mustapa K, Abdul Rahman AZ, Hasliza S, Muhamed Nor AN (eds) Siri simposium berkala III: Ekonomi (Periodic symposia series III: economic). Institut Pemikiran Tun Dr. Mahathir Mohamad, Sintok

Malaysia (1976) Third Malaysia plan, 1976–1980. Jabatan Cetak Kerajaan, Kuala Lumpur

Malaysia (1981) Fourth Malaysia plan, 1981–1985. Jabatan Percetakan Negara, Kuala Lumpur

Malaysia (1986) Fifth Malaysia plan, 1986–1990. Jabatan Percetakan Negara, Kuala Lumpur

Malaysia (1991) Sixth Malaysia plan, 1991–1995. Jabatan Percetakan Negara, Kuala Lumpur

Malaysia (2001) Eight Malaysia plan, 2001–2005. Jabatan Percetakan Nasional Malaysia Berhad, Putrajaya

Malaysia (2006) Ninth Malaysia plan, 2006–2010. Jabatan Percetakan Nasional Malaysia Berhad, Putrajaya

Malaysia (2011) Tenth Malaysia plan, 2011–2015. Jabatan Percetakan Nasional Malaysia Berhad, Putrajaya

Malaysian Department of Statistics (2001) Buku tahunan perangkaan Malaysia 2001 (Year book statistics: Malaysia 2001). Jabatan Perangkaan Malaysia, Putrajaya

Malaysian Department of Statistics (2011) Malaysia yearbook of statistics 2010. Malaysian Department of Statistics, Putrajaya

Mahmudul MA, Chamhuri S, Abdul Hamid J, Basri T, Khairulmaini OS (2013) Agricultural vulnerability and adaptation to climatic changes in Malaysia: Review on paddy sector. Curr World Environ 8(1). Retrieved from http://www.cwejournal.org/vol8no1/agricultural-vulnerability-and-adaptation-to-climatic-changesinmalaysiareview-on-paddy-sector/

Martadha AM, Akmal ZD (2012) The effects of national budgetary on food security policy agenda. Paper presented at Network of Asia Pacific Schools and Institutes of Public Administration and Governance (NAPSIPAG): development challenges in the Asia Pacific: millennium development goals and beyond, Sri Lanka

Maxwell S (1996) Food security: a post-modern perspective. Food Policy 21(2):155–170

Ministry of Agriculture Malaysia (1999) Third national agricultural policy (1998–2010). Ministry of Agriculture Malaysia, Kuala Lumpur

Ministry of Agriculture and Agro-based Industry (2011) National agro-foods policy 2011–2020. Kementerian Pertanian dan Industri Asas Tani, Putrajaya

Ministry of Agriculture and Agro-based Industry (2012) Agro-foods statistics 2011. Kementerian Pertanian dan Industri Asas Tani, Putrajaya
Ministry of Finance Malaysia (2011) Economic report 2011/2012. Ministry of Finance Malaysia, Kuala Lumpur
Ministry of Science, Technology and Innovation (2005) National biotechnology policy. Kementerian Sains, Teknologi dan Inovasi, Putrajaya
Mittal A (2009) The 2008 food price crisis: rethinking food security policies. United Nations, New York
Muhammad Y, Chamhuri S, Shaharudin I (2011) Pengekalan tanah sawah sebagai kawasan pertanian berterusan. Malaysian J Environ Manage 12(2):55–65
Noorfazreen MA, Asmak AR (2011) Perlaksanaan dasar sekuriti makanan di Malaysia: Kajian daripada perspektif ekonomi Islam (The implementation of food security policies in Malaysia: a study of Islamic economic perspective). Jurnal Syariah 19(1):39–62
Padilla M (1997) Food security in African cities: The role of food supply and distribution systems. Paper presented at seminar "Food supply and distribution in francophone African Towns" Food and Agriculture Organization of United Nations, Dakar
Qureshi ME, Hanjra MA, Ward J (2013) Impact of water scarcity in Australia on global food security in an era of climate change. Food Policy 38:136–145
Rasiah R (2011) Overview of Malaysia economy. In: Rasiah R (ed) Malaysian economy: unfolding growth and social change. Oxford Fajar Sdn. Bhd., Shah Alam
Riley L, Legwegoh A (2013) Comparative urban food geographies in Blantyre and Gaborone. African geographical reviews. doi:10.1080/19376812.2013.805148. Retrieved from http://www.tandfonline.com/doi/abs/10.1080/19376812.2013.805148#preview
Ruhaidini AK (2013) Mengajak anak muda menyertai pertanian (Invite young people into agriculture). Mingguan Malaysia, 9 June
Slayton T (2009) Rice crisis forensics: How Asian governments carelessly set the world rice market on fire. Paper presented for Center for Global Development. Retrieved from http://papers.ssrn.com/sol3/papers.cfm?abstract_id=1392418
Sneyd LQ, Legwegoh A, Fraser EDG (2013) Food riots: media perspectives on the causes of food protest in Africa. Food Secur 5:485–497
Sosya I, Gleditsch NP, Gibson M, Sollenberg M (1999) To cultivate peace: agriculture in a world of conflict. Retrieved from http://www.wilsoncenter.org/publication/to-cultivate-peace-agriculture-world-conflict
Stamoulis K, Zezza A (2003) A conceptual framework for national agricultural, rural development and food security strategies and policies. Retrieved from http://www.eldis.org/vfile/upload/1/document/0708/DOC15222.pdf
Tenebaum DJ (2008) Food vs. fuel: diversion of crops could cause more hunger. Environ Health Perspect 116(6):254–257
UNDP (1994) Human development report 1994. Oxford University Press, New York
United States Agency for International Development (2007) Food assistance programming in the context of HIV. United States Agency for International Development, Africa

Part II
Disaster Management

Chapter 6
Bangladesh: Natural Disaster Risk Management

Huong Ha and Akbaruddin Ahmad

6.1 Introduction

6.1.1 Background

Bangladesh, located in the South Asian region, is considered one of the most disaster-prone countries in the world (Ali et al. 2012). Bangladesh has suffered from several natural disasters, such as earthquakes, cyclones, floods and droughts, nearly every year (Asgary and Halim 2011). In addition, the country was also faced with issues associated with disasters, including shortage of food and clean water supply, climate change impact, and land management. Floods, cyclones and earthquakes are regular phenomena that have created a massive impact on the population, land management and socio-economic well-being of the people. Asgary and Halim (2011) explained that the root causes of disasters do not only include flood and cyclone hazards, but also lack of planning and resources, unsafe conditions and many others. Thus, the post-2015 will be a period of great challenge when concerted efforts of the government, the private sector, donors and civil society organizations are required for the timely intervention and mitigation of natural disasters.

Many authors have discussed various issues associated with natural disaster management in Bangladesh. For example, Choudhury (2002) explained the economic losses due to natural disasters and proposed measures to minimise such

H. Ha (✉)
UON Singapore (University of Newcastle, Singapore Campus), Singapore, Singapore
e-mail: artp6075singapore@yahoo.com

A. Ahmad
Policy Research Centre, Dhaka, Bangladesh

A. Ahmad
Darul Ihsan University, Dhaka, Bangladesh

H. Ha (ed.), *Land and Disaster Management Strategies in Asia*,
DOI: 10.1007/978-81-322-1976-7_6

losses. Akter (2009) discussed environmental displacement due to climate change in the country. Khan and Nahar (2014) examined how Bangladesh has addressed natural disasters, and the extent to which the country can adapt to such disasters. Yet, there are insufficient studies on how different sectors in the country have involved in the governance and management of natural disasters. Ha (2012), Ha and Dhakal (2013), Koiman et al. (2008) and Stocker (1998, 2004) asserted that good governance is impossible without the participation and contribution of different groups of sectors in normal conditions. Since disaster management requires resources and efforts from various groups of stakeholders, it is impossible to hold any single group of stakeholders or individuals to be accountable for the whole process of disaster risk management (DRM); this chapter aims to address this gap by discussing national plans and initiatives by various groups of stakeholders to manage disasters in Bangladesh.

6.1.2 Research Objectives

The objectives of this chapter are to (i) revisit key natural disasters and their associated issues in Bangladesh, (ii) examine institutional and regulatory frameworks as well as measures to manage disasters, including pre-disaster planning and post-disaster recovery and (iii) assess the current initiatives to manage disasters against the three-sector governance model, including government, the private sector and civil society. The chapter also includes some policy and technical recommendations, followed by a conclusion.

6.2 Research Method and Framework for Analysis

6.2.1 Research Method

This chapter is conceptual and explorative in nature since it reflects what the authors' views and experience on how disasters have been managed. The authors have gathered secondary data from both academic and non-academic literature, including government reports, non-governmental organisations' (NGOs) reports and international organisations' documents. The authors have also introduced the three-sector governance model which has been adapted to mitigate climate change and manage the environment. Firstly, the authors have reviewed key natural disasters in Bangladesh. Secondly, the authors have discussed various practice and activities to address disaster risk reduction by government, the private sector and civil society, using Bangladesh as a case study. Then, the authors have assessed such activities against the three-sector governance model, and the implications of

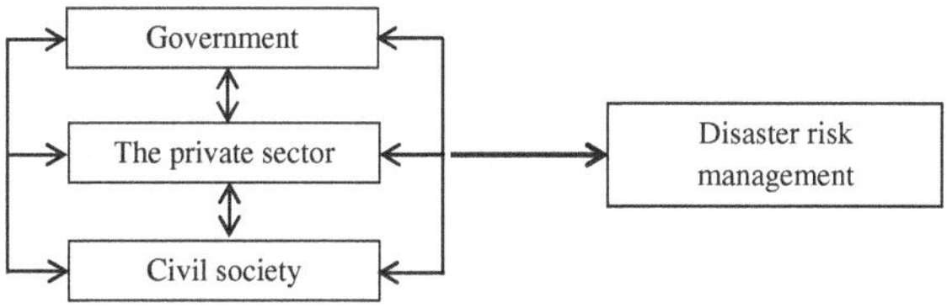

Fig. 6.1 Three-sector governance model and DRM

the current state of DRM in Bangladesh. Finally, the authors have made some policy recommendations to address such gaps.

6.2.2 Framework for Analysis

Governance refers to the structures, mechanisms, institutional arrangements, regulatory and non-regulatory frameworks, etc., to plan, organise, lead and control national affairs (Braithwaite and Drahos 2000; Lidskog and Elander 2010; Fisher and Surminski 2012). In DRM, governance is defined as the instruments, mechanisms, informal and formal institutions in charge of planning, implementation and evaluation of policies and programs to mitigate and adapt to the impact of disasters. It also refers to the allocation of resources and authority, and the coordination among different groups of stakeholders in disaster risk reduction (Ha 2012, 2013; Ha and Dhakal 2013). Disaster management is a challenging task which cannot be addressed by any single sector, any single group or individual. Also, the impact of disasters is long-lasted, far-reached and profound. Thus, the theoretical framework for this chapter has been built based on the three-sector governance model and participation of all groups of stakeholders (see Fig. 6.1).

The three sectors included in this governance model are government (the first sector), the private sector or business (the second sector), and civil society or the third sector. In DRM, government can play the role of a strategic planner, a resource allocator, a policy and regulation maker, a policy implementer and a law enforcer (Braithwaite and Drahos 2000). Government should work closely with other sectors to improve the efficiency and effectiveness of any attempts to prepare for disasters, reduce risk and mitigate the impacts of disasters; government should coordinate the activities of state and non-state actors, and mobile resources from various groups of stakeholders (Koiman 1999; Ha 2012). The private sector and civil society organisations (CSOs) can provide feedback and contribute resources to support programs and initiatives to address political and socio-economic issues associated with disaster management. Government should engage the second and third sectors in any policy discussion and debates since these sectors and the public are the recipients of any policies, and such policies usually affect their welfare and well-beings. In democratic societies, business, CSOs and the public should have equal rights and opportunities to participate in national affairs (Moore and Hartley 2008; Black 2012).

6.3 Natural Disasters and Associated Issues in Bangladesh

6.3.1 Natural Disasters in Bangladesh

Similar to other disaster-prone countries, cyclones, floods, tornadoes, hail storms, landslide and earthquakes are frequent visitors of Bangladesh. This section briefly discusses some major natural disasters in the context of Bangladesh.

6.3.1.1 Cyclones

Severe tropical cyclones and tidal surges during the monsoon period (April–May) are most common. The post-monsoon period of October–November creates equal impact on the population of the coastal belt (Alam and Collins 2010). Low-lying areas in the coastal regions make the residents highly vulnerable causing immense loss to human lives, livestock, food output and overall infrastructure and creating a massive problem for the economic well-being of a large section of the population of Bangladesh (Khan and Nahar 2014).

The cyclone in 1991 hit Bangladesh and killed about 140,000 h (Nehal 2005). In 2007, cyclone "Sidr" swept through the south-western coast of Bangladesh. About 8.9 million people, and 30 out of 64 districts in Bangladesh were affected by this cyclone. The number of casualty reached 3,295 people and over 50,000 people missing or injured (Ministry of Food and Disaster Management 2008). Cyclones usually cause heavy economic loss. For instance, the cyclone Sidr in 2007 damaged assets (e.g. buildings, roads, bridges, etc.), crops, livestock, and fisheries and caused production loss which were totalled about US$437.6 million (Khan and Nahar 2014). Cyclone Aila in 2009 was another severe one in Bangladesh which caused over 7,000 people to be infected by an outbreak of diarrhoea, and about 8,202 people missing (Islam and Chick 2011). Islam and Chick (2011) reported that "about 20 million people were at risk of post-disaster disease due to Aila. The damage caused by Cyclone Alia was estimated to total $US40.7 m" (p. 523).

6.3.1.2 Floods

Floods are frequent but unwelcome visitors of Bangladesh. Severe floods usually occur in July and August. There are four types of floods that occurred in Bangladesh, namely (i) flash floods which are caused by "overflowing of hilly rivers in the eastern and northern" regions in the country, (ii) rain floods which are "caused by drainage congestion and heavy rains", (iii) monsoon floods which are caused by major rivers due to the monsoon and (iv) coastal floods which are caused by "storm surges" (Disaster Management and Relief Division, Disaster Management Bureau 2010, p. 6).

The economic and human losses of floods have been as huge as the losses caused by other natural disasters. For example, more than 60 % of the areas in Bangladesh were affected by the flood in 1988. The flood in 1998, which lasted for more than 2 months, "alone caused 1,100 deaths, rendered 30 million people homeless, damaged 500,000 homes and caused heavy loss to infrastructure" (Disaster Management and Relief Division, Disaster Management Bureau 2010, p. 6). This devastating flood had an enormous impact on the national economy, in addition to causing hardships for people and disrupting people's livelihood in urban and rural areas. The recent flood in 2013 in Bangladesh affected nine northern districts and hundred thousands of people and damaged several properties (Network for Information, Response and Preparedness Activities on Disaster [NIRAPAD] 2013).

Generally, the effects of cyclones and floods are the same as the effects caused by other natural disasters, including loss of life, destruction of assets and property, disruption of economic and sociocultural activities, and change in ecological systems (Pramanik 1991; Ministry of Environment and Forest 2001).

6.3.2 Issues Associated with Natural Disasters

It is impossible not to refer to land and water management, and climate change and other disaster-associated issues when discussing DRM since there is a strong correlation between these above issues and disasters (Schneider 2011).

Land management Bangladesh is one of the most densely populated countries where land management on account of fragmentation and subdivision makes it difficult to obtain economies of scale in agriculture. The total forest area in Bangladesh is 2.53 million hectares that is 17.49 % of the total land area of the country (Choudhury and Hossain 2011). Clearing forest land for pineapple gardens, banana plantations, homestead and agricultural purposes has depleted a great deal of forest resources. Ahmed (2008) explained that

> watersheds that are no longer able to sustain and regulate water flows from rivers and streams. Trees are highly effective in absorbing water quantities, keeping the amount of water in watersheds to a manageable level. The forest also serves as a cover against erosion. Once they are gone, too much water can result to downstream flooding, many of which have caused disasters (p. 17).

Deforestation has been the main cause of soil erosion which in turn has made siltation of rivers and sea level rise as a major challenge for the country to cope with. Bangladesh is overpopulated with a high level of density in the urban areas (Ministry of Environment and Forests 2012). Urbanisation has been considered a major concern for the loss of agricultural land (Naab et al. 2013). A part of the land loss is going for industrialisation, and the other part for real estate development.

Urban dwellers Dhaka, the mega capital city of Bangladesh, has about 15.391 million people (CIA n.d.). This is a major challenge for the city planners to provide safe drinking water, sanitation and a network of communication infrastructure. As of 2009, there were more than 300 "secondary towns (pourashavas)", and each had about 15,000–500,000 people (Asian Development Bank 2009, p. ii). The Asian Development Bank (ADB) also reported that "with their local governments languishing due to lack of budgets and decision-making powers, the economic conditions are not very buoyant and, in many cases, the environmental and social conditions are poor" (p. ii). Apparently, when disasters occur, there would be several problems, including (i) lack of healthcare, food and clean water, (ii) poor hygienic conditions, and (iii) lack of proper shelters which aggravate the effects of such disasters.

Food shortage Although the land of many areas in Bangladesh, e.g. the chars, is very fertile, natural disasters have destroyed many crops and hinder the development process of the local residents (Ali et al. 2012). Floods, cyclone and land erosion do not only destroy several crops, which causes shortage of food to the local residents, but they also wash away residents' shelters, livestock, schools, and hospital and affect any means of making a living (Ali et al. 2012).

Water challenges About 30 million people are exposed to drinking arsenic-contaminated water with 90 % living in rural areas in Bangladesh (Faruque and Alam 2002). Open water ecosystems have been seriously degraded because of pollution, land use changes, impacts of climate change such as prolonged droughts and salinity intrusion. Trans-boundary issue is upstream withdrawal of water that has greatly affected downstream water systems. Water quality and water quantity with grave scarcity during dry season and excess water during wet season have become a major life-threatening issue for Bangladesh (Ministry of Environment and Forest 2001).

Climate change It is unfortunate for Bangladesh to be faced with a situation of a changing climate pattern caused due to the effect of global warming for which the industrialised nations are responsible. It is observed that there has been an increase in the average temperature in Bangladesh. Disaster Management and Relief Division, Disaster Management Bureau (2010), explained that "climatic changes in Bangladesh would likely exacerbate present environmental conditions that give rise to land degradation, shortfalls in food production, rural poverty and urban unrest" (p. 23). The vulnerability of calamities has increased to a level whereby there is a direct bearing on the water supply situation, sustainability of food security and impact on livelihood (Johnston 2014; Ericksen et al. 1993). Since about three-thirds of natural disasters originate from weather–climate extremes, it is imperative to mitigate climate change impact (Disaster Management and Relief Division, Disaster Management Bureau 2010).

Overall, the effects of a natural disaster or a combination of disasters are prolonged and unpredictable. It requires large resources and continuous efforts to address the issues associated with natural disasters and disaster management, including mitigation, recovery and preparedness (Ministry of Environment and Forest 2001). In addition, natural disasters result in environmental degradation which, in turn, poses several threats to the socio-economic development of Bangladesh. Therefore, it is imperative to explore novel and innovative disaster warning and dissemination mechanisms as explained by the Ministry of Environment and Forest (2001) that "a timely and accurate alert system about impending disasters will help reduce the loss of life and property" (p. ii). The next section will discuss how different sectors in Bangladesh have managed disasters.

6.4 Disaster Risk Management in Bangladesh

6.4.1 Activities by Government

6.4.1.1 Disaster Management Model

The government in Bangladesh has adopted a three-step model to manage disasters in the National Plan for Disaster Management 2010–2015. The first step is to define and redefine the risk environment. Relevant government agencies and stakeholders have to scan and analyse the environment, including technical analysis, community risk assessment (CRA), and documentation of risk and hazard factors. Given the correlation between climate and disasters, this step also includes the analysis of "climate change and climate variability impacts" (Disaster Management and Relief Division, Disaster Management Bureau 2010, p. 37). The second step is to manage various types of risk. Respective institutes and stakeholders have to search for different options to reduce risk and assess the costs and benefits of each option to select the optimal alternatives which can help to achieve the preset goals. The last step is to respond to the threat environment. Governmental and non-governmental Institutions need mobile resource and have to implement the selected options. They need to predict potential impact scenarios via the use of risk databases. Importantly, communication and reporting channels and means must be effectively established and sustained. Organisations should also share their expertise and experience and learn from others' experience as well (Disaster Management and Relief Division, Disaster Management Bureau 2010). This model can be compared with the action research model in business strategy since it is a feedback cycle; i.e. feedback gathered during the implementation stage will be disseminated to relevant stakeholders in the first and second stages for evaluation of the system and for further improvement.

6.4.1.2 Disaster Management Institutions

The disaster management system in Bangladesh includes activities of various organisations at five levels, namely national, district, Upazila, union and community levels (see Table 6.1). Several institutes are involved in disaster management in Bangladesh, such as National Disaster Management Council, National Disaster Management Advisory Committee, Inter-Ministerial Disaster Management Coordination Committee and so on. At the national level, the Disaster

Table 6.1 Institutional frameworks at different levels

Level	Institutions
National	National Disaster Management Council (NDMC)
	Inter-Ministerial Disaster Management Co-ordination Committee (IMDMCC)
	National Disaster Management Advisory Committee (NDMAC)
	National Platform for Disaster Risk Reduction (NPDRR)
	Cyclone Preparedness Program Implementation Board (CPPIB)
	Cyclone Preparedness Programme (CPP) Policy Committee
	Focal Point Operation Coordination Group of Disaster Management (FPOCG)
	NGO Coordination Committee on Disaster Management (NGOCC)
	Committee for Speedy Dissemination of Disaster Related Warning/Signals (CSDDWS)
District	District Disaster Management Committee (DDMC)
Upazila	Upazila Disaster Management Committee (UZDMC)
Union	Union Disaster Management Committee (UDMC)
Community	Pourashava Disaster Management Committee (PDMC)
	City Corporation Disaster Management Committee (CCDMC)

Source Information is from Disaster Management and Relief Division, Disaster Management Bureau (2010)

Management and Relief Division (DM&RD), under the Ministry of Food and Disaster Management (MoFDM), is the coordinator of disaster management activities and efforts. In order to respond to the need for better disaster management, the MoFDM issued the Standing Orders on Disaster in 1997 which provides guidance and directions to stakeholders regarding implementation and monitoring of disaster management activities in the country (Disaster Management Bureau, Disaster Management and Relief Division 2010). At the subnational level, several committees have been established to coordinate, review and implement disaster-related activities in their jurisdictions, such as District Disaster Management Committee, Upazila Disaster Management Committee, Union Disaster Management Committee, Pourashava Disaster Management Committee and City Corporation Disaster Management Committee.

One of the action agendas in the National Plan for Disaster Management 2010–2015 is to increase public awareness regarding social mobilisation and disaster reduction via public education, capacity building training for academics in training institutions and engagement of NGOs in disaster management programs and plans. Although the NGO Affairs Bureau and Bangladesh Scouts have been mentioned in the National Plan, they are not included in the list of disaster management institutions in Bangladesh (Disaster Management and Relief Division, Disaster Management Bureau 2010).

6.4.1.3 Disaster Reduction Strategies

Five key strategies to reduce risks and manage disaster have been introduced in the National Plan. The government of Bangladesh does acknowledge that the private sector and civil society are key partners in implementation of these strategies. The first strategy is advocacy which aims to enhance the awareness of government agencies, government officials, media personnel, institutions and relevant groups of stakeholders with regard to their roles in various risk reduction initiatives and the benefits of such risk reduction programs (Disaster Management and Relief Division, Disaster Management Bureau 2010).

The second strategy is to reform policy and planning via reviewing of disaster management policy and planning. This aims to ensure that the planning and policy (i) are consistent with the national plan, (ii) can promote a risk reduction culture at various levels and (iii) facilitate the implementation of the national agenda for disaster management (United Nations, International Strategy for Disaster Reduction Secretariat 2012).

The third strategy is to build capacity of stakeholders at all levels. This strategy has been implemented via the review of the roles and responsibilities of various disaster management committees in order to avoid overlapped functions and waste of resources and efforts. The strategy is also supported by the development of a national training curriculum which can help to build capacity of these committees and enhance their functional capabilities (Disaster Management and Relief Division, Disaster Management Bureau 2010).

The fourth strategy is to develop a planning framework which incorporates the needs and limitations of stakeholders at all levels. The planning frameworks take into account the constraints of different committees (Disaster Management and Relief Division, Disaster Management Bureau 2010).

The final strategy is to provide uniform CRA guidelines in order to ensure consistency in the analysis of community and environment risk identification. This also aims to foster "stronger linkages with scientific analysis information" (Disaster Management and Relief Division, Disaster Management Bureau 2010, p. 40).

6.4.1.4 Disaster Management Regulatory Framework

The government of Bangladesh has enacted a number of legislative and introduced plans with regard to risk reduction and disaster management. These include Disaster Management Act, National disaster Management Policy, Disaster Management Plans and so on. In addition, the National Plan for Disaster Management 2010–2015 and the sixth Disaster Preparedness ECHO (DIPECHO) action plan are comprehensive and include activities and strategies to implement both large- and small-scale adaptive activities in some areas where natural disasters usually occur, such as Galbandha (Ali et al. 2012).

Apart from these above acts and guides, Bangladesh does have plans to address specific natural disasters and their associated issues, such as the earthquake contingency plan, the Cyclone Shelter plan, the Disaster Resilient Cluster Housing, and the Tsunami Response plan. Respective government agencies have also introduced best practice models which are adopted by a number of state and non-state organisations to implement DRM. Some of them are (i) "Disaster Impact and Risk Assessment Guideline, (ii) Local Disaster Risk Reduction Fund Management Guidelines, (iii) CRA Guidelines, (iv) Emergency Response and Information Management Guideline" and many others (Disaster Management and Relief Division, Disaster Management Bureau 2010, p. 45).

6.4.2 Activities by the Private Sector

While there is information about how professional organisations (such as Bangladesh Earthquake Society, South Asian Association for Regional Cooperation (SAARC) Meteorological Research Centre, Network for Information Response and Preparedness Activities on Disaster and International Centre for Climate Change and Development) and civil society (Bangladesh Centre for Advanced Studies) have supported various disaster management initiatives, and shared knowledge and expertise, there is little information about the contribution of the private sector in disaster management. Activities of the private sector to manage risk mainly include donation of money and sponsor risk reduction programs, but business enterprise "tends to lack incentives in funding disaster mitigation initiatives and seems content to leave the job to the public sector and international aid funding through the NGOs" (Matin 2002, p. 13).

Most of the time, interventions by private enterprises are to donate money to disaster victims. The monetary contribution is usually from their employees. For example, it is a common practice in Bangladesh that companies call for donation from their employees in response to the call for relief support after each disaster. Unions' members are usually very active in such campaigns. However, the contribution from private businesses is quite limited. For instance, "during the 4-month period from 9 October 2000 to 12 February 2001, out of 236 entries of donations to the Prime Minister's Fund, 47.9 % were made by various associations and employees' unions that collected contributions of one day's salary by the staff … Only 5.5 per cent were private businesses though it was not clear from the data if the fund was collected from individual subscriptions or came from industrial/business establishments" (Matin 2002, p. 15).

Occasionally, private enterprises do sponsor special programs supporting risk reduction and disaster mitigation. A good example is the case of the Square Pharmaceutical Company which has (i) sponsored the "Dhaka Community Hospital (DCH) for part of its arsenic mitigation activities" and donated vitamins and medicine and (ii) supported "the Health Camps of DCH in rural areas" (Matin 2002, p. 16). Vista Communications Company and WorldSpace Corporation in

Bangladesh have tried to ensure their operational continuity in order to provide robust and effective means of communication, especially during disasters (United Nations, International Strategy for Disaster Reduction [UNISDR] Secretariat 2009, p. 14).

6.4.3 Activities by Civil Society

Activities of CSOs to contribute to disaster management can be classified in three main groups, namely provision of education programs, support relief activities and provision of financial assistance. Many international and national CSOs in Bangladesh have played a critical role in social and economic development, and disaster management, especially in the rural areas in the country. For example, an estimated "32,000 trained volunteers work in the coastal areas under a Cyclone Preparedness Program (CPP) jointly operated by the Red Crescent Society and the Ministry of Disaster Management and Relief" (Khan 2008, p. 667). Key CSOs in provision of public education regarding disaster reduction and management are "Bangladesh Rural Advancement Commission, Proshika, Gono Shahajiyo Shongstha, Dhaka Ahsania Mission, and Disaster Management Forum" (Khan 2008, p. 669). International and national CSOs have contributed to offer formal and non-formal education programs to the public in order to raise their awareness of disaster preparedness and risk reduction. Another example of active NGO is Islamic Relief Worldwide which has established the National Alliance for Risk Reduction and Response Initiatives (NARR) in 2010 in Bangladesh. The aims of these initiatives are to enhance disaster reduction efforts via collaboration and coordination among NGO members (Ali et al. 2012).

CSOs have also contributed to build local capacity regarding disaster reduction via the establishment of disaster management committee, community-based task forces, and rebuilding schools, hospitals, clinics. They have also worked with local government and enterprises to mitigate and adapt to the impacts of disasters and climate vulnerabilities (Ali et al. 2012). Such CSOs have gained support from various communities who have the willingness to improve the living standards of their communities. For instance, as of 2012, there were "625 volunteers in 25 Village Development Communities", working with Islamic Relief Worldwide in Bangladesh (Ali et al. 2012, p. 5).

6.5 Assessment of the Current Activities to Manage Disasters

Bangladesh has strived to implement a comprehensive approach to address disaster-associated issues. The government has emphasised the importance of collaboration and close working relationship with all groups of stakeholders "to build strategic, scientific and implementation partnerships with all relevant government departments and agencies, other key non-government players including NGOs, academic and technical institutions, the private sector and donors" (Disaster Management and Relief Division, Disaster Management Bureau 2010, p. III). A good example of such partnership is the collaboration between the government and non-state actors to disseminate information and good practice through the media and print media, as well as "community-led events" since public education is one of the mechanisms to enhance the public awareness of risk and disaster impact (Ali et al. 2012, p. 5).

In the current framework for disaster governance, the government has included all groups of stakeholders in the planning, development and implementation process of risk reduction. Yet, the roles and responsibilities of such groups have not been clearly and adequately documented in national plans and reports. The government has engaged the private sector and civil society in some of their activities although such opportunities evolve around getting feedback from civil society, partnerships for provision of voluntary work, and resource mobilisation via donation and contribution from different groups.

It is plausible that CSOs are quite active in disaster management in the country. Yet, it is not clear how the private sector and CSOs have been engaged in the national framework for disaster governance. Non-inclusion of all groups of stakeholder in the process of disaster governance will certainly produce negative effects and undermine the effectiveness of any efforts and mechanisms adopted by state and non-state institutes (Ha 2013; Dhakal and Ha 2013; Sealza and Ha 2013). As proposed by Drexhage and Murphy (2010), governments should adopt "partnerships between government, business and civil society to identify and test new approaches, and to scale up promising approaches" in governance (p. 20).

6.6 Policy and Technical Recommendations

Overall, disaster reduction management concerns everyone, every group and nation, and thus, various groups of stakeholders must be involved and engaged in this process (Ishiwatari 2003). Although Bangladesh has shown remarkable resilience in reducing the impact of natural disasters on the poor and marginalized population, the authors propose the following recommendations in order to address issues associated with DRM.

Firstly, the existing disaster preparedness programs and plans, including cyclone and flood preparedness training, evacuation plans, early warning systems and communication, etc., are not inclusive enough since some groups of stakeholders may be excluded in these plans and programs (Akter and Mallick 2013). This can be enhanced by fostering the partnerships between groups of stakeholders, and the recipients of disaster plans and policies.

Secondly, although the government has included "social safety net programs", training programs, shelter programs and so on in the national plan for disaster

Table 6.2 Technical recommendations for disaster risk management in Bangladesh

• Develop and maintain mechanisms to build cooperative land farming/tenure systems as subdivision and fragmentation are a very crucial issue leading to problems associated with land management
• Construct new embankments and increase the height of the existing ones in order to prevent floods
• Continue massive afforestation programs in all 67 districts since upstream deforestation contributes to increasing floods. Deforestation contributes to soil erosion
• Use of surface water from rivers and lakes is the main solution to reverse the trends of the lowering of water table in and around major cities
• Recycling used water in the residential houses, hotels, restaurants and other institutions can also contribute to address the shortage of clean water
• Consider to sign MOUs with neighbouring countries to share water rationally with a fair share of river water distribution for the lower riparian countries for a durable friendship, peace and harmony in the relationship for all parties
• Build more cyclone shelters across the country
• Approach international donors and large private enterprises for financial contribution
• Create a revolving fund to be patronised by the government from the annual budget, and involve the private sector to step up their corporate social responsibility activities to meet challenges of DRM
• Provide public education and orientation to the public regarding disaster preparedness and recovery on an ongoing basis in order to reduce loss of life and properties
• Develop a system of insurance for the coverage of agricultural loss due to natural disasters, and cover life insurance for human loss at a subsidised rate

management, such programs are aimed to respond to disaster impact. (Atker and Mallick 2013, p. 36). They cannot act as "a shield against environmental shocks" mentioned by Atker and Mallick (2013 p. 36). Thus, the government should also introduce proactive initiatives and activities to reduce risk.

Other technical recommendations are discussed in Table 6.2.

6.7 Conclusion

This paper has discussed the key types and associated issues of natural disasters and discussed the three-sector governance model for climate change governance. An all-inclusive governance approach towards DRM should include the public sector (government), the private sector and civil society since they can all contribute to risk deduction and disaster management as well as can affect the outcome of any plans and strategies to DRM. Although the government of Bangladesh has introduced a number of strategies, initiatives, and actions, to mitigate and manage disasters, it is not clear how such initiatives and actions have been implemented and evaluated and how different groups of stakeholders can be more involved in the governance process of DRM. This suggests that further research on disaster governance should focus on stakeholder engagement and participation and mechanisms to evaluate the effectiveness of the current plan and institutional frameworks for DRM.

In conclusion, we acknowledge the fact that it is impossible to completely prevent natural disasters. However, with proper and adequate planning and implementation, effective collaboration and coordination among various groups of stakeholders, damage caused by natural disasters can be mitigated.

References

Ahmed, AIMU (2008) Underlying causes of deforestation and forest degradation in Bangladesh. Global Forest Coalition (GFC), The Netherlands, p 76

Akter T (2009) Climate change and flow of environmental displacement in Bangladesh. Centre for Research and Action on Development, Dhaka

Atker T, Mallick B (2013) An empirical investigation of so I-economic resilience to natural disaster. UFZ economic working paper series 04/13. Helmholtz Centre for Environmental Research, Leipzig

Alam E, Collins AE (2010) Cyclone disaster vulnerability and response experiences in coastal Bangladesh. Disasters 34(4):93–954

Ali SS, Rahman M, Chowdhury NR (2012) Bangladesh: a sustainable and disaster resilient future. Islamic Relief Worldwide-Bangladesh, Dhaka

Asgary A, Halim MA (2011) Measuring people's preferences for cyclone vulnerability reduction measures in Bangladesh. Disaster Prev Manage 20(2):186–198

Asian Development Bank (2009) Urban sector and water supply and sanitation in Bangladesh: an exploratory evaluation of the programs of ADB and other aid agencies. Asian Development Bank, Tokyo

Black J (2012) Paradoxes and failures: 'new governance' techniques and the financial crisis. Mod Law Rev 75(6):1037–1063

Braithwaite J, Drahos P (2000) Global business regulation. Cambridge University Press, Cambridge

Choudhury AM (2002) Managing natural disasters in Bangladesh. Presented at the Dhaka meet on sustainable development in Bangladesh: achievements, opportunities and challenges at Rio + 10 organised by Bangladesh Unnayan Parishad, pp 16–18

Choudhury JK, Hossain MAA (2011) Bangladesh forestry outlook study. Working paper no. APFSOS II/ WP/ 2011/ 33. Bangkok: food and agriculture organization of the United Nations Regional Office for Asia and the Pacific

CIA (n.d.) The World Fact Book: Bangladesh. Available from https://www.cia.gov/library/publications/the-world-factbook/geos/bg.html (Accessed 3 May 2–14)

Dhakal TN, Ha H (2013) Responses to climate change—who is responsible? a conclusion. In Ha H and Dhakal TN (eds) Governance approaches for mitigation and adaptation of climate change in Asia. Palgrave McMillan, London, pp 267–272

Disaster Management and Relief Division Disaster Management Bureau (2010) National plan for disaster management 2010–2015. Government of the People's Republic of Bangladesh, Dhaka

Drexhage J, Murphy D (2010) Sustainable development: from Brundtland to Rio 2012. United Nations, New York

Ericksen NJ, Ahmad QK, Chowdhury AR (1993) Socio-economic implications of climate change for Bangladesh. Bangladesh Unnayan Parishad (BUP), Dhaka

Faruque HSM, Alam S (2002) State of arsenic contamination in Bangladesh. Paper presented at the international water conference, Hanoi, Vietnam, October 14–16

Fisher S, Surminski S (2012) The roles of public and private actors in the governance of adaptation: the case of agricultural insurance in India. Centre for Climate Change Economics and Policy, Munich Re Programme Technical, and Grantham Research Institute on Climate Change and the Environment, UK

Ha H (2012) A Multi-sector governance model for environmental sustainability—Australia case. In: Barker JR, Walters R (eds) New Zealand and Australia in focus: economics, the environment and issues in health care. Nova Science Publishers Inc., USA, pp 35–60

Ha H (2013) Climate change governance—Singapore case. In: Ha H, Dhakal TN (eds) Governance approaches for mitigation and adaptation of climate change in Asia. Palgrave McMillan, London, pp 182–199

Ha H, Dhakal TN (eds) (2013) Governance approaches for mitigation and adaptation to climate change—an Asia focus. PalGrave Macmillan, London

Ishiwatari M (2003) Disaster risk management at the national level. Asian Development Bank Institute, Tokyo

Islam SMT, Chik Z (2011) Disaster in Bangladesh and management with advanced information system. Disaster Prev Manage 20(5):521–530

Johnston I (2014) Disaster management and climate change adaptation: a remote island perspective. Disaster Prev Manage 23(2):123–137

Khan MSA (2008) Disaster preparedness for sustainable development in Bangladesh. Disaster Prev Manage 17(5):662–671

Khan MMH, Nahar N (2014) Natural disasters: socio-economic impacts in Bangladesh. Banglavision 13(1):58–67

Kooiman J (1999) Socio-political governance. Public Manag 1(1):67–92

Koiman J, Bavinck M, Chuenpagdee R, Mahon R, Pullin R (2008) Interactive governance and governability: an introduction. J Transdisciplinary Environ Stud 7(1):1–11

Lidskog R, Elander I (2010) Addressing climate change democratically. multi-level governance, transnational networks and governmental structures. Sustain Dev 18:32–41

Matin N (2002) Corporate social responsibility and natural disaster reduction: insights from Bangladesh. UK Department for International Development, London

Ministry of Environment and Forests (2001) State of environment Bangladesh 2001. Government of People's Republic of Bangladesh, Dhaka

Ministry of Environment and Forests (2012) Rio + 20: May 2012 national report on sustainable development. Government of Peoples' Republic of Bangladesh, Dhaka

Ministry of Food and Disaster Management (2008) Super cyclone sidr 2007: impacts and strategies for interventions. Government of People's Republic of Bangladesh, Dhaka

Moore M, Hartley J (2008) Innovations in governance. Publ Manag Rev 10(1):3–20

Naab FZ, Dinye RD, Kasanga RK (2013) Urbanisation and its impact on agricultural lands in growing cities in developing countries: a case study of Tamale in Ghana. Mod Soc Sci J 2(2):256–287

Nehal K (2005) Cyclonic storms in the coastal areas of Bangladesh: socio-economic impact. Paper presented at the international symposium disaster reduction on coasts scientific-sustainable-holistic-accessible, Monash University, Melbourne, pp 14–16 November 2005

Network for Information, Response and Preparedness Activities on Disaster (NIRAPAD) (2013) Situation report: hazard type: flood. Network for Information, Response and Preparedness Activities on Disaster (NIRAPAD), Dhaka

Pramanik MAH (1991) Natural disasters. Bangladesh Space Research and Remote Sensing Organization (SPARRSO), Dhaka

Schneider RO (2011) Climate change: an emergency management perspective. Disaster Prev Manag 20(1):53–62

Sealza IS, Ha H (2013) Environmental legislations and actions in polity, economy and culture for climate change adaptation: a case study of Misamis Oriental Province, the Philippines. In: Huong H, Dhakal TN (eds) Governance approaches for mitigation and adaptation of climate change in Asia. Palgrave McMillan, London, pp 252–266

Stocker G (1998) Governance as theory: five propositions. UNESCO, Paris

Stocker G (2004) Governing as governance, by Jan Kooiman. Int Publ Manag J 7(3):439–442
United Nations, International Strategy for Disaster Reduction Secretariat (2009) The development of a public partnership framework and action plan for disaster risk reduction (DRR) in Asia. UNISDR Secretariat, Geneva
United Nations, International Strategy for Disaster Reduction Secretariat (2012) Hyogo Framework for Action 2005–2015. Building the resilience of nations and communities to disasters. World conference on disaster reduction. Kobe, Hyogo, Japan. Geneva, Switzerland: United Nations, International Strategy for Disaster Reduction, 18–22 January 2005

Chapter 7
Administrative Planning and Political Response to Hazard: Issues and Concerns

Vikas Saharan

7.1 Introduction

Crisis means an unexpected condition with extreme sudden events, and these events are categorized as disaster, catastrophe, etc. Crisis along with suddenness is accompanied by number of highly unforeseen situation and managing it in terms of preparation and recovery is highly daunting task. Disasters affect people, their livelihoods and environment. Disaster also disrupts community function by damaging physical, economic and social capital. Impact of disaster along with its magnitude and intensity also depends on how much individuals and communities are vulnerable. Disaster whether natural (hurricanes, floods, earthquakes and droughts) or man-made (terror attacks, hazmat spill, sabotage, chemical accidents) always has social, economic and political consequences. Disaster, therefore, threatens general existence of people leading to urgent intervention of government in order to reduce negative consequences of disaster. In response to this, there comes need to prepare, support and rebuild society when disaster occur and this depends on social, economic and local condition in disaster region because disaster relief is a matter of locality, for instance during earthquake, poorer people living in steeper slopes are not only more vulnerable to disaster but it also makes their recovery process quite difficult. Decision-making is also complicated in disaster management, and in order to carry out initial post-disaster relief activities in local communities, it is imperative to work out community-centred disaster preparedness activities on a regular basis prior to a disaster. Along with this, residents of local communities should not only participate and cooperate with local authorities in disaster preparedness activities, but they should also come up with their own disaster mitigation plans which are then ultimately reflected in administrative

V. Saharan (✉)
IAS-STS, Graz, Austria
e-mail: Vikas.sw18@Gmail.com

H. Ha (ed.), *Land and Disaster Management Strategies in Asia*,
DOI: 10.1007/978-81-322-1976-7_7

plans of their local authorities. There is debate that there should be balance between the general public and the needs of citizens in hazard-sensitive area (Helsloot and Ruitenberg 2004).

7.2 Research Question

This study identifies factors which are having systematic and significant effect on administrative capacities to deal with disasters. Study also highlights leadership vision and capabilities towards hazard and linkage between them and administrative bodies. Study examines how better procedures and practices can lead to greater capacity building and able handling of disasters. Study also focuses on how communities and self-organizing processes make a critical difference to the system capacities. Attempt has also been made to highlight the role of disaster management and policy reform during these eventualities. These research questions address extremely important issues for practitioners attempting to formulate appropriate policy interventions.

7.3 Literature Review

There were very few studies on how organization deals with disaster management across the world. Despite an urgent need to learn about disaster administration in other nations and recognizing that they provided a unique opportunity, there had been fewer attempts to study their lessons (McEntire 2006). Disaster management involves preparing, supporting and rebuilding society when natural or man-made disasters occur. It depends highly upon the local economic and social conditions within the disaster region (Gerber 2007). Evidence indicates that low-probability events, such as natural hazards, are systematically misjudged (Faure 2007). Prior to a disaster, it is desired that not only administrative bodies but also residents in local communities come up with community-based disaster preparedness, which are then reflected in administrative plans (Lindell and Perry 2000). Success of an emergency manager is tied with the effectiveness of his/her interactions with other government officials and disaster management community (Waugh and Streib 2006). State and federal agencies play important roles by providing public education, alert and warning systems and evacuation plans, but the tools needed to manage hazards and involvement of non-government organization are most often in mental actors and actors build the capacity of communities to deal with future disasters (Comfort 1999). Importantly, greater vulnerabilities of nations outside the developed world were gaining attention only now, though their significance was recognized early in the field (Dynes 1988). For policy intervention, governments were seen as the only organizations with resources and authority to deal with disasters (Waugh 2000). Reducing vulnerability and affording protection through better preparation, response and recovery were good examples of public goods that needed to be addressed through public policy (Parks and Whitaker 1978). There is a need to undertake comparative issues in disaster research (McEntire and Gregg 2007).

7.4 Theoretical Framework

Response to disaster depends upon the collective effort and participation of all segments of society, i.e. government, community and civil society. Entire process confronts numerous challenges, and it has become imperative to establish a theoretical framework which can serve as base for planning and implementing disaster recovery process. There exist numerous theoretical frameworks which deal with disaster response, but these theoretical frameworks have limitation in addressing issue of disaster with holistic and long-term recovery approach. For example, self-organization theory stresses only on spontaneous reallocation of resources to achieve collective goals in disaster response, but it does not focus on holistic and long-term recovery process related to disaster. Escalation theory on other hand highlights potential related to disaster from conventional base to unprecedented level of complexity and scale. This theory falls short on theoretical guidance on planning and implementation to long-term recovery in response to sudden extreme events. There is another theoretical framework which deals with vertical and horizontal integration, in which network among people in a horizontally integrated community helps mobilization of citizens to actively participate in recovery process. Although this theory stresses on the need for active citizen participation in long-term recovery process, it does not provide necessary framework for the involvement of civil society. There is another theory which is related to strategy of resilience, and this theory highlights community capacity to mobilize and act during disaster. Like other theories, this theory is also focussed on quick response and fails to address long-term mitigation process. One of the main problems with disaster response theories is that it has not been able to highlight long-term disaster mitigation and recovery process which makes communities more vulnerable to sudden events. The main role of policy and theory is to guide practice, and it has been witnessed that many times traditional government policies are inadequate to guide the response to catastrophic events. In fact, bureaucratic rules and regulations instead of facilitating recovery process act as a barrier to recovery.

It is evident from above theories that there is a need for a theoretical framework that highlights long-term disaster mitigation and recovery process. This assumes importance as sudden events create political, social and economic upheaval. Theoretical framework for such events must include a variety of crucial issues, i.e. it should take into account long-term recovery as a continuous, evolving process of inquiry that leads to informed action. Theoretical framework must also strive for broader public dialogue during mitigation and recovery process as interaction foster knowledge, trust and leadership among broader group of people which is crucial for innovative action. This highlights active participation of citizens in the long-term mitigation and recovery process. A theoretical framework for long-term mitigation and disaster recovery must also emphasis on the interaction between government, citizens and civil society because interaction among the participating agents leads to creation of new system which plays an important role in long-term recovery process in form of rebuilding homes, infrastructure, communities, cultures and individual lives. These collaborative networks are more flexible and adaptable as compared to administrative hierarchy of rules and constraints. A theoretical framework which incorporates above-mentioned elements is capable of meeting not only requirement of academic and practitioner communities, but it also provides communities a vital link which is necessary for structuring a successful long-term mitigation and recovery process.

7.5 Research Methodology

Major research designs in social science research included both quantitative analysis and qualitative methods (Phillips 1997). Disaster research domain involved the physical, demographic environment and built-up environments. In the past, disasters provided scholars with opportunities for conducting cross-national and international collaborative research (Britton 2006). Globalization and international efforts for disaster reduction had also increased expectations of growing cross-national research (Peacock 1997). However, systematic efforts to approach disaster management from a comparative framework began late (McEntire 2006). Research had long been dominated by narrow analysis of particular cases (Gilbert 1998). Consequently, there had been little learning about the full impact of these crises (Boin and Hart 2006). A survey of disaster research by McEntire showed that most of the studies had focused on international response, causes and consequences of disaster agents, or about specific subject areas (McEntire 2006). Case studies formed the bulk of empirical scholarly work on disasters, event types and emergency management organizations, although each of them constituted only one case for observation (Drabek 1997). The cases that were explored ranged from a train accident in UK, earthquake in China, nuclear accident at Three Mile Island, earthquake in Iran, plague in Surat, Bangladesh's emergency policies, emergency management in Korea, 9/11 attacks, Hurricane Katrina, etc. (Farazmand 2001).

Nature of the research problem and questions also determines methodology to be used. As this study seeks to explore long-term mitigation and recovery process along with role of government, communities and civil society and leadership skills in handling this issue, qualitative approach is appropriate method for this article as it allowed to establish a general direction and is vital to this study, considering that the long-term recovery effort after disaster. Qualitative method not only provided information but also led to the discovery of additional elements that may be useful for the development of a theoretical framework for long-term disaster recovery programmes.

7.6 Findings and Discussion

Wide array of leadership and management competencies is required to manage routine emergencies and catastrophes, and under these challenging conditions, people see to their leaders for their vision and direction to mitigate impact of disaster, moving them to safer places and ultimately leading to normalcy because efficient leadership can reduce the suffering and minimize the disaster impact while inefficient leadership can turn things in a different way. Emergency and crisis have different meaning, in which former has narrow meaning with focus on context specific, while latter has much more boarder meaning ranging from man-made disaster to social problems. The role of emergency managers becomes important in disaster mitigation as emergency managers make immediate and urgent decision, and when information is unavailable, community completely rely on these officials as a source of information and source of evacuation. Thus, leadership during disaster encompasses strategic and diverse task covering all the stages of disaster management.

Advancement of technology has paved new ways to mitigate and prepare for disaster in better and efficient way by minimizing their impact but technological development has also brought new threats and made human being more vulnerable in form of bio-nuclear terrorism, communication breakdown and other failures, and one of the glaring experience was oil spills devastation that was experienced in the Gulf of Mexico in USA. Therefore, networks of organization from different fields, i.e. public, private and non-profit, along with organization from different level of government shape emergency management body. Emergency manager's success in disaster mitigation is linked with its interaction with disaster management community and different government officials.

In order to develop uniform goals and strategies, emergency managers use networks and relationships, and one of the important tool needed during disaster is efficient communication because all the coordination during disaster to achieve and maintain common goals is accomplished through communication. During disaster, emergency manager's role is important in devising policy and programme implementation as it reduces loss to life and property, promotes environment protection and helps in establishing coordination with other organization. Effective communication during disaster involves few important components, i.e. developing pre-disaster networks with various stakeholders, involving similar means of communications between networking organization, developing common condition to work together and fostering inter-organizational and inter-governmental cooperation, and by adopting these pattern, different stakeholders can communicate efficiently and coordinate and collaborate in better way (Kapucu 2006).

During disaster, emergency manager's involvement becomes much wider which goes beyond conventional operations of providing food, shelter, medical aid, etc. It is the interaction and network with disaster-affected community and related government officials which determine the success of emergency manager during disaster. Numerous organizations are involved in handling the mitigation of disaster with large number of them working voluntary without any affiliation with federal, state or local agencies. Therefore, maintaining linkages, networking, interacting, collaborating and planning with this type of voluntary organization play an important role in strengthening disaster preparedness (Kapucu and Van Wart 2006).

These days, more and more government authorities are adopting and integrating better disaster preparedness methods into their operations because there is more pressure on authorities to deliver to people facing high risk to their life and partly because of legal action they have to face if they become careless and insensitive. Same has been adopted by communities regularly affected by disaster, but financial constraints play an important stumbling block in the rapid adoption and expansion of these facilities (Donahue and Joyce 2001). Recent catastrophic disasters throughout the globe has changed the view regarding natural disaster with pre-disaster mitigation and long-term disaster recovery gaining much more wider acceptance among all the stakeholders, and another important component of this is social and economic challenges in post-disaster phase and connection between economic development to disaster recovery process. This shows that disaster mitigation measures these days are evolving in different manner (Russell et al. 1995). Although government structure plays an important role during disasters, at the same time, it is also dependent on community economic, social–psychological and political resources. This involvement of communities plays an important role in creating a sense of bonding, which in turn gives rise to efficiency to entire process. Disasters always demand some kind of reaction in form of response, and this happens only with some kind of collaboration, and this collaboration especially at local level depends on involvement of community and voluntary work of different organization. It is the work of voluntary organization which provides network and linkage to different local communities as effective prevention and mitigation of disaster is largely dependent on involvement of native communities. In local areas, it is the local government which plays an important role in regulating land use and adopting and enforcing building codes and buildings' standards to mitigate natural and man-made hazard. There has been increased awareness among local community to collaborate with the local government officials especially in terms of enforcement of land and building code; in other words, there is more convergence among different stakeholders at local level to mitigate the impact of disaster.

Managing emergency represents complex situations because on one side, managing emergency is spontaneous, while on other side, it requires very fine organization and planning. Those involved in process of managing emergency always work on to innovate, adapt and improvise new things to adjust to local condition because pre-decided plans not always work accordingly and making new plans to work in these condition is quite mammoth task. This has led to the evolution of experts in emergency management which by their consultancy agencies offers wide range of solution. Due to delegation of emergency services to these experts, there is blurring of boundary between government and non-governmental activities. During emergency situation, managers after developing emergency operations and mitigation plans move from one community to other community and coordinate all the exercise by ensuring collaboration between government agencies and communities. Management of emergency is not like routine functions of government offices and agencies, but it is much more meticulous in function involving planning from ground level because it is evident that vulnerability of communities is increased when external resources are unavailable and internal capacity is unable to cope it, and in these situation, relief measures of community and neighbourhood have to work on their own as it take some time for external help to arrive, and as a result of this, there is general understanding among different communities for mutual coordination. It has also led to the formation of umbrella type of organization by involving different organization. There is belief that emergency management programme effectiveness depends not on technical skills of emergency managers but on their interpersonal skills because coordination of emergency operations by these emergency managers is done by maintaining systematic communications. Therefore, management of emergency represents a different challenge because there is not only close linkage between disaster response, disaster recovery, disaster preparedness and hazard mitigation, but it is more closely intertwined. In these situations, there is need of a more liberal, flexible, effective approach with broader perspective which helps in solving puzzling issue; this prompts emergency manager to know the situation and adopt collaborative measures with other organization because collaboration is one of the measures to get job done especially during disaster. During disaster, there is a feeling among public that somebody responsible should take charge of managing emergency and may held accountable for this but this is contrary to basis of emergency fields and especially to collaboration because during emergency, it is the pressure to deliver for community forces emergency managers for collaboration, network and security. Thus, management of emergency both in theory and practice has been shaped by contemporary happening.

Whenever, scale of emergency is small, officials working at the local level are able to handle it very easily. As the scale of emergency increases i.e. when emergency assumes the form of disaster causing destruction to life and property, things become difficult for local officials and community to manage and mitigate it and when emergency assumes extreme form, i.e. catastrophic, during such situations things become complex and government is compelled to involved in multiple level to mitigate this as these events happen at unpredictable level. In these situations, decision-making and collaboration with other partners are directly related to magnitude and impact of disaster. It is the decision-making ability of leader which decides the level of relief measures taken at particular place. Any lapses on this part are related to decision-making of leader, and in most of cases, failure of relief measures is always related to decision-making of leader. Challenging condition during these circumstances depends upon initial decision quality which determines final outcome, capability of decision-making of leaders which ensures quality outcomes, decision-making on part of policy making, and constraints in decision-making of leadership even when leaders are handling important issues. Leadership success in managing disaster depends on assessing and adapting to local situation, fixing all the communications in proper way, capability and willingness to make all type of decision, and facilitating coordination among various stakeholders to mitigate disaster. In order to manage disaster, certain routine functionalities are necessary, i.e. a well-devised plan and system in place, use of latest means of information technology, willingness and ability of leader to restore proper communication between different organization as it is most vital element for smooth functioning of relief measures, ensuring pre-arranged decision-making processes, ensuring cooperation, coordination and proper delimitation of boundaries among respective organization, i.e. government, private and non-profit sectors, and flexibility in decision-making of leaders (Waugh and Streib 2006). With the development and advancement of technology, governments and other agencies are equipped with more better and sophisticated way to deal with all type of emergencies. During the time of crisis, it is the legitimacy and the trust of the people on which leaders rely more which help them in managing and running system smoothly along with this; it is also important that leaders during this situation should show urgency in decision-making, thereby avoiding the development of chaotic situation. Management of these emergencies depends on few things, i.e. negation of anything which is on pattern and assumed lines; focusing more on dynamism with stress on adaptability and flexibility; knowing different pre-condition to manage surprising and chaotic; going beyond conventional working of administration and governance by stressing on attitude, skills and cutting-edge knowledge; and disciplined and extraordinary leadership in order to control resources (McEntire and Gregg 2007).

In order to manage the unexpected, it is important to have pre-plans regarding failure, general avoidance regarding simplification of interpretations, becoming aware of un-expectation and developing skills to contain these types of unexpectedness with commitment to resilience. History has been testimony to the fact that many leaders have been able to transform this type of crisis into prosperity, while others have failed in this attempt. According to (Lester and Krejci 2007) in order to avoid crisis induced by reform, crisis leader should adopt few strategy, i.e. formulation of crisis management philosophy by leaders which helps in negotiating the inherent dilemma of reparation and reform, taking into consideration the opposite arguments by leaders, when reform is pursued because ignorance of this will led to gagging up of all the opposition at time when their performance is under the process of scrutiny, these types of reform paves the way to meet long-term challenges. Important area that is pivotal for making emergency experts competent is building team along with leadership, coordinating along networking with other organization and looking things from bureaucratic, social and political context.

During crisis along with creativeness, willingness and firmness of emergency managers, it is the collaboration which provides manpower, funding, energy and time because sometime during emergencies, there is conflict between officials of central, federal and local level over jurisdiction, which gives felling of competition among agencies. In these conditions, it is the team work with common vision and mission along with dynamic leadership which binds all the stakeholders and transforms existing situation. Efficient leadership is first and foremost component needed to bring required change at intra- and inter-organizational level, and there are many factors needed to bring change in organization among this important factor are willingness within the organization to bring change and developing common goal among organization to bring transformation. Important thing, which is too noted, is that leadership in organization is not concentrated at one place but it is widely scattered throughout the organization pattern. It is this distribution of leadership which plays an important role in achieving the transformation of organization along with wider societal good. In order for leadership to be successful during emergency, it is necessary to have following characteristic: sense making among leaders which help them in realizing the future crises and handling the mitigating process in order to avoid any eventualities; decision-making ability among leaders, so that leaders have capability in making decision at appropriate time and making coordination and networking among other stakeholder in order that decision reaches to wider community; meaning making in which leaders not only gives direction to public but also motivate community in getting out of this situation and ensuring proper line of communication; tracking which makes leaders efficient in keeping eye on effected; and finally learning which makes leader to evaluate the entire happening and come up with its own recommendation based on the success as well as failure (Slovic et al. 1974).

Studies have shown that during crises and emergency, leadership has administrative and political aspects as leaders should establish proper line of communications between public and officials both elected and un-elected. Leaders are seen with the hope by all to control situation before, during and after disaster; along with this, leaders should also be able to command their respective organization and supervise the entire completion process. Administrative leaders during emergencies should focus on mitigation, preparedness, response and recovery (Kapucu and Van Wart 2008). When the government goes for relief measures, it is prone to commit two types of errors; one error is related to under cautiousness which is more visible, more overt with more chances of open outburst from different section of society, e.g. if rescue workers are allowed by the government to enter hazard zone and during relief measures, they are hit by leakage of poisonous gas, then people will blame the authorities for being insensitive to life of workers, while other is related to over cautiousness effort of government which is less visible and more covert, and have less impact in form of public anger, e.g. if government responses very late and slowly to any type of relief measures, it will be blamed for not responding fast. Both of these errors cause damage to people, although latter will have lesser negative campaign. Therefore, for any disaster relief, it is important to have exact information and this is challenge among all the authorities who are involved in disaster relief measures that how to make coordination with all by making use of scattered information. It is not the unawareness regarding the potentiality of disaster which is responsible for slack in preparation among the government. It has always been part of policy negligence among higher echelons of government for not spending the present resources for future because current leadership is always focused more on current gain as compared to future perspective.

Given the impact and scale of hazard, it is important for authorities to devise some disaster protection measures which are capable of saving the livelihoods and assets of individual from calamities. Kobe conference in 2005 came with declaration which emphasized that main responsibility of protecting the people and their assets lies with the existing government and government should adopt policy measures which give priority to reduce disaster according to their resources and capability. Integration of development and humanitarian programme is one of the measures to reduce the impact of disaster and ensuring livelihood security because long-term sustainable development is possible only by knowing and adopting measures to impact of disaster. Over the period of time, number of disasters has increased and these have become largest source of human being casualty as poor people especially those living in developing countries are easily affected. Adoption of proper mitigation measures not only decreases vulnerability of individuals but also paves the way for sustainability.

7.7 Conclusion

Disaster, whether natural or man-made, not only causes loss to life and property but it also impedes development process. Adoption of proper mitigation plans along with coordination and networking with different stakeholders assumes very crucial and significant role in this entire process. These measures along with proper monitoring and evaluation have been adopted by authorities throughout the globe. Role of leadership assumes important role in this entire process as decision-making during time of disaster plays an important role in providing direction to relief measures. Strengthening and improving organizational structural arrangement and measure (effectiveness, practices and procedures), augmenting financial capabilities, improving human resources, building technical competence, enhancing training, and raising performance standards along with creation of disaster management authorities, new statutes, changing from a relief orientation to a comprehensive developmental outlook, setting up emergency support functions, developing fail-safe communications, decentralization of responsibilities (institutional strengthening through local self-government as active and responsive local government is seen to be a better explanation for superior outcomes, and local institutions are recognized as effective voices of local requirements as local communities are the first to reach the scene) and involving local communities and non-state actors; along with this, there is need to promote organizational learning, an inter-governmental context of shared governance by improving coordination between different units and existing agencies, economic resources, adoption of technological change, organization design, subjective abilities of leadership and experience of calamitous events are the measures that can be adopted and implemented in order to prevent any disasters.

References

Boin A, Hart P (2006) The crisis approach? In: Rodriguez H, Quarantelli EL, Dynes RR (eds) Handbook of disaster research. Springer, New York, pp 42–54

Britton N (2006) National planning and response: national systems. In: Rodriguez H, Quarantelli EL, Dynes RR (eds) Handbook of disaster research. Springer, New York, pp 347–367

Comfort LK (1999) Shared risk: complex systems in seismic response. Pergamon/Elsevier Science, Oxford

Donahue A, Joyce P (2001) A framework for analyzing emergency management with an application to federal budgeting. Public Adm Rev 61(6):728–740

Drabek TE (1997) Following some dreams: recognizing opportunities, posing interesting questions, and implementing alternative methods. Int J Mass Emerg Disasters 15(1):21–46

Dynes RR (1988) "Cross-cultural international research: sociology and disaster. Int J Mass Emerg Disasters 6(2):101–129

Farazmand A (2001) Handbook of crisis and emergency management. Marcel Dekker, New York

Faure M (2007) Financial compensation for victims of catastrophes: a law and economics perspective. Law Policy 29(3):339–367

Gerber BJ (2007) Disaster management in the United States: examining key political and policy challenges. Policy Stud J 35(2):227–238

Gilbert C (1998) Studying disaster: changes in the main conceptual tools. In: Quarantelli EL (ed) What is a disaster? Perspectives on the question. Routledge, London, pp 11–18

Helsloot I, Ruitenberg A (2004) Citizen response to disasters: a survey of literature and some practical implications. J Contingencies Crisis Manag 12:98–111

Kapucu N (2006) Interagency communication networks during emergencies: boundary spanners in multi-agency coordination. Am Rev Public Adm 36(2):207–225

Kapucu N, Van Wart M (2006) The evolving role of the public sector in managing extreme events: lessons learned. Adm Soc 38(3):279–308

Kapucu N, Van Wart M (2008) Making matters worse: an anatomy of leadership failures in managing catastrophic events. Adm Soc 40(7):711–740

Lester W, Krejci D (2007) Business "Not" as usual: the national incident management system, federalism, and leadership. Public Adm Rev 67(s1):583–591

Lindell MK, Perry RW (2000) Household adjustment to earthquake hazard: a review of research. Environ Behav 32:461–501

McEntire DA (2006) Local emergency management organizations. In: Rodriguez H, Quarantelli EL, Dynes RR (eds) Handbook of disaster research. Springer, New York, pp 168–182

McEntire DA, Gregg D (2007) The intergovernmental context. In: Waugh WL Jr, Tierney K (eds) Emergency management: principles and practice for local government. ICMA, Washington, pp 57–70

Parks RB, Whitaker GP (1978) Patterns of metropolitan policing. Ballinger Publishing Company, Cambridge

Peacock WG (1997) Cross-national and comparative disaster research. Int J Mass Emerg Disasters 15(1):117–133

Phillips B (1997) Qualitative methods and disaster research. Int J Mass Emerg Disasters 15(1):179–196

Russell L, Goltz JD, Bourque LB (1995) Preparedness and hazard mitigation actions before and after two earthquakes. Environ Behav 27:744–770

Slovic P, Kunreuther H, White GF (1974) Decision processes rationality and adjustment to natural hazards. In: White F (ed) Natural hazards, local, national, and global. Oxford University Press, New York, pp 187–205

Waugh WL Jr (2000) Living with hazards, dealing with disasters: an introduction to emergency management. Sharpe Publishers, Armonk, M.E

Waugh WL Jr, Streib G (2006) Collaboration and leadership for effective emergency management. Public Adm Rev 66(s1):131–140

Chapter 8
Deficiency of Disability Issue in Disaster Risk Reduction Strategy: Implications on Human Security and Social Cohesion

Md Shafiqul Islam

8.1 Introduction

Worldwide 15 % of the population is living with some form of disability (World Health Organization and World Bank 2011). Prevalence of disability is little higher in high income countries but being poor the PWDs of low income countries are more vulnerable to disasters. The wide ranging vulnerabilities of climate change (CC) disproportionately impacts persons with disability (PWD) (World Health Organization 2005). The man-made disasters also impact the PWDs in similar way. This is evident from the study of Hurricane Katrina (USA, 2005), Cyclone Sidr (Bangladesh 2007), Haiti Earthquake (2010) and Rana Plaza Building Collapse (Bangladesh 2013).

PWDs are among the most adversely affected during conflict situations or when natural disaster strike. They experience higher mortality rates, have fewer available resources and overlooked during evacuation, relief, recovery and rehabilitation efforts (Mitchell 2014). During natural disaster, PWDs not only suffer the impacts of natural disaster as the general population but are also less able to cope with the deterioration of environment. They are not able to flee from the impact area and likely to be left behind or abandoned during evacuation. These instances result due to a lack of prevention, planning and lack of accessible services. PWDs also experience greater loss of autonomy due to loss of impairments or mobility aids left behind. As per the report of the International Federation Of Red Cross And Red Crescent Societies (IFRC), PWDs are vulnerable to disasters, both on the account of impairments and poverty. They are often excluded at all levels of disaster preparedness, mitigation and interventions (IFRC 2007). The impacts of disasters on PWDs are also recognized and addressed by Article 11 of the United

M.S. Islam (✉)
Centre for the Rehabilitation of the Paralysed, Savar, Dhaka, Bangladesh
e-mail: shafiq022002@yahoo.com

H. Ha (ed.), *Land and Disaster Management Strategies in Asia*,
DOI: 10.1007/978-81-322-1976-7_8

Nations (UN) convention on the rights of persons with disabilities (CRPD). As one of the world's most vulnerable and disaster affected countries, the issue is very significant for human security and social cohesion in Bangladesh. Every year, lots of people are affected by natural and human induced disasters in Bangladesh. Besides casualties, disasters physically disable many people creating a vicious cycle of more disability, increased poverty and social exclusion.

Over the years, Bangladesh has significantly improved its disaster management capacity both at national and local levels. In line with the constitutional and United Nation International Strategy for Risk Reduction (UNISDR) requirements, the government has enacted relevant acts and developed policies. But surprisingly, PWDs and other marginalized groups are yet to be addressed in disaster risk reduction (DRR) policy instruments particularly in implementation plan. This not only violates the constitutional and CRPD commitment but also hinders poverty reduction, disability sensitive development, human security and social cohesion.

8.2 Research Objectives

The objectives of this paper were to (i) highlight how the disability issues are neglected in DRR strategy and are affecting the human security and social cohesion with particular reference to Bangladesh. (ii) Identify approaches required for the mitigation of such threats and challenges.

8.3 Research Method

8.3.1 A Brief Baseline of Literature

Disability and Disaster are cross-cutting issues having impacts on every aspect of human life like health, economy, poverty, education, security, social life and citizen's affairs. There are wide ranging policy and published documents which links the disaster and disability issues. At the international context, UN took the pioneering role of sensitizing the issues by under taking Universal Declaration of Human Rights, UN Conventions on the Rights of the Child, The World Conference on Special Needs Education (Salamanca Declaration 1994), CRPD (2006), (Commission on Human Security 2003). Intergovernmental Panel on CC Report (2007), the UN International Decade for Natural Disaster (1990–1999), International Strategy for Disaster Reduction and Hyogo Framework for Action (2005–2015). Following the UN initiatives, nation state prepared its own Act, Policy, Strategy and Standing orders to manage the disasters and face the adverse effects of CC. Bangladesh government enacted Disaster Management Act (2012), Disaster Management Rules (2013), Bangladesh CC Strategy and Action Plan 2009 were introduced. There were also many orders and instructions. But these do not mention the special need and support for the disabled. Even very few studies were conducted to determine the prevalence of disability in Bangladesh.

In the absence of a national census or empirical study mitigation planning became difficult. To ameliorate the situation, sporadic studies were conducted on purposive sampling method. One such study conducted by Unnyan Onneshan mentioned disability prevalence in the range of 8–6 % which is far below the estimated rate (Onneshan 2005). This issue has been analyzed by many social scientists and many articles have been published. For example, 'How to include disability Issue in Disaster Management' by Handicap International, Disability in Bangladesh: A Situational Analysis' by the Danish Bilhaziasis Laboratory for the World Bank (2004), 'Disaster Disability and Rehabilitation' by WHO (2005). The linkages between poverty and disability have been mentioned in many publications (Alam et al. 2005). However, other writers have different view to articulate poverty as a consequence of precise terms (Misra and Tripathi 2001). In this study, these will be systematically analyzed in line with the objective of the study.

8.3.2 Methodology

A mixed method design is applied in this study. The contextual framework of the paper is based on secondary published sources and empirical research literature of co-related topics. The focus is on a systematic secondary document analysis from the wide variety of published and online search articles relevant to the theme of the study. The other method includes field level experiences in some serious post-disaster incidents. It may be noted that the author himself was involved in the day to day management of a huge man-made disaster i.e., Rana Plaza Tragedy of 2013. All these sources helped in generating the ideas expressed in this paper for the improvements of disability issue in DRR strategy of Bangladesh.

8.4 Defining Disability, Human Security and Social Cohesion

8.4.1 Disability

The disability concept has changed over the decades. Earlier, disability used to be defined as an individual problem based on the concept of charity and medical model. During 1980s came the social model that defines disability as long-term physical, mental, intellectual or sensory impairments. These impairments interaction with various barriers may hinder people's full and effective participation in the society on equal basis with others (CRPD 2006). Another definition is given by WHO on the concept of bio-psychosocial model. This definition explains how people are disabled through stigmatizing social interactions, environmental barriers and other social phenomena (Njelesani et al. 2012).

8.4.2 Human Security

Human security is broadly understood as freedom from fear and want. It also means protecting and empowering the world's most vulnerable people. It encompasses safety from chronic threats, such as hunger, disease and political repression, and protection from sudden and hurtful disruptions, such as war, violence and disaster. In other words, human security is understood as security of survival (mortality/injury and health), security of livelihood (food, water, energy, environmental needs, shelter and economic security) and dignity (basic human rights, capacity and participation). UN Commission on Human Security defined human security as creating political, social, environmental, economic, military and cultural systems that together give people the building blocks of survival, livelihood and dignity (Commission on Human Security 2003). In accepting these definitions, the question arises, are PWDs secure especially from disasters?

8.4.3 Social Cohesion

A sense of belonging is one main aspect of social cohesion. It is important for disabled people to belong, to be part of shared values, as much as it is for everyone else. But the global, national and social reality is different which is reflected in the research findings of CRPD and also that of many others. Social inclusion and respect for human rights are essential for peaceful societies. The term social cohesion was first coined by Emile Durkheim in his 1983 book, *The Diversion of Labour in Society*. Today, it has various definition narrated by many scholars. Some of the attributed characteristics of social cohesion are belonging, shared values, shared challenges, equal opportunities, sense of trust, hope, capacity to live together in some harmony and a sense of mutual commitment. Social cohesion can further be elaborated as everyone having equal access to establish basic social relationship in society i.e., work participation, family life, political participation and activities in civil society. Social cohesion is directly linked to indexes, indicators and various measures of well-being and to various facets of human security.

8.5 The Vicious Cycle of Disaster, Disability, Poverty and Human Security

It is known fact that disaster not only impacts disabled people disproportionately but also results in increase in the number of PWDs. This happened in the case of both man-made and natural disaster. Again, disability and poverty are inextricably linked. In developed and developing countries alike, people with disabilities are more likely to be poor than their nondisabled counterparts. The 2006 Human Development Report (UNDP 2006) states that 'people with disabilities are among the most vulnerable members of society and among the poorest'. A vicious cycle links disability and chronic poverty: if you are poor you are more likely be disabled, and if you are disabled you are more likely to be poor. This poverty factor of the disabled, infringe their human and livelihood rights leading to human insecurity. The cycle goes on expanding unless it is checked by disabled inclusive disaster management program to turn the cycle into a virtuous one. For each permanent disability resulting from disaster, a family is tied down to daily care. They face significant barriers when seeking employments and block their rights to employment (Nuri et al. 2012). They are virtually excluded from normal productive life leading to poverty and human insecurity. Disabled inclusive and sensitive disaster management and DRR strategy can arrest the situation.

8.6 Bangladesh Case Study

8.6.1 Prevalence of Disability

There is no authentic data on disability in Bangladesh. Currently, a nationwide survey is in progress but its accuracy and methodology are questionable. It is estimated that disabled people represent about 10 % of the population meaning approximately 15 million disabled people are there in Bangladesh. This is a huge number and no development discourse can achieve its goal excluding this huge population. Unfortunately, these people face discrimination, which has an influence on all aspect of their life. They often have less access to education, employment, information or to essential goods. Thus, they become part of the most disadvantaged and marginalized group. The problem is more acute due to inaccessible infrastructure and transport system, non-inclusive culture and insensitive social customs. A concerted effort by the government, NGO (non-governmental organizations), civil society and community can easily introduce an inclusive culture in all spheres including DRR and development discourse.

8.6.2 Rationale for Disability Inclusion in DRR Strategy and Implementation Plan

People living with disabilities are vulnerable to disasters in many ways. Due to their disabilities, they have limited movement and access to communication, education, health care, shelter, food and work. Family members often try to hide disabled persons from the society and as a result they become invisible and in most cases their names never get included in the government data base. It is already mentioned that disaster disproportionately affects disabled than the able persons. Moreover their requirements in all phases of disaster is different and special from the able bodied men and women. Due to their limited ability, they often remain invisible and excluded from accessing emergency support and essential services such as food, medical care, shelter, water, and sanitation and hygiene facilities. Information on disaster preparedness and disaster warnings is often in inaccessible format. Environmental, social and attitudinal barriers also prevent their inclusion to essential services resulting in human insecurity and social exclusion. On the contrary, disability inclusive DRR process reduces morbidity and mortality.

8.6.3 An Analysis of Bangladesh's Acts, Rules, Strategy, Standing Orders and Policies on Disaster Management and Disability

During the past decades, Bangladesh has significantly improved its disaster management capacity both at national and local levels. In line with the constitution, CRPD and United Nations International Strategy for Disaster Risk Reduction (UNISDR) requirements the government has enacted relevant acts and developed policies like National Policy on Disability 1995, Disable Persons Right and Protection Act (2013), Disaster Management Act (2012), CC strategy and action plan 2009, Revised standing order for disaster (2010), National plan for disaster management (2010–2015), South Asian Association for Regional Cooperation (SAARC) and Agreement on rapid response to disaster. But surprisingly PWDs and other marginalized groups are yet to be addressed in DRR policy instruments particularly in implementation plan. This has resulted in government actions, development, preparedness and plan which are insensitive, inaccessible and non-inclusive for the disabled (Anam et al. 2002). For example, the cyclone shelters are inaccessible and the warning and information system is unreachable. The volunteers and government machinery is unskilled in assessing the needs and handling the disabled during disaster. Various committees and planning processes are non-inclusive for the disabled. This not only violates the constitutional and CRPD commitment but also hinders poverty reduction, disability sensitive development, human security and social cohesion. Without including everyone at all levels of the disaster management cycle, it is not possible to build maximum resilience for all communities. The current disaster management practice, its planning and implementation are not disabling sensitive. The socio-economic development model of Bangladesh and its related policies do not include disability, whereas as per CRPD, it should be disable sensitive.

In Bangladesh, social service department (SSD) and some NGOs conduct some disable inclusive program considering their special needs. But there is no mention of disable people in the pre-disaster and post-disaster action. Even there is no mention about the special considerations and need of the PWDs for the disaster preparedness and response activities. The Disaster management committees at district, upazilla and union level are not disabling inclusive and disabled are not consulted and have no voice in these committees. As a result government officials, volunteers and relief workers are not aware of the need and do not have adequate skill to take care of the disabled at different phases of disaster management. As a result, very few disaster response officials are trained to deal effectively with these people and very few institutions have been developed to cater for the rescue and rehabilitation of the disabled. All these gaps have amplified the sufferings of the disabled in the man-made and natural disaster.

This was also reflected during post Rana Plaza care of the victims. An institutional mapping conducted by Action Aid showed that very few organizations (only 4 out of 32) provided some rehabilitation service to the disabled which is grossly inadequate. Only recently government has enacted Disabled persons' right and protection Act which adequately addressed disability issue and their representation at all level and all level decision making process. But no rule and policy has been developed based on this Act. So its implementation is going to take long time resolving the conflicts arising from existing Acts, rules and policies. Moreover, two important articles (Articles 31 and 36) regarding registration and identity card and forbidding discrimination and paying compensation has been put on hold for indefinite period. Without the implementation of these two important articles, disability issue is not going to be addressed positively for building an all-inclusive society.

Therefore, active and meaningful participation of PWDs in the planning process, implementation and monitoring of DRR actions must be ensured by enhancing the capacity building programs at grass root levels. All risk assessment and planning process should be disability inclusive and the database of persons with disabilities should be available at upazilla and district levels so that they do not become invisible or get excluded during an emergency. A report of Handicap International estimates that in the districts of Gaibandha and Sirajganj approximately 60 % of PWDs were overlooked in flood emergency programmer in 2007. PWDs are often part of the poorest groups of people. And the most alarming thing is that the disasters create new impairments and new disabilities. In the recent incidence of Rana Plaza, about 2,500 people were injured and many of them have been disabled permanently.

8.6.4 Case Study Sidr

A large cyclone named Sidr hit southwest part of Bangladesh on November 11, 2007. It killed 3268 people and injured around 55,000. Research conducted by Centre for Injury Prevention and Research Bangladesh (CIPRB) revealed that 11 % of the injured persons sustained permanent disability. It was further revealed that approximate half million disabled people were affected by Sidr and only one third were evacuated to the cyclone shelter which were disable unfriendly. It was further revealed that most of the disabled were left alone in the home where most of the abled family members had to run to safer places. The volunteers, social mobilizers, relief workers and rescuers were not equipped with knowledge, skill and resources to cater for the need of the disabled.

8.6.5 Case Study: Rana Plaza Building Collapse

On April 24, 2013, the deadliest man-made industrial accident occurred in Bangladesh, which resulted into approximate 1,200 deaths and 2,500 rescued alive with various injuries. The study of rescue, search recovery and post-disaster activities revealed the inadequacies at all levels to handle the special needs of those who became permanently disabled. The deficiency was noticed in handling the spinal cord injury and limb amputee cases. The overall shortage of orthotics and prosthesis professional was very critical and there is no training and education facility on this important disability-related profession. The garments factories housing this building were inaccessible to the disable workers. So it is presumed that there were very few disabled persons who faced this disaster. But it is after effect created new generations of PWD.

8.6.6 Impacts on Human Security and on Social Cohesion

It is mentioned that approximate 10 % of Bangladesh population is suffering from different form of disability. It is also revealed that every natural and man-made disaster results in newer generation disability. It is also revealed that Bangladesh is at the hot spot of CC disasters and the DRR strategy is insensitive to the need of the disabled. Being insensitive to the need of such a huge population is detrimental for inclusive society and violates human rights and other basic right of the disabled. Violation of human rights of the disabled puts human security of Bangladesh at risk. Violation of an individual's rights puts that person into a situation of risk to his well-being. When people's rights are violated because they belong to a particular social group it puts the well-being of entire group into risk. When this violation becomes systematic due to the strategic failure like the DRR strategy the human rights, social cohesion and human security of the entire population of the state are risked.

8.7 Conclusion

PWD rights are often unmet due to their exclusion from disaster management initiatives. This also results into high rates of mortality and morbidity, deterioration of health conditions and loss of autonomy which ultimately impacts on human security and social cohesion. The situation is same in Bangladesh also. Though Bangladesh has developed a well-integrated disaster management strategy, but it lacks disability inclusion particularly at implementation phase. Being at the hot spot of CC, this has disproportionate impacts on PWDs which is revealed from the study of different natural and man-made disasters. The situation can be reversed through inclusion of PWDs at every stage DRR which will ultimately ensure security of all human being and help in building an inclusive society. CC, disaster and disability studies in terms of the themes of human security and social cohesion are significant. For too long PWDs have been marginalized from mainstream society .The absence of disability issue in the discourse of CC and disaster management is echoing the same marginalization of the PWDs. Unless their social, material and security issues are understood in the context of disaster, these peoples live will not be secured which has ultimate impact on social cohesion.

References

Alam KJ, Bari N, Khan AHM (2005) Community based rehabilitations practices and alleviation of poverty of peoples with disabilities in Bangladesh. In: Workshop report on community based rehabilitation and poverty alleviation of persons with disabilities, Bangkok, 5 Jul 2005

Anam N, Khan AHM, Bari N, Alam KJ (2002) Unveiling darkness, the situation analysis on disaster and disability issues in the coastal belt of Bangladesh. In: 2nd ed., Centre for Services and Information on Disability (CSID), Dhaka, Bangladesh

Commission on Human Security (2003) Human Security Now, Retrieve from: http://www.humansecurity-chs.org/finalreport//English/FinalReport.pdf

Disable Persons Right and Protection Act (2013) Government of the peoples' Republic of Bangladesh

Disaster Management Act (2012) Government of the peoples' Republic of Bangladesh

Disaster Management Rule (2013) Government of the peoples, Republic of Bangladesh

International Federation of Red Cross and Red Crescent Societies. IFRC (2007) World disaster report 2007: focus on discrimination, international federation of red cross and red crescent societies. International Federation of Red Cross and Red Crescent Societies, Bern, Switzerland

Misra G, Tripathi KN (2001) Psychological dimensions of poverty and deprivation. In Pandy J (ed), Psychology in India revisited: developments in the discipline: vol 3. Applied social psychology and organizational psychology. Sage Publications, New Delhi, India

Mitchell D (2014) Crises, conflict and disability: ensuring equality. In: Karr V (ed). ISBN 13: 9780415811651, Routledge (advances in disability studies)

Njelesani J, Cleaver S, Tataryn M, Nixon S (2012) Using a human right based approach to disability in disaster management initiatives. In: Cheval S (ed) Natural disaster, ISBN: 978-953-51-0188-8, In Tech. Available from: http://www.intechhopen.com/books/natural-disaster/using-ahuman-right-based-approach-to-disabilty-in-disaster-management-initiatives

Nuri RPM, Hoque MT, Akand MMK, Waldron MS (2012) Impact assessment of vocational training programme for persons with disabilities in Bangladesh. J Asia Pac Disabil Rehab 23(3):2012

Onneshan U (2005) Disability in Bangladesh: prevalence, knowledge, attitudes and practices. Unnyan Onneshan, Dhaka

Standing Orders on Disaster (2010) Ministry of food and disaster management, peoples' Republic of Bangladesh, Apr 2010

UNDP (2006) Human development report 2006. UNDP, New York, USA

United Nations Convention on the Rights of Persons with Disabilities (CRPD) (2006)

World Health Organization (WHO) (2005) Disasters, disability and rehabilitation. Available at http://www.who.int/violence_injury_prevention/other_injury/en/disaster_disaability.pdf

World Health Organization (WHO) and World Bank (2011) World report on disability 2011, World Health Organization, ISBN 978-92-4-068521-5. Switzerland, Geneva

Chapter 9
Geopolitics and Community Vulnerability: A Case of Diu Island

Vinita Yadav

9.1 Introduction

9.1.1 GeoPolitical Linkages: Its Impact on an Island

Geopolitics involves the study of the influence of political and economic geography on the politics, foreign policy as well as between geography, politics and power (Hafeznia 2006). It affects the trade and relations between the states. Vulnerability is a mix of structure and processes. Structures are economic, social, cultural, environmental and institutional and process related to poverty and risk (Brouwer et al. 2007; Kelly and Adger 2000). For India, not only the south-west dialogue is important but also south–south and south-east so as to establish strong political and economic linkages with neighbouring nations. India's Look East policy has helped to establish economic and strategic linkages in East Asia. Diu Island is at the southern tip of Gujarat. It is surrounded by Arabian Sea and shares international water boundaries with Pakistan.

9.1.2 Research Objectives

The research aims at understanding the Coastal Regulation Zone (CRZ),[1] the impact of disaster and suggesting ways to bring socio-economic changes in Diu Island. The objectives are to (1) understand the geopolitics and conflicting

[1] The coastal stretched influenced by tidal action up to 500 m from high tide line and land between low and high tide line are declared as CRZ (MoEF 1991).

V. Yadav (✉)
Department of Regional Planning, School of Planning and Architecture, New Delhi, India
e-mail: v.yadav@spa.ac.in

H. Ha (ed.), *Land and Disaster Management Strategies in Asia*,
DOI: 10.1007/978-81-322-1976-7_9

situation; (2) identify the areas falling under different type of coastal zones under CRZ and analyse its impact on people; and (3) identify issues and suggest strategies to minimize the economic and human loss.

9.1.3 Structure

The paper is divided into four sections. The first section introduces and states the objectives of the study. The second section explains the conceptual framework followed by a section on Research Methodology. The last section explains in detail about the findings of the research regarding the areas under different CRZs, floods, earthquake, cyclone faced by Diu Island and fisherman, technologies used and their stories. It is followed by conclusion and strategies to minimize the losses.

9.2 Literature Review

The geopolitics refers to a reciprocal relationship between politics, geography, demography and economies and interactions among them (Hafeznia 2006). It requires a study of the state, international relations and localities and their interest focused on an area (Toncea 2006). Critical geopolitics is part of contemporary geography but is anti-geopolitics, anti-cartographic and anti-environmental (Haverluk et al. 2014). Geopolitics in a climate change context becomes significant due to power equations between not only super powers but also emerging economies (Maxmillian and Depledge 2013). The disaster creating unanticipated damage only gets attention of institutions, whereas less emphasis is laid on reducing the occurrence of disasters. The vulnerable population ventures into sea despite being aware of impending disaster due to poverty and fragile livelihoods (Gaillard et al. 2009). The occupational pluralism is found less in fishermen due to their dependence on sea for livelihood (Badjeck 2008). For fisheries sectors, a normative approach, of removal of fishing vessels to safer areas and compensating the affected fishermen for the damage to gear and vessels, has inadequately addressed the issue (Delaney et al. 2003). An urgency to adopt 3R (reduce risks, rebuild communities and restore permanence) to show results may lead to an adverse socio-economic and environmental impact (Ingram et al. 2006). Vulnerability actively shapes their coping strategies and has a degree of dynamic resilience in the face of shocks than merely being passive recipients of policy and other forms, but socially, vulnerable groups at times fail in their approach (Bohle 2007). Myhr and Nordstrom (2007) concluded from their study on Tanzanian fishermen that mobile phones empower people, give knowledge about market opportunities and decrease risk.

9.3 Theoretical Framework

The conceptual framework used in the research is to understand the geopolitics and locational specificities impacting the population. An island is an interesting case to understand the climate change, disaster risk and population engaged in subsistence activities getting impacted by such change. The climate changes in terms of an increase in temperature, precipitation and sea water level, proneness to cyclone and resulting floods and earthquake influence negatively the adjoining population. The inter-country conflicts arise both at land and sea, but for an island, the conflicts need to be understood more in the light of adversity faced by its occupants.

The coastal zone regulations decide the land utilization, and its violation can be disastrous in the ecologically sensitive zone. In an island, vulnerability is understood by studying natural hazards, exposure and sensitivity on one hand and coastal zone regulations and its impact in geopolitical sensitive area on the other. The island population is dependent on sea, and its fury results in loss of human lives and economic loss. In a Union Territory (UT), which is divided into two parts Daman and Diu, administrative capital is Daman district located at a distance of 768 km away from the Diu district.

9.4 Research Methodology

The literature review was undertaken to understand the geopolitics and its affect, vulnerabilities and assessment, coastal zones and regulations impacting the lives of residents. The secondary data related to elevation, survey boundaries from regional plan and land ownership detail were collected from the department of land revenue survey, and plain table maps were collected from the Town and Country Planning Department (TCPD) at Diu. The elevation data were extracted through plotting the contour lines with an interval of 5 m. The survey was conducted by TCPD for rural areas whose boundaries were geo-referenced and digitized. For urban areas, plane table survey maps were stitched together, geo-referenced and digitized for survey boundaries only. The rural and urban maps were then combined to get the survey number boundaries for the entire district. The CRZ boundaries data were geo-referenced and digitised to extract the areas falling under different CRZ zones. The errors, if any, in the digitized maps were removed so as to create polygon for extracting data. The primary data were collected with the help of interviews with stakeholders, i.e. officials, fisherman, households and industry owners engaged in allied activities. and focus group discussions with fisherman associations, panchayat heads were held to understand the issues associated with the disasters and community vulnerability in geopolitical sensitive zone.

9.5 Diu an Island: A Case Study

Diu is part of Daman and Diu Union Territory in India (TCS 2002). This is second smallest UT in India. It was liberated along with Goa on 19 December 1961 after 5 decades of Portuguese colonial rule. Since then, Goa, Daman and Diu were single UT. In 1987, Goa was given statehood and Daman and Diu became separate UT. Daman and Diu, with an area of 112 km^2 and population of 243,247 [an increase of 53 % (158,204) from 2001 population], are two separate districts governed by central government in India (MoIB 2008). Daman and Diu (219 %) have the highest decadal growth rate in urban population during 2001–2011. Due to proximity to Gujarat and Maharashtra state, it has mixed culture.

Diu Island is located at 20°41′ North Latitude and 70°58′60″ East Longitude and situated in the southern side of Gujarat near the Viraval port. The coastline is of 21 km out of which north–south length is 4.6 km and east to west width is 13.8 km. Diu is a famous tourist destination due to Nagoa and Jalandhar beaches, Diu Fort, Portuguese churches and Gangeshwar temple. It has limited connectivity with mainland through National Highway 8E, Kesariya Diu road and Una Diu road connecting Diu with Somnath Gir district in the north. The nearest railway station is Delwada at a distance of 9 km and Vapi located at a distance of 12 km. However, Veraval is the town connected by major trains at a distance of 90 km. Valsad junction is another major station nearby. Diu is connected by Jet Airways flight via Mumbai.

Diu district has a population of 52,074 spread over an area of 40 km^2 in 2011. The population has increased from 44,110 in 2001 with a decadal growth rate of 18 %. The population density is 1,342 persons per square kilometre, but in habituated area, it increases up to 7,234 persons per square kilometre if we consider only the built-up area. The reason for which is the large area being covered by wetland, forests, agriculture and shrimp farming. The sex ratio is 1,066 females per 1,000 male, which is more than the national average.

Diu district has five settlements out of which Diu Municipality is an urban settlement, and Vanakbara, Saudwadi, Bhucharwada and Zolawadi are four village panchayats rural settlements (Census of India 2011). Diu Municipality has an area of 17.7 km^2 with a population of 23,991. One of the settlements in Diu Municipality, i.e. Ghoghla is located in main land. The topography of Ghoghla consists of flat plains. A small part of Diu named Simbor is located in the Gujarat state at a distance of 25 km from Diu. The central region comprises of airport, forest and agricultural farms of Bucharwada and Zolawadi panchayats. Most of the rural settlements located in the central region are along the forest and major road. Till 1996, Chassi River used to flow along the northern part of the Diu region. However, the river dried up after the construction of a dam in Gujarat. This area has become creek now. It is known as Chassi creek as per government records.

Diu district has four minor ports, i.e. Diu-Ghoghla, Simbor, Vanakbara and Malala declared by government notifications in 1967, 1972, 1984 and 1997, respectively, under the section 4 of Indian Port Act, 1908. Diu was once a sea port and naval base with strategic location for trade of salt, dry and fresh fish and ivory work in fourteenth to sixteenth century but lost its importance to Gujarat ports due to their accessibility.

The large chunk of population is dependent on fishing for their livelihood due to higher level of productivity. In primary sector, contribution of fishing in the employment is 96 %. The various kinds of fish found in Diu are shark, pomfret, prawns, dara and eel. The export of marine product through the Diu port has also gained prominence. Diu's economy is interdependent on Gujarat's economy especially of coastal areas. The coastal erosion is another area of concern for the authority and an area of 516 ha covered with casuarinas, and other tree species has been proposed as reserved forest area. The coastal stretch plantation leads to restricting coastal erosion.

9.6 Findings and Discussion

9.6.1 Elevation and Its Impact on Disaster

Diu's elevation has an important role in understanding the impact of the disaster and vulnerability. Its elevation ranges from 1 to 25 km^2 above mean sea level. Around 45 % of the Diu region has an elevation of less than 5 m from mean sea level, which is basically the low-lying region (Table 9.1). These low-lying areas are mainly located in the northern part along the creek. At the time of high tide, many of the areas get submerged and remain as marshy land.

Diu town is located within elevation levels ranging from 5 to 15 m above mean sea level. The south-western part of Diu district stretches from Vanakbara to Nagoa beach along the coast. It has sandy beaches and sand dunes having height of 5–10 m above mean sea level. The south-eastern part of the region spreads from Nagoa to the southern tip of Diu town and along the coast has rocky area Malala and Fudam. The rocky area contains limestone and acts as natural barriers against high tides. North-western part of Ghoghla is under the influence of tidal inflows and outflows, whereas eastern part is protected by retaining wall. Southern side of Ghoghla is sandy beaches, which faced maximum damage during cyclone.

9.6.2 Costal Regulation Zones and Diu

In 1991, Ministry of Environment and Forests (MoEF), Government of India had declared coastal stretches of seas, bays, estuaries, creeks, rivers and backwaters, which are influenced by tidal action (in the landward side) up to 500 m from the high tide line (HTL) and the land between the low tide line (LTL) and HTL, as CRZs (MOEF 1991, 2009). However, in the case of stretches along rivers, creeks and backwaters distance from the HTL, there can be variation for preparing

Table 9.1 Topography of Diu

Elevation (in metres)	Area	
	In (km^2)	(%)
<5	17.5	45
5–10	13.6	35
10–15	6.2	16
15–25	1.6	4
Total	38.9	

Source extracted from ASTER GDEM Worldwide Elevation Data (2011)

Coastal Zone Management Plan (CZMP), but this distance shall not be less than 100 m or the width of the creek, rivers or backwater whichever is less. The CRZs are meant for regulating activities after bringing out a notification under Section 3(1) and Section 3(2) (v) of Environment Protection Act, 1986 and Rule 5(3) (d) of Environment Protection Rules, 1986. The CRZ regulation was the beginning of state's involvement in protecting the coastal area (Nandakumar and Muralikrishna 1998).

Tide line of landward side is classified into four categories, namely CRZ-I, CRZ-II, CRZ-III and CRZ-IV. The Administrator, Union Territory of Daman and Diu notified the CRZs according to CZMP. CRZ-I is the ecologically sensitive areas between HTL and LTL where no development activity can take place. CRZ-II is already developed areas, and construction is allowed on the landward side of the road as per the existing building bylaws of TCPD. CRZ-III is undisturbed area other than CRZ-I and CRZ-II, which are falling in rural area and legally designated urban area without substantial built-up. No construction activities are allowed on land up to 200 m from HTL areas, and construction is permissible on land between 200 and 500 m from HTL, but with only 33 % of ground coverage, two floors and the maximum height of building can be 9 m (Table 9.2). A special permission is required for hotels and resorts from MoEF. For the areas other than CRZ-I, CRZ-II and CRZ-III, which are categorized as CRZ-IV, construction can be done on land between 60 and 500 m from HTL with 50 % ground coverage, two floors and maximum height of building up to 9 m.

The total area covered under CRZ is 23.71 km^2, i.e. 61 % of the total area of Diu, and different kind of restrictions are in place for the construction activities (Table 9.2). The areas covered under CRZ at sea side and creek side vary as different set of regulations is in place. Out of total CRZ affected area, 8.62 km^2 (38 %) is at the sea side and rest 14.69 km^2 (62 %) is at the creek side. More than 35 % of total area falling under CRZ-I make the people more vulnerable during disasters.

Due to their coastal location, they face severe impact of any disaster whether it is cyclone or earthquake triggered due to cyclone or floods due to sea level rise. The impact of disaster is determined by type of hazards and locations of the

Table 9.2 CRZ zone in Diu

CRZ	Area	
	In (km^2)	(%)
CRZ-I	13.60	35
CRZ-II	1.73	4
CRZ-III	3.90	10
CRZ-IV	4.48	12
Total CRZ	**23.71**	**61**
Non CRZ	**15.19**	**39**
Total Area	**38.90**	**100**

Source Administration of UT of Daman and Diu (1996). Coastal Zone Management Plan

vulnerable group including persons living with disability (PwD). Diu has faced all kinds of disaster during last decade. The severity of disaster and its impact varied on its nature and location of settlements. The different kind of disasters and their impact have been explained in next section.

9.6.3 Impact of Disasters on Diu Island

9.6.3.1 Cyclones and Its Impact in Diu

Diu being a coastal area has faced many severe cyclones periodically causing loss of life and property in the area. The details of severe cyclones are listed in Table 9.3.

In the June 1996 severe cyclone, 2,082 cattle and 2,472 people were affected including 33 casualties. Areas affected by storm surges in Diu district are Gomtimata temple to Vadi Sheri in coastal strip of western side, Jalaram Society in northern side and Saudwadi in northern side in Vanakbara settlement, Kankaimata temple northern side in Bucharwada, Gandhipara northern side and Raberi road coastal side in Diu and Koliwada southern side Mithabava area eastern side in Ghoghla.

Table 9.3 Cyclones in Diu

Sr. No.	Type of cyclone	Month	Year
1	Hurricane cyclone	November	1982
2	Medium cyclone	June	1983
3	Severe cyclone	June	1996
4	Medium cyclone	October	1996
5	Medium cyclone	December	1998
6	Severe cyclone	May	1999

Source Disaster Management Diu District (2012, p. 26)

Table 9.4 Details of damage caused by 1996 severe cyclone

Sr. no.	Name of industries/office/ department	Approximate damages (rupees in lakh)
1	Salt industries	21.34
2	Industries	7.73
3	Agriculture	45.00
4	Property of Diu district	11.54
5	PWD roads and government property	26.63
6	Electricity	12.56
7	Tourism	2.21
8	Port (including fishing boats, nets and government owned boat)	34.23
9	Range forest office (RFO)	45.00
10	Health	1.50
	Total loss (in Rs.)	207.74

Source Disaster Management Diu District (2012, p. 28)

In 1996 cyclone, Diu's salt pans, industries, agriculture, roads, electricity poles, fishing vessels, etc., were severely damaged creating a loss of rupees 207 lakhs (Table 9.4). The maximum loss was reported by agriculture and forest office followed by port authority. During cyclone, the roads between Diu-Vanakbara road via Bucharwada and Nagoa got blocked due to falling of trees and heavy influx of sea water. The felled trees overtopped roads in several areas causing numerous breaches and erosion of side slopes and caused heavy damages to surface. Approach road to bridges especially Diu-Ghoghla was completely washed away. Water supply and electricity supply in Diu district were badly affected.

During 1999 severe cyclone, one foreign going ship was hit at the Nagoa area and got completely damaged, and 141 fishermen of Diu fishing in Okha region (in Gujarat) died and 82 fishing boats (14 totally damaged, 43 damaged, 25 lost) faced different kinds of damage.

Cyclones and Ghogla Coast

Ghogla is a peninsular area in the north-east of Diu, inhabited by fishermen for centuries. Ghogla is one of the low-lying areas of Diu, and its altitude is less than 5 m from the mean sea level. It is at high risk of inundation during cyclones, yet a strong cultural affinity to sea and the economic dependence on it force the Ghogla fishing community to remain close to the sea. Ghogla coast is lined by hundreds of small and large fishing boats, and the smell of fishes permeates the air throughout the day. It is a densely populated town, and the poor condition of its buildings reflects the socio-economic status of its residents. According to one of the resident of Ghoghla,

> He is one among the thousand fishermen in Ghoghla. He owns a small boat, which he operates with a partner and splits the profit in half. He has been fishing for 25 years, ever since he was a young boy, and proclaims himself to be a Sagarputra, or son of the sea. His community respects the sea, as the provider of their lives, and is not afraid to venture into it. Even during cyclones or when the sea is under turbulence, the residents of Ghoghla often refuse to move to safer areas (fisherman (2013): personal interactions, Ghoghla, 20th August 2013).

Dissemination of Information During Disaster

Collectorate, port office and police office receive the cyclone warnings from metrological department located at Ahmadabad. The control room at collectorate office further transmits the warning to municipal council and gram panchayats. In municipal areas, police and collectorate use loud speakers. In rural areas, gram panchayat has got loud speakers installed at varied places to announce any such information. The port office disseminates the warning through signals from Bandar Chowk area.

9.6.3.2 Floods

Natural calamities due to heavy rains or influx of water due to high tides causing floods are a periodical phenomenon. Depending upon elevations, some parts are more prone to occurrence of floods than others. Vanakbara, Bucharwada and Ghoghla are the most affected areas due to their low elevation. The chance of flood occurence in Diu has become minimal on account of construction of Kodinar Dam at Chassy River, Kodinar in Gujarat. The UT Administration has opened a flood control room in collectorate, which receives information regarding discharge of water from Kodinar dam from time to time.

9.6.3.3 Earthquake

The UT of Diu is geographically part of Gujarat in Saurashtra region on the Arabian Sea Coast. According to earthquake hazard map of India, Diu is located in the moderate damage risk zone with probable earthquakes of 5.0–6.0 magnitudes on Richter scale. During last 200 years, it recorded nine earthquakes of moderate to severe intensity in the years 1819, 1845, 1847, 1848, 1864, 1903, 1938, 1956 and 2001. During the last earthquake on 26th January 2001, around 10 houses collapsed and walls of several houses got cracked. It sounded like underground explosions and heavy rumblings and continued for some seconds.

9.6.3.4 Fire

As per the information given by Diu Fire Service, the average number of fire incidents in Diu is around 50 per annum. Out of these incidents, nearly about 5–10 % fire incidents are of serious nature. But there is no loss of human life or property damages in any incidents of fire.

9.6.3.5 Coastal Erosion

The sea water is eroding the limestone rocks. The Jalandhar beach, Malala, Diu fort and Gangeshwar temple area are continuously getting eroded due to sea waves. This slowly reduces the area available for development. In order to restrict the coastal erosion, an area of 516 ha covered with casuarinas and other tree species has been proposed as reserved forest area. The coastal stretch plantation leads to restricting coastal erosion.

Thus, disasters and Diu goes simultaneously. There is not even a single month when the coastal community does not feel vulnerable nor there a family in Diu, which has not faced adverse impact of disasters in terms of loss of a family member or economic loss.

9.6.4 Disasters and Community Vulnerability

Fishermen in Diu go out for fishing expeditions lasting 12–15 days at a stretch and travel to distant areas in search of better fish catch. Their trawlers have been installed with latest technology global positioning system (GPS) devices and very high-frequency (VHF) device/radio. The GPS helps in navigation and VHF radio is used for transmitting messages about cyclones as well as helps in sending other messages to crew members so that they are able to efficiently communicate with each other. The Indian government has introduced subsidies up to 75 % on GPS and VHF and made it mandatory for all trawlers to install them. This helps in tracking their position at international borders. The government has also made it mandatory for them to register with the port office at Bandar Chawk before each trip to the sea. This helps in identifying the crew member at the time of eventuality and gives the compensation to their family.

Whenever the district authorities receive a cyclone warning, it is immediately signalled to all marine vehicles. These cyclone warnings are generally received 24 h before the cyclone strikes. It impacts differently to the people falling in two categories: one who are residing along the coastal areas or are about to leave the harbour and second and one who is deep inside the sea when warning is being sent.

As a result of warning, the people residing in the coastal areas are able to evacuate their houses and move to safer locations as a precautionary measure. However, during site visit, it was observed that there are not many choices of such places on the coastal areas. Boats scheduled to leave the harbour are also not allowed to leave and the trip gets postponed. This leads to economic loss as most of the crewmen are daily wageworkers. However, hundreds of boats and their crew members, which are already in sea, are left vulnerable to the disaster. They are often 250–300 km away from the coast and are unable to reach back within 24 h. Even if they received the cyclone warning, they were helpless and stuck at sea. There have been earlier instances of 25–30 boats with 300 people overturning at the same time during cyclones. As narrated by one of the women construction worker in Vanakbara village from Vanakbara panchayat of Diu,

> Her husband was a fisherman and used to work on trawlers for a fixed wage (workers on trawlers used to get a salary of 6000-8000 rupees per month). He was killed in a fishing accident on sea, and since then her financial condition has deteriorated. The village panchayat and fishing community gave her a one-time grant of Rs. 10,000 but a regular source of income is still missing. Savitaben works for Rs. 150 per day as construction labour together with her brother. Often she is unable to find work regularly, and is forced to seek his financial support (Construction Worker (2013): personal communication, Vanakbara Panchayat, 19 August 2013).

Another Ghoghla fisherman stated that *fishermen also refuse to return from sea, saying they have a better idea about cyclones* (personal communication, Ghoghla, 20 August 2013).

This reflects that vulnerability is both at the seashore and in the sea at the time of severe cyclones.

Diu port has a very small jetty which functions only during high tide. It does not have a 24-h port (unlike the Gujarat ports). Sometimes during cyclones, the boats are forced to wait just outside the port for 10–12 h, leaving the fishermen susceptible to dangers of life and property and increasing their vulnerability. Earlier insurance company used to give insurance coverage for the boats to the fisherman but now even that has been discontinued for past 8–10 years.

Vanakbara port has a mismatch between its carrying capacity and number of trawlers of different sizes. This leads to an increased competition near the coast, and many of the boats have to go to deeper seas to find the catch. Due to silt getting deposited on its bank and de-siltation process is not a continuous activity, it becomes difficult to take out the trawlers during most of the day. Only during the high tide, fishermen are able to take their boats out of the minor port. There is also no constructed space to keep the boats so even if they are parked in the seashore; they have to face the fury leading to damage at stationary position.

9.6.5 Vulnerable Fisherman Across International Borders

The geography and tidal movement of the waves in the region is such that larger fish reserves are found very close to the international border with Pakistan. They always desire to catch maximum fish and those varieties of fish, which has higher economic value. Therefore, fishermen are forced to go out to longer distances over a period of 10–15 days at a time in search of a larger catch and often crossover to the Pakistani territory in search for better profits. One such location is Okha where plenty of good variety fish is available. It is located in Gujarat at a distance of 350 km from Diu and 20 km from the international border with Pakistan. The fishermen calculate the distance travelled on the basis of time taken. They have relied on their time–distance estimation and navigational skills for centuries. Despite installation of GPS devices for navigation, they found it difficult to adapt to these technological advanced equipment due to lower educational level and lack of training to operate such devices. Many times, water flows at considerably higher speed making them (the fishermen) travel longer distance at shorter time frame. At such occasions, the distance of 20 h to India–Pakistan line of control (LoC) gets covered in just 14 h. The Pakistani coastal guards seize their boats along with captain, crew member and the fish catch immediately after crossing the international boundary. At times, other than captain, crew members are sent back on a single marine vehicle. It takes effort from Indian Consulate side to get engaged in interaction with their Pakistani counterparts for release of the captive fishermen. The last such release was on 23rd August 2013 when 62 Diu's fishermen were released by the Pakistani authorities at Wagah border. They were in Pakistani prison for more than 2 years. In these years, the family members of the captured fishermen were left with no source of income and assistance from governments. Moreover, the captured boats are never returned. According to the President of Vanakbara fishermen association of Diu, "In past 12 years, 700 boats have been retained by Pakistan and it resulted into a heavy financial loss to the fishermen as each boat costs 40–50 lakhs" (excerpts from Focus Groups Discussion, August, 2013).

There is another side of the story, which is contradictory to above-explained argument, as narrated by one of the trawler crew member. According to him,

> the fish catch has decreased in the proximity to Diu due to higher pollution level as well as increased competition. The higher pollution is due to increase in number of trawlers using diesels and industries located towards Una side discharging their waste in the sea. The increase in number of fishermen and trawlers has not been proportionate to fish catch. The trawler requires huge inbuilt cost of body, transportation and machinery with a minimum of 30-50 lakh rupees per boat. Other than this, each trip costs 50,000 to 1,00,000 rs. With limited catch availability in the vicinity and large investment, boat owner's emphasis is not only on quicker recovery of the investment but also to maximize their profits. The boats have been provided with GPS devices, but they prefer to switch it off while crossing over to international waters. This is to show their ignorance about the route if caught by the Pakistan Coastal Guards. They do not want GPS to be used as evidence that they were having knowledge about the sea route and crossing the boundaries. The fishermen doing so

are fully aware of their act, yet are forced to take a risk of such magnitude. Whenever they got caught, they lose the entire catch, are forced to spend a minimum of 6 months in Pakistani prisons and their release is also uncertain. This is more of a compulsion for them than a wishful act of disobedience (Fisherman (2013): Personal Interactions, Vanakbara Panchayat, 19th August 2013).

9.6.6 Community Associations and Reduced Vulnerabilities

Diu district has three hierarchies of fishermen associations to fight for the rights of fishermen whether the fight is for allowing/disallowing the extra number of trawlers or construction of space to park the trawlers on the shore or release of prisoner fishermen by the Pakistan's coastal guard or seeking subsidy for different equipment. At district level, Diu district fishermen association is apex body followed by Vanakbara fishermen association and Ghoghla fishermen association, at second level for these two specific settlements. Vanakbara fishermen association has five smaller associations based on castes group, i.e. Kharwa, Koli, Vadiseri Koli, Vadi Vistar, and Timbasari. Most of the fishermen are members of these associations and meet regularly to discuss the issues faced by them. On one hand, fishermen associations help their own community by mobilizing financial resources for widows, through collection from community. On the other, they send the letter to collector and port office for demanding subsidy or fixation of number of boats or for getting the crew members registered.

9.6.7 Positive and Negative Impacts of Neighbouring State

Diu depends upon on Gujarat for its socio-economic upliftment. The patients in Diu are referred to Gujarat for specialized treatment, student travels to Gujarat for higher education and workers go in search of job opportunity. The fish catch is also sent to ports in Gujarat for export to China, Europe and other countries. The agricultural produce is also sold in the Gujarat Mandi's. People from Gujarat also travel to Diu for tourism purposes as well as for liquor consumption as Gujarat has been declared an alcohol-free state. The latter increases the revenue of Diu district but also leads to negative influence in terms of road accidents and pilfering of solid waste in and around tourist places.

The difference of being state, i.e. Gujarat state being administered by the Chief Minister and UT, i.e. Daman and Diu under direct control of Centre also adds to further complicacy. The State and Centre control leads to further complicacy, as there is no coordination for integrated planning.

9.7 Conclusion

9.7.1 Summary of Key Points and Implications

Coastal Zones Regulations are in place in Diu, but cases of its violation have been legally fought in the courts. Diu has a high vulnerability to disasters especially cyclones and resulting flood due to its being an island and difference in levels of elevation. The community engaged in fishing activities is most vulnerable at the time of cyclone. The people associated with fishing activities, who stay in low-lying areas, does not have any government provided spaces to take refuse at the time of disaster. The retaining walls have also not been created to stop the influence of floods in coastal areas. The crew members of trawlers are uneducated and lack understanding of usage of GPS and VHF radio. They are also not covered under insurance policies and belong to poor households. In case of casualties, the compensation given is too less to sustain a family, which has lost its only earning member. There are often inordinate delays in disbursement of the funds from the government authority to the beneficiaries.

The warning system fails to help them when they are deep in the sea water, and despite receiving the warning, they are unable to return on the seashore. Crossing the international boundaries in water also lands them in prison in Pakistan for several years. In the absence of a signed treaty between two nations, it takes a long time to get the poor men released from the prison.

9.7.2 Recommendations

The CRZ regulations shall be strictly followed, and any violation in this regard shall be dealt with according to land of the law. It shall be publicized through different bill boards in important tourist destinations to spread awareness. The official inspection shall be a regular feature to observe any violation regarding CRZ regulations and notice to the violators shall be issued.

The facility on higher grounds shall be created to evacuate the affected population in case of a warning regarding severe cyclone. The existing forests help in environment upgradation and preventing coastal erosion, but need to be conserved well and declared as reserve forest. The retaining wall shall be created, wherever possible, to minimize the impact of disasters.

The Union Government shall frame insurance rules and policies specially catering to fishermen and other vulnerable population. The policy shall be inclusive and state shall give collateral for such policies, as poorer population cannot afford it.

The adequate compensation shall be given to the affected families not only from natural disasters but also to those who are being caught by bordering country till the time their release is ensured.

Current warning system only intimates the fishermen before 24 h. An early warning system shall be devised so as to give adequate time for trawlers to reach to the safe seashore.

The awareness campaign and capacity building programmes to educate the fishermen about GPS, VHF radio and other devices as well as different government schemes are the need of the hour. The fishermen associations shall be engaged in disseminating the information, and their representations shall be given priority in governance.

The diplomatic channels shall be streamlined for early release of all the fishermen currently in the prison of neighbouring country as well as to stop such capturing in future. The steps shall also be initiated to get the vessel released by their coastal guard.

There shall be a high-level coordination committee with officials from Centre and State for coordinating the economic activities surrounding the Diu region. This is to ensure that there shall not be any adverse impact of such activities on Diu's ecological sensitive zone.

An integrated planning approach by inclusion of fisheries sector in national planning for disaster reduction can reduce their vulnerabilities.

9.7.3 Limitations

As no written account of human and economic loss because of different kind of natural and man-made disasters was found, it becomes difficult to study the trend data of such losses. Due to small number of population and an insignificant vote bank for politicians, Diu has not received the required attention in planning and policy making. Many of Gujarat's development impact negatively the Diu's environment. The evidence in this effect has come up during personal interactions with stakeholders in Diu. However, as the study was focused on Diu, it was beyond the purview of the research to understand the implications of such developments.

9.7.4 Further Research

The need is to undertake an in-depth analysis of the families who faced the loss of family member or served economic losses at the time of disasters. The detailed study of critical geopolitics and geopolitics is required to understand the social, economic, political and environmental consequences of interrelationship between different actors. To understand the issue and its solutions holistically, the feasibility of setting up a joint committee of Diu and Gujarat administration shall be carried out.

Acknowledgments The primary and secondary data were collected during the exercise undertaken for preparation of Regional Plan for Diu in 2013. I am grateful to faculty (permanent and visiting) from School of Planning and Architecture, New Delhi and officials from Town and Country Planning Organisation (TCPO), New Delhi for their valuable inputs and suggestions during discussions and students especially Deepika, Mithun, Pankaj, Suman and Poonam for their help in collation of data.

References

Administration of UT of Daman and Diu (1996) Coastal regulation zone map. Daman and Diu Coastal Zone Management Authority. Administration of UT of Daman and Diu, Daman and Diu

Badjeck MC (2008) Vulnerability of coastal fishing communities to climate variability and change: implications for fisheries livelihoods and management in Peru. Bremen, Mai, dissertation, Universitat Bremen and ZMT Bremen. Available from World Wide Web: http://www.d-nb.info/989897052/34. Accessed 10 Oct 2013

Bohle HG (2007) Living with vulnerability: livelihoods and human security in risky environments. Inter-sections interdisciplinary security connections, United Nations University Institute for Environment and Human Security (UNU-EHS) and Munich Re-foundation from knowledge to action, no. 6/2007, pp 1–32. Available from World Wide Web: http://www.microinsuranceconference2005.com/dms/MRS/Documents/InterSection2007_Bohle_Vulnerability.pdf. Accessed 10 Oct 2013

Brouwer R, Akter S, Brander L, Haque E (2007) Socioeconomic vulnerability and adaptation to environmental risk: a case study of climate change and flooding in Bangladesh. Risk Anal 27(2):313–326

Census of India (2011) Provisional population totals paper 1 of 2011: Daman and Diu. Accessed from: http://www.censusindia.gov.in/2011-prov-results/prov_data_products_daman.html

Daac LP (2011) ASTER GDEM worldwide elevation data: 1.5 arc-second resolution, produced by METI and NASA, maintained by the NASA Land Processes Distributed Active Archive Center (LP DAAC). Available from: http://www.gdex.cr.usgs.gov/gdex/on. Accessed 14 Aug 2013

Delaney R, Michael F, Murray PA (2003) Planning for disaster management and vulnerability reduction in the fisheries sector of Caribbean Island states in 54th Gulf and Caribbean Fisheries Institute, pp 127–135. Available from World Wide Web: http://procs.gcfi.org/pdf/gcfi_54-10.pdf. Accessed 10 Oct 2013

Disaster Management Diu District (2012) Available from http://www.diu.gov.in/Others/Collectorate/Disaster-Management-Plan.pdf. Accessed 10 Oct 2013

Gaillard JC, Maceda EA, Stasiak E, Le Berre I, Victoria M, Espaldon O (2009) Sustainable livelihoods and people's vulnerability in the face of coastal hazards. J Coast Conserv 13(2–3):119–129

Hafeznia MR (2006) Principles and concepts of geopolitics. Popoli Publications, Iran, pp 37–39

Haverluk TW, Beauchemin KM, Muller BA (2014) The three critical flaws of critical geopolitics: towards a neo-classical geopolitics. Geopolitics 19(1):19–39

Ingram JC, Franco G, Rumbaitis del Rio C, Khazai B (2006) Post-disaster recovery dilemmas: challenges in balancing short-term and long-term needs for vulnerability reduction. Environ Sci Policy XXX:1–7. doi:10.1016/j.envsci.2006.07.006

Kelly PM, Adger WN (2000) Theory and practice in assessing vulnerability to climate change and facilitating adaptation. Clim Change 47:325–352

Maxmillian T, Depledge J (2013) Great-power politics, order transition, and climate governance: insights from international relations theory. Clim Policy 13(3):572–588

Ministry of Environment and Forests (MoEF) (1991) Department of environment, Forests and Wildlife 1991, Notification under Section 3(1) and Section 3(2) (V) of the environment (protection) Act, 1986 and Rule 5(3) (D) of the environment (protection) rules, 1986 declaring coastal stretches as coastal regulation zone (CRZ) and regulating activities in the CRZ. New Delhi, the 19 Feb 1991 (as Amended up to 3 Oct 2001)

Ministry of Environment and Forests (MoEF) (2009) Order, New Delhi, the 21 Dec 2009, SO 3250(E)

Ministry of Information and Broadcasting (MoIB) (2008) Programmes for the people: four years of UPA Government, Daman and Diu 2004–08, Directorate of Advertising and Visual Publicity, Government of India

Myhr J, Nordstrom L (2007) Livelihood changes enabled by mobile phones—the case of Tanzanian fishermen, students/researchers. Uppsala University, Uppsala. (online). Available from World Wide Web: http://www.wideopenaccess.net/files/webform/submission/Myhr%20Nordstrom%20Livelihood%20changes%20enabled%20by%20mobile%20phones.pdf. Accessed 10 Oct 2013

Nandakumar D, Muralikrishna M (1998) Mapping the extent of coastal regulation zone violations of the Indian coast, Report for National Fish workers Forum, Valiathura, Thiruvananthapuram

Tata Consultancy Services (TCS) (2002) Final report on 20-Year Perspective Plan for Daman and Diu, Government of India, Ministry of Tourism and Culture, Department of Tourism, Market Research Division

Toncea V (2006) Geopolitical evolution of borders in Danube Basin. PhD

Chapter 10
Lack of Integrity Moving the Goals Away

Vinay Sharma, Rajat Agrawal and Kumkum Bharti

10.1 Introduction

A paradox in today's era in the form of a major question has emerged, when on the one hand, achievement of goals such as Millennium Development Goals (MDGs) is associated with the survival of mankind itself and, on the other hand, the goals are moving away because of lack of fundamental commitment amongst the implementing agencies as if a large number of people involved in the process are adamant on losing out on their integrity required to sustain commitment. India has gone through tumultuous times recently when scams and corruption incidents involving gross and deliberate misconduct have been witnessed with reference to schemes those were supposed to benefit the masses and achieve MDGs. Had it not been like that those goals always seemed achievable.

On the other side, presumably because of the dominance of desire for the tangible realization of value, social restructuring and hypothetical deductive approach in the past century value generation and realization has usually been expressed in constructions, attributes (Prahalad and Ramaswamy 2002) and factorial terms. Though thinkers like Gandhi, Drucker, Mandalla, Lincoln and several others did not deviate from their thoughts which inevitably due to their radical strength have supported economies in different ways. For example, Mahatma Gandhi's favourite hymn begins with the words,

Vaishnav jan to tene kahiye je
Peer paraayi jaane re
Dukh dooje ka harkar bhi
Abhiman na mann me aane re…

V. Sharma (✉) · R. Agrawal · K. Bharti
Indian Institute of Technology Roorkee, Roorkee, Uttarakhand, India
e-mail: vinayfdm@iitr.ernet.in

H. Ha (ed.), *Land and Disaster Management Strategies in Asia*,
DOI: 10.1007/978-81-322-1976-7_10

That translates in English as

A godlike man is one,
Who feels another's pain
Who shares another's sorrow,
And pride does disdain...[1]

Value co-creation along with its definitional theorization seems to have a backdrop of participants, especially the beneficiaries or customers or target group as the agents of change themselves (Spohrer and Maglio 2008). This view explained aptly by Sen (2000) supports elemental categorization of value co-creation as a process.

Once every aspect of the preceding initiation of goal setting and Gandhi's approach to understand people's plight is taken into consideration, one imagines an explicit requirement of such a thought process and implementation zeal along with the involvement of people especially in case of disaster management. A huge calamity in Uttarakhand in regions on Holy Shrine of Kedaranath Temple in Himalayas occurred in mid of June, 2013, wherein thousands of people died but the major aspect was that several people could not be saved not only because of the ill-preparedness for such a calamity but also because of the simple reason this paper highlights about the lack of integrity. How integrity can be achieved, co-creation of value can bring in supportive change is the context of this paper wherein the examples from diverse fields highlight the same aspect of Gandhian thought of getting associated with the pain and plight of people.

Further, it is pertinent to understand that co-creation can be understood as a process resulting in value or may be seen as an end and means by itself to propel value co-creation. However, if one visualizes value derivation as innate and never ending, connoting and connected to values while observing the relationship of freedom centric view with value co-creation, one must acknowledge means as ends and vice versa (Sen 2000). Xie et al. (2008) propose that values are usually embedded in history, culture and economic development of a society; however, the formation of values is influenced by the personal, social, economic, historical and cultural development.

10.2 Goals

Goals are associated with time, organization, direction, implementable steps (methodology), sustenance and improvement but intensity and will are important that is why who is doing what is the deciding factor. Goals are themselves motivators as the literature says, but studies do endorse the key elements of challenge,

[1] Translated by noted Indian author and columnist Mr. Khushwant Singh. In a column written for The Hindustan Times, Mr. Singh published his English rendering of Bapu Gandhi's favourite hymn.

self-efficacy and commitment, which are intrinsic, have to be present for the achievement of goals. Ordóñez et al. (2009), in their article say that "Hundreds of studies conducted in numerous countries and contexts have consistently demonstrated that setting specific, challenging goals can powerfully drive behaviour and boost performance". Locke (1968) wrote that "specific and challenging goals lead to better performance as opposed to vague and non-quantitative goals or no goals at all". In fact, Miner (2003) endorsed Locke and said that "out of 73 management theories, goal-setting theory" (Locke 1968) is ranked as the most important one by organizational behaviour scholars. Further, Locke and Latham (2006) claimed "so long as a person is committed to the goal, has the requisite ability to attain it, and does not have conflicting goals, there is a positive, linear relationship between goal difficulty and task performance". Additionally, "empirical studies have found that people who have higher levels of self-efficacy tend to achieve better performance" (Bandura 1991; Gist and Mitchell 1992).

Goals bear greater magnanimity if are larger and encompassing in nature. This entails bigger effort and greater, cumulative will, integrated sensitivity, sustenance hence integrity. The MDGs of eradicating extreme poverty and hunger, achieving universal primary education, promoting gender equality and empowering women, reducing child mortality rates, improving maternal health, combating HIV/AIDS, malaria, and other diseases, ensuring environmental sustainability, and developing a global partnership for development (United Nations Millennium Development Goals (United Nations Millennium Development Goals 2014) website, http://www.un.org/millenniumgoals/bkgd.shtml) are suggestive of magnanimity.

On fulfilment part, data and reports are suggestive of shortfalls and analysis not only is suggestive of lack of efficiency and effectiveness but also diversion of resources allocated for the achievement of these goals for the benefit of the people responsible and entrusted for realizing those.

10.2.1 Major Indian Government Schemes, Value Co-creation and Models

People when start organizing and implementing policies, models emerge. Structure gets developed, monitoring begins. The precursor for policy formulations and implementation is goal formation. Goals are associated with time, organization, direction, implementable steps (methodology), sustenance and improvement but intensity and will are important that is why who is doing what is the deciding factor. Goals are themselves motivators as the literature says, but studies do endorse the key elements of challenge, self-efficacy and commitment, which are intrinsic, have to be present for the achievement of goals (Sharma 2009a, b, c). Locke and Latham (2006) claims "so long as a person is committed to the goal, has the requisite ability to attain it, and does not have conflicting goals, there is a positive, linear relationship between goal difficulty and task performance". But, lack of integrity hinders the accomplishment of goals but can be built by raising the level of human will commitment and spirituality (Sharma 2009a, b, c).

In India, involving the rural population in community development activities is not new. In Indian villages, the Gram Sabha and the Gram Panchayats have provided a platform to the villagers where the issues related to development of the village are raised, discussed and decisions are taken unanimously. The role of Gram Sabha is, thus, to identify the need and areas that are to be developed in the village. Gram Sabha takes the decision in consonance with the Gram Panchayat and offers the solution to the villagers with the financial help offered by the state government. Therefore, to unleash the latent needs of the villagers, intense and continual involvement has to be made (Bharti et al. 2013). This entire process of identification of problem and solution development is an example of co-creation taken through the "community of creation" (Sawhney and Prandelli 2004).

For instance, India has seen many schemes run by the Government and has learned through the experiences of international agencies or developed here itself for health care provision, educational and poverty alleviation through employment generation as stated in Table 10.1.

For example, co-creation of value is applicable to Mahatma Gandhi's National Rural Employment Guarantee Scheme (MGNREGS) that guarantee work for 100 days in one financial year to the poor[2]. The scheme is believed to build capacity, create public awareness and effective management in the country and reduce poverty by increasing the participation of the poor in the economy development.

Moreover, many internationally acclaimed models have been found to be very successful like,

- Grameen Bank Model of affordable microfinance services initiated by Muhammad Yunus of Bangladesh and recipient of Nobel Peace prize.
- Fortune at the Bottom of the Pyramid Model of identifying approximately 4 billion population as a promising market by C. K. Prahalad
- Narayan Hrudayalaya Model of affordable health care by founder Dr. Devi Prasad Shetty
- E-choupal model of connecting Indian farmers to ITC by ITC (India)

And even a theoretical model that I proposed on the basis of what has been happening and what may happen called "Affordability for the Poor and Profitability for the Provider" (wherein few of the aspects have now been followed not because of I proposed one must say but because of the utmost necessity associated with those developmental schemes and steps to be taken) (Sharma 2009a, b, c), and such models can be sustained and proposed for health for all through affordability and profitability with public spiritual partnership (Sharma 2009a, b, c).

[2] www.nrega.nic.in/rajaswa.pdf.

Table 10.1 Projects to alleviate poverty

Development of Women and Children in Rural Areas (DWCRA) scheme based on the model of development of children through women (mothers)
Swarnjayanti Swarozgar Yojana based on the model of development of self-employment amongst unemployed (especially youth)
Mahatma Gandhi National Rural Employment Guarantee Scheme (MGNREGS) based on assurance of employment amongst rural period for fixed number of days
National Rural Health Mission for health care provision
Mid-Day meal scheme which was based on the model of increasing the attendance of children in schools through provision of mid-day meals to the children. It is quite successful so far

10.2.2 Lack of Integrity Moving the Goals Away

Human will, integrity, commitment, goals, spirituality *all such key words and the related concepts* though are subjective and interpretative in nature but are simultaneously explicit. Every human being understands the meaning and the meaningfulness associated with these terms in almost universal terms. There are strong references available proving the relevance of such concepts but it cannot be categorically said that the text provides us any clue on who and when used these terms for the first time and then when were these uniquely interpreted so much so that they gained special meaning. The entirety of these natural characteristics of human beings being present in all of us has several times been referred to in terms of attributes and constructs but that interprets and strengthens the aspects of gaining (actually strengthening) these characteristics and then elaborates on the results those may emerge. Therefore, this paper chooses to not to specify the framework of the usage and interpretation of these terms but to exemplify them with the help of relevant examples from real life and is suggestive of how sustenance of these characteristics may be strengthened on the basis of the first-hand observation, interviews and discussions with the people who have been practising this art of development and sustenance of these characteristics. One more thing explicit from the research conducted in the due course of understanding these characteristics is that the people who possess these without exception have a meaningful life, meaningful in terms of fellow human beings and anything supposed to be achieved through these people is achieved.

Another aspect that emerged in relation to the research itself is that to understand these characteristics; the possessors of these characteristics have to be understood in relation to their actions exemplifying these characteristics; therefore, though decided in the earliest most part of the research conducted, no descriptive research design could be carried forward, and an associative conclusion also emerged saying that these traits are to be found in everyone, wherein the individual himself has to find and strengthen these characteristics within himself, but of course people who have done so especially in public life and in the positions associated with implementation of said policies and achievement of said goals have to be looked upon and not only to be exemplified but also to be followed.

This paper while acknowledging
The existence of development goals,
Refers to few models for achieving such goals,
Suggests the role of human will for this achievement
through strong commitment levels by choosing few
examples exemplifying such commitment and will
and specifically focuses and elaborates on the
requirement of integrity for the said commitment and
will
Along with the key factor which is required for the
sustenance of integrity called "Spirituality".
The paper once again chooses real life examples to
explain spirituality and its role in strengthening and
sustaining integrity for achieving development for all.

10.2.3 Human Will, the Deciding Factor

Models get followed or generated as it has been observed but an utmost desire and "will" becomes one of the most important reasons for them to happen as may be seen from the following very important examples, such as

- Assam's "Boat Clinic" carries health services and hope is a very interesting, thoughtful and zealous effort to provide health care to people (Health.India.com 2012).
- A man in his mid-50s helped grow a huge forest on a sandbar in the middle of the mighty Brahmaputra in Assam's Jorhat district, which has caught attention of the government, tourists and film-makers (Times of India 2012). The 30-year-long effort of Jadav Payeng, known amongst local people as "Mulai", to grow the woods, stretching over an area of 550 hectares, has been hailed by the Assam forest department as "exemplary". Mulai began work on the forest in 1980 when the social forestry division of Golaghat district launched a scheme of tree plantation on 200 hectares at Aruna Chapori situated at a distance of 5 km from Kokilamukh in Jorhat district. "This is perhaps the biggest forest in the middle of a river"; not only tourists are flocking to the woods in droves, a famous British film-maker Tom Robert went there 2 years back to shoot one of his films. The forest, known in Assamese as "Mulai Kathoni" or Mulai forest, houses around four tigers, three rhinoceros, over a hundred deer and rabbits besides apes and innumerable varieties of birds, including a large number of vultures. It has several thousand trees amongst which are *valcol, arjun, ejar, goldmohur, koroi, moj* and *himolu*. There are bamboo trees too, covering an area of over 300 hectares.
- Radical improvement in Delhi Government School system (2008): An initiative of Directorate of Education, Government of NCT, and this IT project was initiated in the year 2003 to improve the education system in Delhi Government School. In-house development of the software, grading of teachers in three categories red, green and yellow-based upon their performance for Class

X and XII, merger of two single-sex schools running in two shifts in the same building, introduction of online attendance for teachers, generation of online marks for evaluation of performance teacher-wise, class-wise and zone-wise, district-wise of Class VI onwards, the onus of taking admission was shifted from student parents to Heads of Schools of Parent and Feeder Schools are some of the major features of the project. To cover important topics from Class I to V, a special bridge course was developed. A choice-based computerized transfer system combined with GIS-based posting of new recruits was introduced. MIS Mail and Online Feedback system for the citizens has been introduced. The major objectives of the initiatives were universal enrolment, reduce dropout, improve quality of education, all-round development of child and joyful learning. This initiative brought out an immense improvement in the administration efficiency, transparency, accountability and monitoring capability of the system. As an outcome, result increased by 11.7 % in Class X in year 2006 and 17.39 % in 2007, results of the merged went up by almost three times from 27.65 % in 2005 to 50.94 % in 2006 to 77.02 % in 2007, verifiability of attendance of teachers at any given point of time, substantial improvement in the pass percentage of all the classes.

- The former deputy commissioner of Nawanshahar, Kishan Kumar is a pioneer of the Nawanshahar model[3] now being implemented in other districts of the country. This model transformed Nawanshahar,[4] which was once famous for female foeticide. As of today, all major departments in the district have women heads. He brought about the change by doing "what he thought needs to be done". Kumar improved the ratio from 774 to 910 per 1,000 in just one year. Some of the measures employed were developing special computer software to monitor pregnancies, cracking the whip on scanning centres, nursing homes, doctors and corrupt district health officials and sending decoy patients to suspect clinics. He also introduced a unique e-governance model known as *Suvidha* centre that has facilitated the citizens in a number of services from issuing of licences to booking tickets (Sareen 2012).
- Arunachalam Muruganandam is the man behind the invention of a sanitary napkin vending machine, which could produce 1,000 pieces of napkins in mere eight at a cost of ₹1 per piece. Many self-help groups and nationalized banks acquired the machine and started selling napkins at ₹2 per piece. This invention has transformed the lives of large number of girls in the rural areas of Tamil Nadu not only in terms of ease of use but also on grounds of personal hygiene. To add to it, the napkins are environment friendly and the self-help groups earn their livelihood out of selling napkins. Professor Thomas M. Stove of the famous Massachusetts Institute of Technology (MIT), Mechanical department,

---

[3] http://punjabnewsexpress.com/news/1293-Nawanshahr-once-infamous-for-female-foeticide-now-witness-the-woman-empowerment.aspx (Accessed September 1, 2012).

[4] Punjab News Express (2012), "Nawanshahar, once infamous for female foeticide, now witnesses the woman empowerment", available at: OutlookIndia.com (2007), available at: http://m.outlookindia.com/story.aspx?sid=4&aid=234467 (Accessed September 1, 2012).

offered a contract between MIT and Muruganandam for buying the machine after being impressed by his work (Kalagu.wordpress.com 2009).

- Ram Babu Bhatt is a Jaipur-based researcher and social worker. He is a well-known name in the field of social development and is deeply committed to the issues of women, children, tribal and *dalits* (Purva 2012).
- Dr Uma Tuli is the founder of Amar Jyoti Rehab centre in Delhi. It is an organization that works for the holistic development of the disabled through inclusive education, vocational training and self-employment. (http://www.satyamevjayate.in/).
- Jeroo Billimoria is the founder of CHILDLINE, the country's first toll-free tele-helpline for street children in distress. Initiated in June 1996 as an experimental project of the department of Family and Child Welfare of TISS, CHILDLINE has come a long way today, to becoming a nationwide emergency helpline for children in distress. As of March 2011, total of 21 million calls since inception have been serviced by CHILDLINE service and operates in 210 cities/districts in 30 States and UTs through its network of 415 partner organizations across India. (http://www.satyamevjayate.in/).
- Dr. Rajat Mishra: Clinical psychologist Dr Rajat Mitra is the Director of Swanchetan Society for Mental Health, a non-profit organization dedicated to providing evidence-based care and psychological support to victims of heinous crimes. He has been working for the past 20 years in Delhi with prisons, law enforcement and judiciary in different capacities to understand the criminal justice system and work towards developing better methods of care within it. (http://www.satyamevjayate.in/).
- Dr Samit Sharma is an Indian Administrative Services (IAS) Officer, Managing Director, Rajasthan Medical Services Corporation, Jaipur, is a paediatrician and practice in Rajasthan before joining the Indian Administrative Service. He has undertaken pioneer and innovative work in the generic medicine project, with the aim of providing affordable health care and medicine to everyone regardless of their income level. (http://www.satyamevjayate.in/).
- Winnie Singh graduated in Psychology and Philosophy from Miranda House, Delhi University, and is a Ph.D. scholar in Public Health at Jodhpur National University. She is a committed social worker with over two decades of experience at national and grassroots level in public health care, with special emphasis in the area of health, elderly and gender equity. Her lifelong interest in humanitarian relief, dynamic nature and action has been the driving force that led her to found Maitri India, a non-profit organization committed to poverty alleviation, education and health. (http://www.satyamevjayate.in/).

10.2.4 Motivation Absorbs Magnitude

We have personally participated in three *Kumbh Melas*, a Hindu festival that is actually, World's largest congregation of human beings. One of the largest of these Kumbh's took place in Allahabad in the State of Uttar Pradesh, on the confluence of the River's Ganga, Yamuna and Saraswati (now not present physically as per the mythological evidences) in the year 2001 on the occasion of *Mauni Amavasya*, where more than 30 million people on a single day and over 70 million participated over a few weeks.

We were intrigued by the methodology that was followed by Government Machinery to manage the occasion, especially health care system, wherein with very limited resources such as only 200 doctors and about 6,000 support staff, things such as sanitation, cleanliness and health care had to be taken care of. I stayed there and interviewed officials especially Dr. G. R. Sharma, the head of the health care services there and realized that methodology and models were normally which have been used with improvisation related to time and tests but the most important thing was the level of commitment, purposefulness, transparency, efficiency, efficient administration and good governance in the delivery of health care services. Finally, we reached to the conclusion that it is a matter of spirituality-driven integrity that makes such things happen. For detailed description, readers may access my paper on the same, Sharma (2009a, b, c). Motivation absorbs Magnitude: An analysis of health care services of Kumbh Mela published by Chinese Public Administration Review in volume 6 issue 1.

As we tried to put up that goals are followed, models emerge and implementation of models is quite dependent on the role of the "will" of the implementing agencies so much so that their motivation absorbs the magnitude of the tasks.

10.3 How Sustainable the Commitment Is?

The report on "Reaching MDGs in India, by Centre for Legislative Research and Advocacy" (2009) reported that the next 5-year term from 2010 to 2014 marks the final opportunity India has of reaching the MDGs and unless urgent action is taken, India will fail to reach the MDG targets relating to health, nutrition, gender and sanitation. India is on track for some of the MDGs but the progress is not so encouraging for the other MDGs.

Two-thirds of the way to the 2015 finishing line for achieving the eight globally agreed MDGs, India is at a crucial turning point with a few successes and some failures. Persistent inequalities, ineffective delivery of public services, weak accountability systems and gaps in the implementation of pro-poor policies are the major bottlenecks to progress said experts (CSO 2009).

On the one hand, sustenance of commitment and will is the solution, but on the other hand, it has to address as a question in itself first because of evidence associated with corruption marring it. One of the most important lessons that has always been learnt in pursuance of goals is that if those are not achieved well within time, the resource burden that they create kills the relevance of the achievement itself and the other aspect is that even if the goals are achieved well within stipulated time, the reach and the resources allocated have to be properly monitored because of the effect that over allocation may create on other goals.

Corruption in India is a major issue and adversely affects its economy (Singh 2010). A 2005 study conducted by Transparency International in India found that more than 62 % of Indians had first-hand experience of paying bribes or influence peddling to get jobs done in public offices successfully. Indian media has widely published allegations of corrupt Indian citizens stashing trillions of dollars in Swiss banks (Agencies 2012).

10.3.1 Sustained Integrity: A Possible Solution

For over three decades, integrity has been widely studied and reported on in both the industrial/organizational psychology and human resources management literature (Martin 2011). Yukl and Van Fleet (1992) stated "Integrity means that a person's behaviour is consistent with espoused values and that the person is honest and trustworthy". Integrity is soundness of and adherence to moral principle and character, uprightness and honesty (MerriamWebster 2012). According to Socrates, "when reason rules but also respects the proper roles of emotion and desire, then the person will never commit sacrilege or theft nor treat others treacherously, neither will he ever break faith where there have been oaths or agreements". This is a classical definition of integrity (Ralph 2008). The most salient manner in which integrity expresses is through trustworthiness. As mentioned above, when we imagine a person or institution with integrity, one of the first concrete examples that come to mind is that of keeping promises, of maintaining our covenants, and of nobly and honourably caring for any trust we are responsible to safeguard (Ralph 2008). The person who is trustworthy in very small matters is also trustworthy in great ones and the person who is dishonest in very small matters is also dishonest in great ones (Ralph 2008). Commitment is the state or quality of being dedicated to a cause, activity, etc. It is also known as a pledge or an undertaking. In management literature, it can be defined as an emotional bond with one's organization (Meyer and Allen 1997) or relative strength of an individual's identity within particular organization (Steers 1977).

10.3.2 Sustained Integrity Is Led by Spirituality

MerriamWebster (2012) dictionary defines spirituality as, "of, relating to, consisting of or affecting the spirit; of relating to sacred matters; concerned with religious values; of, related to, or joint in spirit". According to Legere (1984), spirituality can be defined as " ... the attempt to give ultimate meaning to things, it is the ultimate context for humanity to understand itself. It is that interior quest for meaning in life which expresses itself in both contemplation and action and through which its divine presence is felt and understood ...". Elkins et al. (1988) declared spirituality as "state of being and experiencing which comes through awareness of transcendent aspects which is reflected by particular identifiable values with respect to self, other, nature, life and whatever somebody considers to be the ultimate". Conger (1994) stated spirituality as "the experience when someone literally transcends oneself, going beyond the limits of self-interest, for example love or social justice without self-interest or when someone is able to transform one's vision and feelings beyond the ordinary to discern extraordinarily godly presence in one's life and universe". Freshman's (1999) qualitative research, "An exploratory analysis of definitions and applications of spirituality in the workplace", found that "there are many possible ways to understand such a complex and diverse area as spirituality". Spiritual individuals in the workplace are more likely to demonstrate enhanced teamwork (Mitroff and Denton 1999; Neck and Milliman 1994), greater kindness and fairness (Biberman and Whitty 1997), increased awareness of other employee's needs (Cash and Gray 2000), increased honesty and trust within their organizations (Brown 2003; Krishnakumar and Neck 2002), higher incidences of organizational citizenship behaviour (Nur and Organ 2006) and express more servant–leader behaviour (Beazley and Gemmill 2006). Spirituality leads to commitment; Jurkiewicz and Giacalone (2004) asserted that spirituality "creates an environment where integration of the personal and professional selves is possible, engaging the whole person in the work process". Some authors suggest that spirituality and religion are related (Fabricatore et al. 2000) with others claim that spirituality and religion are ultimately not the same (Duffy 2006).

10.3.3 Practicing Spirituality

Willingness, commitment and sustained integrity has a basis of spiritual orientation, which was validated through the recent research that was conducted on understanding the perspective of management of spiritually oriented organizations; the methodology was based on reflexive approach (Alvesson 2003) and an amalgamation of in-depth interview and focused group discussions and the outcomes are suggestive of the fact that every task which especially has association with the well-being of humans should be delivered with spiritual orientation.

The study focussed on spiritually oriented organizations because of the reason that these organizations not only have scaled sustainable heights but also have sustained those heights through principles that are beyond contemporary management approach and reach and have shown exceptional results as depicted in Table 10.2.

Table 10.2 Principles in practicing spirituality

Every member of such organizations merges his identity with the objectives and identity of the organizations concerned
Fundamentals were extremely clear and interpreted universally in the same form
All for one and one for all
Work with complete capacity and compensation as per the basic needs
Service to all the humans with true sense and submission
Trust and complete acceptance
Respect for the law of the land
Transparency
The respondents had a common chord as far as the discussion and their contribution is concerned and that was service of others before self, the concept of legacy, association of legacy with management procedure evolution, regular introspection for managers and leaders, strict discipline, character and display of character, concept of precedence and maintenance of purity
Looking at so many aspects and so much of expectation one finds it extremely difficult to be followed but this has been the most enduring part of this research that one could find all these characteristics in single personalities and a very large number of such personalities and the immediate thought process which emerges is that if all the managements of all the organizations in the world could have few such people the culture of all the organizations would be associated with utmost and the sustained development of its members and customers. But is it really possible?

Yes, if some of the following principles are reiterated and followed as they do:

- Means are as important as ends;
- Organizations must emphasize on processes along with outcomes, whereas processes must integrate the human resource growth and satisfaction factor;
- Your time is the only thing that is useful for others because actually "your time is for others";
- One of the most important organizational management principals that discuss was that sensitivity and sensibility are the basis of almost everything, whether it is motherhood, our physical being, spirituality, family, organizations, society and whole of the world and thus derived out of evolving knowledge- and sensitivity-based static wisdom, action change and processes get improved upon.

10.4 What If Integrity Is not Sustained?

India has been rocked by scams and there is a long list that is suggestive of such a dark picture that one wonders whether anything would ever be ideally achieved leave aside the development goals.

There is an exhaustive list of scams that took place in recent times in India and the magnitude in terms of money involved, people involved and the people effected is huge so much so that if a systematic study may be conducted in terms of that if such scams and instances of corruption would not have happened what would be the positive implications those would have taken place.

Further examples of humans deceiving humans can be fetched from many stories along with a television-based movement generated recently called *Satyamev Jayate* meaning "Truth is Victorious" (Satyamev Jayate 2012). An example from *Satyamev Jayate*: women of Kowdipally village of Andhra Pradesh have been misguided by doctors over the years. Without any need, they are asked to get their uterus removed. The operation costs the poor farmers so much that they sell their land for it and even took loans.

10.5 Conclusion

Integrity has to be sustained is the observed and proposed solution, but before being a solution it is a question in itself. The answer to this question could only be found through observing and understanding the individuals and organizations who have been sustaining their integrity leading to sustained commitment and will. A study of such individuals and organizations requires qualitative research, and the most appropriate is reflexive methodology along with detailed observation of the examples. Spirituality emerges as a key basis, and the reason for sustenance of integrity which ultimately becomes the reason for achieving goals.

The paper highlights how disaster-like situations can be met by suggesting the examples in daily life and the aspects of integrity achieved through spirituality responsible for such achievements, wherein the most important thing is to understand the pain and plight of the people through Gandhian thought that has imbibed characteristics of spirituality. Co-creation and its pertinence in were highlighted in the process to involve people to solve their problems and to achieve larger goals for the good of all.

References

Agencies (2012) Money held by Indians in Swiss banks stands at Rs. 12, 740 cr. The Indian Express. http://www.indianexpress.com/news/moneyheldbyindiansinswissbanksstandsatrs12740cr/962007/.June14. Accessed 26 Sept 2012

Alvesson M (2003) Beyond neopositivists, romantics, and localists: a reflexive approach to interviews in organizational research. Acad Manag Rev 28(1):13–33

Amin (2012) Assam's 'Boat Clinic' carries health services and hope. Health.india.com. http://health.india.com/diseases-conditions/assams-boat-clinic-carries-health-services-and-hope/. July11. Accessed 11 June 2012

Bandura A (1991) Social cognitive theory of self-regulation. Organ Behav Human Decis Process 50(2):248–287
Beazley D, Gemmill G (2006) Spirituality and entrepreneurship. J Manage Spirituality Relig 3(3):258–270
Bharti K, Agrawal R, Sharma V (2013) Will value co-creation help bottom of the pyramid development? Indian Manage 52(7):26–29
Biberman J, Whitty M (1997) A postmodern spiritual future for work. J Organ Change Manage 10(2):130–138
Brown RB (2003) Organizational spirituality: the sceptic's version. Organization 10(2):393–400
Cash KC, Gray GR (2000) A framework for accommodating religion and spirituality in the workplace. Acad Manage Executive 14(3):124–134
Conger JA (1994) Spirit at work: discovering spirituality in leadership. Jossey-Bass, San Francisco
CSO (2009) Millennium development goals-India country report 2009. http://mospi.nic.in/rept%20_%20pubn/ftest.asp?rept_id=ssd04_2009&typeNSSO. Accessed 26 Sept 2012
Duffy RD (2006) Spirituality, religion, and career development: current status and future directions. Career Dev Q 55(1):52–63
Elkins DN, Hedstrom LJ, Hughes LL, Leaf JA, Saunders C (1988) Toward a humanistic-phenomenological spirituality: definition, description and measurement. J Humanistic Psychol 28(4):5–18
Fabricatore AN, Handal PJ, Fenzel LM (2000) Personal spirituality as a moderator of the relationship between stressors and subjective well-being. J Psychol Theol 28(3):221–228
Freshman B (1999) An exploratory analysis of definitions and applications of spirituality in the workplace. J Organ Change Manage 12(4):318–325
Gist ME, Mitchell TR (1992) Self-efficacy: a theoretical analysis of its determinants and malleability. Acad Manage Rev 17(2):183–211
Jurkiewicz CL, Giacalone RA (2004) A values framework for measuring the impact of workplace spirituality on organizational performance. J Bus Ethics 49(2):129–142
Kalagu.wordpress.com (2009) Sanitary napkins for poor women. http://kalugu.wordpress.com/2009/05/20/sanitary-napkins-for-poor-women/. Accessed 3 Sept 2012
Krishnakumar S, Neck CP (2002) The what, why and how of spirituality in the workplace. J Manag Psychol 17(3):153–164
Legere TE (1984) A spirituality for today. Stud Formative Spirituality 5(3):338–375
Locke EA (1968) Toward a theory of task motivation and incentives. Organ Behav Human Perform 3:157–189
Locke EA, Latham GP (2006) New directions in goal-setting theory. Curr Dir Psychol Sci 15(5):265–268
Martin DE (2011) Whistle blowing, religiosity, spirituality and integrity: understanding the impact of social dominance orientation and environmental context. J Moral Occup Psychol, Forthcoming. Available at SSRN: http://ssrn.com/abstract=1970970
MerriamWebster (2012) Integrity. http://www.merriamwebster.com/dictionary/integrity. Accessed 26 Sept 2012
Meyer JP, Allen NJ (1991) A three-component conceptualization of organizational commitment. Hum Res Manage Rev 7(1):61–89
Miner John B (2003) The rated importance, Scientific validity, and practical usefulness of organizational behaviour theories: a quantitative review. Acad Manage Learn Educ 2:250–268
Mitroff II, Denton EA (1999) A study of spirituality in the workplace. Sloan Manage Rev Summer pp 83–92
Neck CP, Milliman JF (1994) Thought self-leadership: Finding spiritual fulfilment in organizational life. J Manage Psychol 9(6):9–17
Nirvikar Singh (2010) The trillion-dollar question. The Financial Express. December 19. http://www.financialexpress.com/news/the-trilliondollar-question/726482/0. Accessed 9 Oct 2012

Nur YA, Organ DW (2006) Selected organizational outcome correlates of spirituality in the workplace. Psychol Rep 98(1):111–120

Ordóñez LD, Schweitzer ME, Galinsky AD, Bazerman MH (2009) Goals gone wild: the systematic side effects of overprescribing goal setting. Acad Manage Perspect 23(1):6–16

OutlookIndia.com, Kumar K (2007) Creating a successful model for checking female foeticide, running a district that is a model of efficient governance. http://m.outlookindia.com/story.aspx?sid=4&aid=234467. Accessed 23 Apr 2007

Oxfam India (2009) Centre for legislative research and advocacy. Reaching MDGs in India. http://www.clraindia.org/include/MDGs.pdf. Accessed 26 Sept 2012

Prahalad CK, Ramaswamy V (2002) The co-creation connection. Strategy Bus, pp 50–61

Press Information Bureau, Government of India (2008) Radical improvement in Delhi Government school system. http://pib.nic.in/newsite/erelease.aspx?relid=38364. Accessed on 13 Sept 2012

Purva D (2012) Satyamev Jayate hits the heart! The Times of India. May 07. http://articles.timesofindia.indiatimes.com/2012-05-07/tv/31609772_1_aamir-khan-female-feticide-ultrasound. Accessed 1 Sept 2012

PTI (2012) Man creates forest single-handedly on Brahmaputra sand bar. Times of India. http://articles.timesofindia.indiatimes.com/2012-03-25/the-good-earth/31236015_1_forest-gunin-saikia-plantation. Accessed 25 March 2012

Ralph D (2008) Integrity-psychological, moral and spiritual. http://www.lcsc.edu/diessner/pdf/Integrity%20paper%20web%20version%202007june20pdf. Accessed 13 Oct 2012

Sareen V (2012) Nawanshahr, once infamous for female foeticide, now witnesses the woman empowerment. Punjab News Express. http://punjabnewsexpress.com/news/1293-Nawanshahr-once-infamous-for-female-foeticide-now-witness-the-woman-powerment.aspx. Accessed 26 April 2012

Satyamev Jayate (2012) Satyamev Jayate—Every life is precious. May 6. http://www.satyamevjayate.in/. Accessed 8 Feb 2014

Sawhney M, Prandelli E (2004) In: Starkey K, Tempest S, Mac Kinlay A. (Eds.). How organizations learn: managing the search for knowledge. CengageBrain.com

Sen A (2000) Development as freedom. Oxford University Press, New Delhi

Sharma V (2009a) Affordability for the poor and profitability for the provider: provision of health for the rural and the poor population of the world with special reference to Uttar Pradesh in India. VDM Verlag ,Germany

Sharma V (2009b) Health for all through affordability and profitability with public spiritual partnership. http://www.adbi.org/files/2010.12.11.cpp.sess2.5.sharma.paper.health.affordability.pdf. Accessed 6 Sept 2012

Sharma V (2009c) Motivation absorbs magnitude: an analysis of health care services of KUMBH MELA. Chin Publ Adm Rev 6(1):1–8

Spohrer J, Maglio PP (2008) The emergence of service science: toward systematic service innovations to accelerate co-creation of value. Prod Oper Manage 17(3):238–246

Steers RM (1977) Antecedents and outcomes of organizational commitment. Adm Sci Q 22:46–56

United Nations Millennium Development Goals (2014) http://www.un.org/millenniumgoals/bkgd.shtml. Accessed 11 April 2014

Vargo SL, Maglio PP, Akaka MA (2008) On value and value co-creation: a service systems and service logic perspective. Eur Manag J 26(3):145–152

Xie C, Bagozzi RP, Troye SV (2008) Trying to presume: toward a theory of consumers as co-creators of value. J Acad Mark Sci 36(1):109–122

Yukl GA, Van Fleet DD (1992) Theory and research on leadership in organizations. In: Hough LM, Dunnette MD, Handbook of industrial and organizational psychology, 2nd ed. Consulting Psychologists Press, Palo Alto, pp 147–197

Chapter 11
Anticipating and Managing an Unwelcome Guest: Climate Smart Approach in Disaster Management

Ashok Kumar Singha, Saroj Kumar Nayak, N.C. Kar, Swapna Tripathy and Barendra Krushna Sahoo

11.1 Introduction

Enhanced concentration of greenhouse gases causes global warming. The link between global warming and extreme weather events such as cyclone, cloud burst and enhanced return period of drought have become increasingly clearer. This requires urgent attention of all stakeholders, precise early warning system, adequate preparedness and swift response to minimise loss to life, property and livelihoods. As per the Inter Government Panel on Climate Change (IPCC 2013) Fifth Assessment Report, climate change impacts weather-related hazards and may lead to more number of intense extreme weather events more frequently in the coming decades. The effect is catastrophic; due to geophysical (Tsunami, Earth Quake, etc.) and hydrometeorological (flood, cyclone and drought) events globally more than 226 million people are affected every year. Weather-related disasters comprise about 81 % of all events, causing 72 % of all economic losses and 23 % of fatalities.[1]

According to the International disaster database, the disaster profile of India showed there were 52 extreme temperature-related events that killed 14,808 people; floods that were the result of excess precipitation killed 66,500 people in 263 events. Out of which, 7,419 died in 22 flash flood events and 569 in 4 storm surge/coastal events. But the number of people affected in these events was 7,419

[1] According to EM-DAT/CRED statistics that tracks disaster and extreme weather events globally.

A.K. Singha (✉) · S.K. Nayak · S. Tripathy · B.K. Sahoo
CTRAN Consulting, Bhubaneswar, Odisha, India
e-mail: ashoksingha@gmail.com

N.C. Kar
Utkal University, Bhubaneswar, Odisha, India

H. Ha (ed.), *Land and Disaster Management Strategies in Asia*,
DOI: 10.1007/978-81-322-1976-7_11

and 569, respectively, while in 133 general flood events, 326 million people were impacted.

The objective of this paper is to examine the response of two state governments in India to two different disasters and to examine their response through the climate smart disaster management framework for validation. Uttarakhand and Odisha responded to these two disasters differently (without attribution of motive of good or bad governance)—one applied the Climate Smart Disaster Management (CSDRM) approach and one without—and we examine the outcome ex-post.

The key research question the paper tries to answer is to see that irrespective of the location and context, the application of CSDRM principles result in better response and leads to resilience in handling such extreme events better. The paper is divided into the following components: a background followed by a detailed literature review on the subject. A theoretical framework developed on CSDRM has been included to familiarise the readers, and subsequently, the research methodology highlights a case-based approach and participant observation method as tools. This is followed by a discussion and conclusion.

11.2 Literature Review

Climate change will significantly impact the future extreme weather events in Indian subcontinent in terms of frequency and intensity. For the Indian subcontinent, the multi-model mean (RCP 8.5) for warm spells will be longer and may extend to 150–200 consecutive days beyond the 90th percentile (Sillmann et al. 2013). In terms of thermodynamics, a warmer atmosphere will carry more moisture than the cooler one, thus releasing more projected increases of moisture approximately 2.3 % per degree of warming for summer monsoon rainfall (Menon et al. 2013). However, this will be with weaker circulation and uneven spread. Other projected changes include increased temperature in land areas in northern India, relatively longer winter and post-monsoon seasons and increase in annual average temperature by 3–6 °C. There are also reported findings to suggest significant changes in glaciers in the Himalayan ecosystem and resultant water balance change in the river basins, especially, in the north-eastern region and Indo-Gangetic plains. Tropical cyclones hit India time and again with varied intensity. Globally, about 80 cyclones form in a year. As per Indian Meteorological Department (IMD), the average annual frequency of tropical cyclones in the north Indian Ocean (Bay of Bengal and Arabian Sea) is about five. The Bay of Bengal is four times more vulnerable to tropical cyclone than the Arabian Sea, the ratio being 4:1. The monthly frequency of tropical cyclones in the north of Indian Ocean displays a bimodal characteristic with a primary peak in November and secondary peak in May. The months of May–June and October–November are known to produce cyclones of severe intensity (in Odisha, most cyclones of severe intensity have been in October).

There are several approaches to respond to a disaster. Various risk management instruments have been used by the Governments and civil society, aid agencies in different circumstances, emergency, conflict, prolonged weather events such as drought or sudden events like Tsunami to protect the livelihood and also devise protocols to safeguard communities in future (Ewbank 2010). There is also argument about equity and climate justice, responsible development (UNISDR 2005) and respect to property rights. Hyogo framework envisages to engage all stakeholders (local, national, regional and international) to act responsibly in reducing loss of lives and social, economic and environmental assets when hazards strike. It is also noticed that global warming enhances disaster both in intensity and frequency, so careful analysis of development planning in fragile zones, blending appropriate adaptation and mitigation action for low carbon development (Urban et al. 2010) would help, and this has to be adopted by industrial, social, economic and environmental networks. There has to be prioritisation of initiatives in business as usual scenario and emergencies to identify vulnerabilities in planning and standard operating procedures. Continuous learning and reflection is a smart initiative that enhances resilience (Kobasa et al. 1985).

In this document, attempt has been to analyse two distinct weather events in two different locations in hill and coast, both fragile ecosystems. In Odisha, population impacted is largely fixed even though it is a festive season and more people at home location; in Uttarakhand, it is mixed population (surge in floating tourist population due to a religious event like *chardham yatra* and local residents). Idea in this paper is not to compare the extent of damage as the baselines differ but to understand the genesis of the extreme events due to climate change and proper understanding and interpretation of this change through the scientific data for handling such events smartly. Climate change in India has aided the increase in water balance and anthropogenic activities in the glacial features in north leading to cloud burst and flood. In coastal areas, the warming of sea surface aided the intensification of tropical cyclone in the Bay of Bengal. In both the cases, global warming is likely to have resulted in two distinct extreme weather events putting in test the disaster management machinery of the state and community in response, rescue, recovery and relief.

11.3 Theoretical Framework

CSDRM approach has been developed based on the expert input of over five hundred researchers, community leaders, nongovernment organisation (NGO) workers and government officials from ten disaster-prone countries. CSDRM framework has three distinct focuses: (1) assessment of climate risks and uncertainties and collaboration across diverse stakeholders and using their knowledge as a tool of management; (2) strengthening the adaptive capacity of the people by strengthening networks, policy implementation, reflection and learning with requisite flexibility; and (3) reducing the vulnerability and addressing endemic structural issues relating to poverty and vulnerability through equity, transparency and accountability in a collaborative arrangement with diverse stakeholders (Government, NGO, the private sector and international agencies). The framework [2] has three distinct pillars: *Pillar 1*: tackle changing disaster risks and uncertainties (three subcomponents under this pillar are collaboration, assessment of risk and uncertainties, and integration of risk and uncertainties with policy, plan and programmes to reduce the exposure and vulnerability); *Pillar 2*: enhancement of adaptive capacity (three subcomponents under this are networking, learning and reflection and flexibility in policy and programs to incorporate feedback and lessons); *Pillar 3*: addressing poverty, vulnerability and structural causes that reduce adaptive capacity (subcomponents include equity and social justice, partnerships that ensure access and just entitlement and empowerment of local community, local authorities, NGOs and CBOs to influence both decisions of the government).

11.4 Research Methodology

In this paper, we have tried to analyse two cases using the Climate Smart Disaster Risk Management Framework (CSDRM). The data have been collected through participant observation method, focus group discussions and key informant interviews. Scientific data have been gathered from satellites (both for the cyclone and the flash flood event) and other secondary sources including media reports. The sole objective of the paper is to validate the framework based on the developed cases and in no way reflects on the efficacy of one government or the other. However, the hypothesis is that a government which follows the climate smart approach is in a position to create a better adaptive capacity in the state that they govern to respond to climatic events and stress.

We have used participant observation method by accurately following two major climate events to validate the above framework. De WALT and DeWALT (2002) describe in their guidance note that participant observation is a method to develop a holistic understanding of the phenomena under study that is as objective and accurate as possible given the limitations of the method (p. 92). They suggest that participant observation can be used as a way to increase the validity of the study, as observations may help the researcher have a better understanding of the context and phenomenon under study. Validity can be stronger with the use of additional strategies used with observation, such as interviewing key informants, document analysis, media reportage or other more quantitative methods like sample survey. In this instant case, the authors directly followed the events in Uttarakhand for flash flood and severe cyclone Phailin in Odisha from satellite

[2] Mitchell et al. (2010) Climate Smart Disaster Risk Management, Strengthening Climate Resilience, Institute of Development Studies: Brighton, UK.

image, media reports, direct interaction with key informants, affected people and other stakeholders.

For both the areas, 100 key informants from army, civil authorities, tour operators, journalists, scientists, members of the community, NGOs, bi- and multilateral institutions were consulted. The media report i.e., visual, print and social media were tracked three days around the occurrence of the actual event and up to a week after the event. The discourses after that have not been included for this analysis as they turn spurious due to political economic considerations. Scientific imagery from satellites is in the public domain and has been downloaded during the period cited above.

11.5 Findings and Discussion

Two cases have been listed below which are analysed using the principles of participant observation.

11.5.1 Case on Flash Flood in Uttarakhand

Uttarakhand in the Himalayan ecosystem has 90 % of its area under hills. Extreme climatic events from 15 June 2013 to 17 June 2013 wreaked havoc in this state which coincided with religious festival Char-Dham yatra. Alaknanda River and the Mandakini, both tributaries of the Ganga originating upstream in Himalayas, received heavy precipitation (4 times more than normal). This was due to the fusion of westerlies cloud system and additional water balance available through the melting of glaciers in the heat waves scorching the area in May. The area geomorphologically is fragile with high erosive capacity. During Late Quaternary period, the towering topography in upper reaches of Himalaya had witnessed extreme climate fluctuations with recurrent episodes of glaciation and warming, synchronous with the global events that affected the fluvial discharge and thereby the gradational processes (Pandey et al. 2014). In 1999, the area had also suffered mild earth quake resulting in loosening of rock structures. Anthropogenic activities (Dharmadhikary 2008) compounding the scale of the disaster included unabated expansion of hydropower projects (70 such projects) and construction of roads to accommodate ever-increasing tourism; the registered vehicles in the state grew from 83,000 in 2005–2006 to nearly 180,000 in 2012–2013. There has been de-notification of fragile areas and unplanned construction of hotels and rest houses to accommodate tourist inflow and thus enhanced the vulnerability of the area. On 16 and 17 of June 2013, heavy rainfall triggered a landslide in the Kedar Valley and debris from the landslide and water from the lake travelled down the slope, channelled into the glacier, and came down to Kedarnath town. Unfortunately, due to years of unplanned development, debris deposits created a lake-like feature not allowing smooth run-off downstream.

The lake boundary could not withstand the extreme pressure due to simultaneous downpour in the region, and water gushed out with powerful velocity, creating a new stream in addition to the existing two streams and swept away anything that came on the way[3] (hotels, shops, roads, structures and tourists, local inhabitants and wild life, etc.). As per the Government of Uttarakhand, 4,364 people lost their lives due to the natural disaster caused by heavy rainfall, cloudbursts, flash floods and landslides in June 2013 in Uttarakhand including 4,119 missing persons, who are feared to be no more.[4] The overall evacuation as per National Disaster Mitigation Agency stood upwards of 105,000 people from the affected region. The assessment of forest cover loss conducted by National Remote Sensing Centre in the affected area using IRS LISS IV data (spatial resolution 5.8 m) shows vegetation/forest cover change at 50 locations (grassland—29 patches of about 124.9 ha and forest—21 patches of about 46 ha).

The heavy loss of life and property and long-term damage to ecosystem can be analysed in the lenses of CSDRM. The warning from Indian Meteorological Department was given at least 72 h ahead of the tragedy. This was scientifically correct, textual and not specific. The moderate to heavy and very heavy rain fall would have significantly different effect on hills and plains. This was not clear enough signals for the administration to interpret and act. On the other hand, Meteorological and Oceanographic Satellite Data Archival Centre (MODSAC) of ISRO had issued a warning[5] on 16 June 2013 with specific locations and impact with visuals.

In precise, it shows that the vulnerability points had been identified by some agencies close to 24–48 h ahead of its actual incidence. The state too had a disaster management agency (since 2007) but did not have any disaster management plan or standard operating procedure.[6] The state was also almost a year into preparing the strategic climate change action plan by an UN agency. It had invested in the vulnerability assessment supported by a prestigious climate change network but could not make adequate preparation. Its proximity to the National Capital Region

[3] ISRO's preliminary estimates show that a total of 745 landslides occurred along the river valleys of Mandakini, Mandani, Kali and Madhyamaheshwar.

[4] Written reply of Minister of State for Home, Mr M Ramachandran in Parliament of India (Lok Sabha).

[5] Prediction given by the model included (1). Possibility of high rainfall event at 28.75 N , 77.52 E with radius of influence of 28.89 km. (2). Possibility of high rainfall event at 28.90 N , 80.70 E with radius of influence of 5.66 kms. (3). Possibility of high rainfall event at 29.30 N , 78.50 E with radius of influence of 5.66 km. (4). Possibility of high rainfall event at 30.44 N , 78.69 with radius of influence of 58.71 kms. Possibility of cloud burst event in which major cities to be affected are BARKOT, KIRTINAGAR, MUNIKIRETI, NARENDRANAGAR, PAURI, RAIWALA, RISHIKESH, RUDRAPRAYAG, SRINAGAR TEHRI, UTTARKASHI, (5). Possibility of high rainfall event at 31.33 N , 78.63 with radius of influence of 10.80 km. (6). Possibility of high rainfall event at 31.60 N , 78.80 E with radius of influence of 5.66 km.

[6] Times of India report, 21 June 2013 based on Comptroller and Auditor General Report on 23rd of April 2013.

and presence of army also could not reduce the impact of tragedy. After the initial shock, army and air force[7] with local tour operators with knowledge of the terrain did a commendable job in extremely inclement weather to save lives and reach with essential supplies to people stuck in different patches of mountain and forest.

If we look at pillar 2, different agencies of the government could not work in tandem and many had mandates with conflicting objectives of revenue generation (from tourists, power generation from hydro-resources and transport, etc.) with ecosystem preservation. There was no systematic assessment (CAG 2013) of long-term damage to ecosystem, forest and biodiversity loss and soil conservation.[8] The support networks were not involved even though there was a vibrant civil society movement in the state. The lessons from the past on such events (as recent as four events in 2009) were also not analysed and standard operating procedures not developed. This left the line administration in the quandary, and top administration turned defensive. Under pressure, the innovation was a casualty, decision-making was not orderly, and the actions were chaotic leading to further failure in delivery of basic services and emergency services.

If we look at the pillar 3, there seemed to be gross abuse of property rights, entitlement failure and also elite capture in taking decisions which are not in the interest of the poor and vulnerable. In fact, assiduous attempt was made to circumvent green laws by successive governments by allowing construction in ecofragile areas. No attempt has been made by any stakeholders to break a common ground to promote sustainable development.

11.5.2 Case on Tropical Cyclone Phailin in Odisha

Odisha has 480-km coast line and is a victim of cyclone and flood intermittently. Before Phailin, Odisha suffered extensive damage during the super cyclone in 1999. Very severe cyclonic storm Phailin originated in Andaman Sea on 6 October 2013 and intensified into a vortex with intensity T1.0 and centre 12.0°N/98.5°E at 0600 UTC on 7 October in a shear shape. Since then, its path was constantly monitored by indigenous Kalpana-1 satellite, Doppler radar in Vizag and IMD station in Bhubaneswar. Due to the climate change, sea surface temperature has remained significantly hot in the southern coast (Odisha and Andhra Pradesh) and this coupled with ocean thermal energy led to rapid intensification from 10

[7] Operation Surya Kiran (Hope) was commanded by Lieutenant General Anil Chait, General Officer Commanding in Chief (GOC in C) Central Command (now retired) in which 10,000 troops participated. It was conducted in tandem with the response by Indian Air Force (IAF) (Operation Rahat), the Border Road Organisation, National Disaster Response Force (NDRF), Indo-Tibetan Border Police (ITBP) and other para military forces under the Ministry of Home apart from the state police and disaster management team.

[8] CAG Report of 2013 highlights that Geological Survey of India had identified 101 villages as 'vulnerable' in June 2008 in Uttarakhand. The state government did not take any measures for their rehabilitation till date.

October 2013. Shy of the experience in Uttarakhand, IMD this time issued specific and visually coded pictures and interpretation to Odisha and Andhra Pradesh. It updated bulletins every half-hour and Red Message—the highest alert from IMD was concise and specific in terms of track, likely landfall route, intensity in terms of wind speeds and rainfall forecasts.

Odisha Government responded to the situation with adequate alacrity. Being a festive season (Dussehra is one the biggest festival with long holiday span in eastern region), many officials had planned their vacations. First, leaves of officers of the affected areas were cancelled and they were ordered to report to their respective departments (almost all major departments of the government); later, leaves of officials of all the districts were cancelled to keep them on the standby. Chief Minister led from the front taking constant review of the preparedness. Chief Secretary led the interdepartmental coordination effort and interfaced with various central agencies to provide logistical backup (air force, navy and National Disaster Response Force columns). Control rooms were set up in Ganjam (where the likely landfall was expected) and other nine districts. Mobile phone numbers of various agencies, officers and even suppliers of essential commodities were updated and verified. Special relief commissioner was the nodal officer for rescue and relief. He was in constant touch with district collectors, and they in turn divided areas into high, medium and low priority for evacuation. The block development officers visited cyclone shelters and arranged for dry and cooked foods which were assembled in a stadium in Bhubaneswar and despatched to various locations. Energy secretary reviewed the preparedness of the power system as it was likely to be worst hit. The labour availability to restore is a problem during the cyclone. Therefore, the labourers were mobilised from the neighbouring states and kept in the grids. More than a million people were evacuated (about 0.6 million in Ganjam alone) in one of the strongest coordinated effort seen in recent history. The state had set a zero-casualty target. After rescue, the focus shifted to relief where Indian Air Force started dropping food packets, water and rations to affected area. In total, 151 blocks covering 18,117 villages and 43 urban local bodies and more than 10 million people were directly impacted with 21 casualty due to cyclone and 17 due to the flood in the aftermath. More than 0.6 million ha of crop area was affected; 0.9 million people were evacuated from the cyclone-affected areas and 0.1 million from flood-affected areas.[9] It is important to note that the human casually due to cyclone and flood is less than 50.

If we see the entire Phailin management in CSDRM lenses, in pillar 1, the vulnerable areas had been identified almost a week ahead of actual incidence and were constantly monitored. The state leadership and executives identified and prioritised areas for response and coordinated with central agencies like NDMA, defence crisis management group, railways, tourism, transport, civil society organisations, private enterprises and specialists in the state with a firm focus but requisite flexibility. Information was available in plenty to drown the unnecessary rumours.

[9] *Source* Government of India, Report on Very Severe Cyclone Phailin, p.21.

In pillar 2, the state has learnt its lessons from the tragedy of the past super cyclone in 1999 and developed institutional framework, policy and standard operating procedures. Orissa State Disaster Mitigation Authority is an autonomous organisation created in the aftermath of the super cyclone in 1999. The state has developed disaster management plans, has identified multi-hazard zones and conducted several capacity-building programmes with the support of NDMA and agencies like UNDP. Civil society networks are actively engaged too. This has enhanced the adaptive capacity of the state (Twigg 2007). The administration has used policy, innovative approaches to respond to the situations (e.g. a district collector declared a hotel as hospital to accommodate expectant mothers, and doctors rushed to take care of the newborn during the incident).

If we look at the pillar 3, it is important to note that Odisha despite its recent growth is home to a large number of poor and economy is agrarian. These incidents create irreparable damage to livelihood and economic assets. The disaster management plans have been prepared not only to tackle the disaster but also to prepare the ground for the aftermath. The disaster management plans start from the village/ward level and are consolidated through similar planning at the panchayat, block, district and urban local body's levels in the selected districts. A cadre of village volunteers has been created to carry out the village-based natural disaster risk management programmes (Hedger et al. 2010). While it is premature to say that the just entitlements have been provided and developments in the state have been sustainable, the management of disaster as compared to the past is far better. The management of ecosystem due to the higher industrialisation would still be a challenge and may bring forth more unwanted guests like Phailin that the administration would have to handle.

The validity of this framework is tested using the different contexts. The robustness of the framework is based on the hypothesis that a CSDRM approach can better address a disaster event, be it cyclone, flood, GLOF, heat wave, etc. This is not to be interpreted as an impact assessment framework rather a management framework. The activities laid out in the pillars lead to a better outcome than the control situation (where such approaches were not followed). We have discussed the two scenarios pillar by pillar in two case studies (Table 11.1). The content analysis shows statewide details. The case studies give two different pictures under the same framework.

If we do a content analysis in terms of salient indicators, we find the following results.

11.6 Limitations

The main limitation in this study is to effectively engage with affected on site. While this was possible in Odisha for a longer period of time, for Uttarakhand, the period was short. The heavy casualty and chaos had impact on getting quality information.

Table 11.1 Comparison between Odisha and Uttarakhand

Parameter	Odisha	Uttarakhand
1a. Interagency collaboration starting from the warning	Odisha had developed a clear SOP based on the lessons of super cyclone of 1999 and subsequently institutional development. It included a multiagency protocol steered at state and district level involving government agencies, NGOs, emergency repose support team and volunteers. The impact was very high during this event. It helped reducing pressure on any single stakeholder during the emergency and ensured orderly response	Uttarakhand did not have any such prior experience of large-scale devastation; the frequency of occurrence was low. However, they had weather events beforehand and did not manage to develop any support network as envisaged in their own policy. They did have mandated government agencies, good conservation NGO network but could not build the collaborative network. The only difference is they had to deal with floating population (though they are all supposed to be registered for the festival). Therefore, it created pressure on every stakeholder and led to chaos. It persisted till the Army was put in charge
1b. Early warning system and vulnerability assessment	Vulnerability assessment is periodic not episodic, whereas an early warning system is largely episodic. Odisha has been able to use early warning system and does periodic assessment of vulnerability in flood, cyclone and heat waves but not much on slow onsetting events like drought. In Phailin, precise, timely warning led to proper identification and prioritisation of vulnerable areas (many were known); the system was user-friendly. Here, Odisha too benefitted due to the Uttarakhand event. The scientific message which administrators could easily understand colour-coded images helped. Response was vigorous and engaging	There has been several assessment of vulnerability in the state as it is an ecofragile area. However, this has always been locked into academic debates while politicians continued to overexploit the available resources even though from a maximisation point of view. They ignored red zones. In the recent event, the warning too came too close to the actual disaster and cannot be termed timely. The messages were not user-friendly and only can be interpreted by trained personnel who also chose to remain passive. Response was muted and passive
1c. Risk assessment and redesign of policy and programmes	State is an early mover, and lessons of super cyclone have been absorbed to develop policies and plans. Local administrations also tried various innovations as well as situational response. It enhanced localisation and was effective	The state has not learnt from past events, and the institutional frameworks for policy and programmes are weak. No standard operating procedures were quickly activated

(continued)

Table 11.1 (continued)

Parameter	Odisha	Uttarakhand
2a. Strengthening networks for experimentation and innovation	Odisha had systematically conducted mock drills, invested heavily in different networks supported by bilateral and multilateral agencies as well as own resources. Therefore, during crisis, these existing networks were functional and empowered to innovate and experiment with options	As described earlier, the state had a large number of conservation-based networks, but no investment and engagement by the state made them dysfunctional with notable exception of local tourist guides' body playing a commendable role in evacuation
2b. Learning and reflection	This is a key element of resilience and development practice into adaptive policies. There is some evidence of that in case of Odisha when the administration leant a lot form 1999 super cyclone, but even recent learning from Uttaranchal, early warning and response shows states approach to this. This many not be true for all occasions but for extreme events certainly	In case of Uttaranchal, it will be premature to comment on this as they did not have a crisis of the magnitude of Odisha. It is presumed that if the recent learning is factored in SOP, they could be better prepared. Operationally, Indian Army and air force not only followed time-tested methods but also did a few innovations in manoeuvres to get quick result
2c. Flexibility	In some sense, large-scale evacuation is procedure driven, but it is also important that the local administration get enough flexibility in procurement, deployment. In this case, district administration of affected areas had that; now there is a plan to tweak laws and policies that act as barriers	It was not ordered flexibility. Since the state operated in a chaos, the various agencies operated even though flexibly but not orderly. However, whether Odisha or Uttrakhand, a lot more needs to be done by state in building flexibility in programming
3a. Equity and justice	This aspect is not fully established; however, there was a forced evacuation of everyone in an affected area; the better-off ones may have stock foods and water, but the relief was for all. However, some influential members may have better access-definitely ruling party had upper hand. In terms of policy, the intent is equity and justice	The state had abused property rights, and it has been well documented, and it has affected one and all, rich and poor. However, its policies and documents reflect good intent (equity, justice)
3b. Empowerment	This was reasonably good; though a lot more work is needed to achieve	The community is empowered, but was numb with the magnitude of the tragedy. This is a systematic process—to take decisions; the right has been largely usurped by the state

(continued)

Table 11.1 (continued)

Parameter	Odisha	Uttarakhand
3c. Influence	The communities do influence decision-making especially long-term planning	The communities do have influence
Damage	CSDRM in a way ensures low casualty and if focused on adaptive capacity can address structural inequity and long-term sustainability. The human casualty has been minimal	High human casualty and long-term damage. Many tourists vowed not to return to the state and this might lead to a vicious cycle
Political cost	Strong leadership during the emergency has a positive political pay-off to the current political leadership	Heavy political cost, the chief minister was replaced and several leaders stare at reversal of political fortune in their constituencies

The above table clearly shows the state following CSDRM has better adaptive capacity (based on the parameters gathered through participant observation method and FGDs) to respond to extreme weather events. The hypothesis has been tested in two different events and locations without any prejudice using participant observation technique, and the idea is to test its robustness which is validated

11.7 Conclusion

It is concluded that climate change that leads to global warming can play havoc in several ecosystems. The intensity and frequency of disaster would increase in near term. A government using CSDRM approach is better prepared to manage disasters. It strengthens the collaboration between government agencies, private sector, and civil society and provides requisite flexibility and innovation required during emergency. This enhances the adaptive capacity and reduces the vulnerability. However, having a CSDRM policy in place does not automatically ensure the structural issues relating to climate change, disaster risk and poverty.

One of the key things in CSDRM approach is to actively engage all stakeholders to understand and interpret the information and knowledge about disaster, adaptation and mitigation. This includes converting scientific analysis to actionable information, guideline and protocol.

Second important aspect is to identify vulnerability (current and future) on a sustained basis and device strategy, governance arrangement accordingly (e.g. planation in buffer zone in mining areas, proper landfill, zonal planning in hills, coastal regulation zone guidelines).

Third important aspect is to plan mitigation actions that have cobenefits. The new market mechanism and existing mechanisms like CDM ensure both domestic and international support to reduce the impact of global warming. Forest and biodiversity conservation should be rewarded under common but differential responsibility framework.

Existing legal framework needs to be strengthened to arrest ecosystem damage.

Balanced approach is needed in developing social, economic and technical solutions (e.g. storage dams where hydropower is generated, shelter belt planation and mangroves along the coast). Engagement of all sections of the community (residents in upper reaches and residents of meadows; riverine settlers and farmers and fishers along the coast; tour operators respecting and adopting responsible tourism in fragile areas) is necessary to achieve sustained resilience. It is important to review the approaches that succeeded and that failed after the event to develop a plan of action for the future.

Use of modern technology (computation, prediction, navigation, communication, materials that sustain) also helps in enhancing adaptive capacity, and the state needs to invest in such technology, infrastructure and processes.

Trained human resources at the local level in emergency are the most effective solutions to speedy recovery, building such skills in educational curriculum, vocations, would enhance resilience.

Financial protection strategy will include design of appropriate insurance and savings products, investment in infrastructure that reduce/withstand hazards (e.g. cyclone shelters, savings cum relief schemes, weather insurance products, etc.).

References

Comptroller and Auditor General, Government of India (2013) Performance Audit of Disaster Preparedness in India, Report No. 5

Dharmadhikary S (2008) Mountains of concrete: dam building in the Himalayas, International Rivers, December, http://www.internationalrivers.org/fi les/IR_ Himalayas_rev.pdf

DeWalt Kathleen M, DeWalt Billie R (2002) Participant observation: a guide for fieldworkers. AltaMira Press, Walnut Creek

Ewbank R (2010) Integrating climate change adaptation into secure livelihoods, toolkit: framework and approach. Christian Aid, London

Hedger M, Singha A, Reddy M(2010) Building climate resilience at state level: disaster risk management and rural livelihoods in Orissa in climate smart disaster risk management, strengthening climate resilience, Discussion Paper Series, IDS, Brighton

IPCC (2013) Climate Change 2013: physical science basis. https://www.ipcc.ch/report/ar5/wg1/. Accessed 20th May 2014

Kobasa SC, Maddi SR, Puccetti MC, Zola MA (1985) Effectiveness of hardiness, exercise and social support resources against illness. J Psychosom Res 29:525–533

Menon A, Levermann A, Schewe J (2013) Enhanced future variability during India's rainy season. GRL, Under revi. http://www.pik-potsdam.de/~anders/publications/menon_ levermann13.pdf

Mitchell T, Ibrahim M, Harris K, Hedger M, Polack E, Ahmed A, Hall N, Hawrylyshyn K, Nightingale K, Onyango M, Adow M, Sajjad Mohammed S (2010) Climate smart disaster risk management. Strengthening Climate Resilience, Brighton

Pandey AK, Pandey P, Singh GD, Juyal N (2014) Climate footprints in the Late Quaternary–Holocene landforms of Dun Valley, NW Himalaya, India. Curr Sci 106(2):245

Sillmann J, Kharin VV, Zwiers FW, Zhang X, Bronaugh D (2013) Climate extreme indices in the CMIP5 multi-model ensemble: Part 2. Future climate projections. J Geophys Res. doi:10.1002/jgrd.50188

Twigg J (2007) Characteristics of a disaster-resilient community: a guidance note. DRR Interagency Coordination Group, London

Urban F, Mitchell T, Silva Villanueva P (2010) Greening disaster risk management: issues at the interface of disaster risk management and low carbon development, in Strengthening Climate Resilience Discussion Paper 3, IDS, Brighton

UNISDR (2005) Living with risk: a global review of disaster reduction initiatives, UNISDR, Geneva, Switzerland UN-ISDR (2005), Hygo framework for action 2005–2015, http://www.unisdr.org/we/coordinate/hfa

Part III
Public Education and Other Issues Related to Disaster Management

Chapter 12
Disaster Risk Management and Public Education in Singapore

Huong Ha and Suriani Jamil

12.1 Introduction

12.1.1 Background

Similar to the impact of any other environmental problems, natural disasters do not respect any race, any groups of stakeholders, any income groups and any countries. The impacts of both natural and man-made disasters and environmental problems, such as the cross-national Tsunami in 2004, the Tsunami in 2011 in Japan, the recent haze incident in Singapore due to the forest fires in Indonesia in 2013, the Haiyan typhoon in the Philippines in 2013 and many others, are tangible and intensive posing threats on several aspects, including socio-economic, human and ecosystem development (Ha 2013).

Although the impacts of natural disasters are cross-bordered and difficult to be controlled and contained within a country or a region, public awareness of disaster management is important to increase self-resilience before, during and after disasters. Public awareness is defined as "the extent of common knowledge about disaster risks, the factors that lead to disasters and the actions that can be taken, individually and collectively, to reduce exposure and vulnerability to hazards" (International Federation of Red Cross and Red Crescent Societies 2011, p. 5). One of the measures to enhance public awareness of disaster risk management (DRM) is to provide educational programs to the public. Nielsen and Lidstone (1998) also explained that an increase in disasters has entailed higher expectations of public safety. Thus, the public have expected governments to provide them with

H. Ha (✉)
UON Singapore (University of Newcastle, Singapore Campus), Singapore, Singapore
e-mail: artp6075singapore@yahoo.com

S. Jamil
Postgraduate student, University of Newcastle, Newcastle, Australia

H. Ha (ed.), *Land and Disaster Management Strategies in Asia*,
DOI: 10.1007/978-81-322-1976-7_12

(1) information about how to prepare for and avoid disasters (assuming that risk can be controlled and managed), and (2) what to do during and after disasters (Goldstein 1990).

12.1.2 Research Objectives

This chapter aims to discuss (1) different educational programs conducted in different countries, and (2) public education by the public sector, the private sector and civil society to educate the public about disaster preparedness and management, using Singapore as a case study. It also (1) evaluates the effectiveness of such programs, using the five element of the *Wisconsin Model of Public education* as the assessment criteria, and (2) analyses the implications of public education in DRM. The chapter will end with some policy recommendations, followed by the conclusion.

12.1.3 Significance of the Study

Since Southeast Asia is one of the most vulnerable regions to natural disasters, this chapter is significant as it provides information and evaluation about the public education initiatives for further research in public education and disaster management. Interdisciplinary nature is also a significant aspect of this study since this chapter discusses cross-disciplinary issues, including disaster management, strategies and education.

12.2 Literature Review

12.2.1 Public Education and Disaster Management: Concept and Importance

Community refers to a group of stakeholders, comprising the general public members such as students, academicians, environmentalists, neighbourhood committees, residents, and non-governmental bodies such as the Red Cross (Buccholtz and Carroll 2012). Community is important because it represents the social identity within the spectrum of people whereby their uniqueness is cultivated in their cultures, practices and perspectives.

According to Ahmadi and Laei (2012), the main objective of public education is to create public awareness about how risk can be reduced, and disaster can be managed. Different countries have applied different approaches to offer public education. The question is, whether public education should be made compulsory and free? In Iran, the people consider the overall concept of education as a key point in enhancing knowledge due to the shift from an agricultural-based economy to a modern-industrial lifestyle (Parsizadeh et al. 2010). It is emphasised by Ahmadi and Laei (2012) that education is a tool for the society to facilitate learning to respond to various types of incidents. In fact, many governments provide funds to support public education to promote the society with more knowledge towards the changing needs in information technology, science and technology in order to respond to risk (Ahmadi and Laei 2012, p. 869).

According to Pitoska and Lazarides (2013), the importance of public education is to ensure strong integration in values and practice among all individuals towards achieving the common goals. This will encourage disaster managers and the community to work together on "looking at situations, problems and opportunities" (Mumford 1995, p. 4). The following section will discuss a model of public education given the importance of public education in disaster management, and the need for cooperation and collaboration in managing such public education initiatives.

12.2.2 The Wisconsin Model of Public Education

In support of public education regarding DRM, Dr. George Kliminski and Dr. Eric Smith introduced a framework termed as *The Five Components of the Wisconsin Model of Public Education*. The model includes the following elements: (1) citizen involvement (2) needs assessment and planning, (3) extended use of public education facilities, (4) interagency co-ordination and co-operation, and (5) leadership and accountability (Wisconsin Department of Public Instruction 2014).

Citizen Involvement: Citizen involvement is defined as "public service bodies empower citizens" (Goulding 2009, p. 38), whereby more individuals are found to be participating in activities organised by the committee members in the neighbourhood (Goulding 2009). Dobos and Jenei (2013) explained that "openness and transparency of communication" (p. 1089) will allow disaster managers to identify their roles and responsibilities and to "adopt new methods to facilitate" the process (p. 1089). Indeed, Salehi and Gholtash (2011) also explained that active involvement of the citizens will lead to less ambiguity towards their given roles. When there is commitment within the community, there will tend to be "positive influence" and loyalty in their work (p. 310). This prepares individuals towards dealing with challenges during pre- and post-disasters (Fakhri and Talebzadeh 2011).

Needs Assessment and Planning: The community's learning and information needs have to be analysed so that community leaders are able to plan and design suitable public education programs. This will further promote a positive learning culture and create a greater sense of community bonding (Wisconsin Department of Public Instruction 2014). In assessing the needs of the community, Whipple and Nyce (2007) focused on "ethnography" and suggested that community analysis should be conducted before any public education programs are designed (p. 695).

This aims to obtain greater insights of the "characteristics of the people and the environment they lived in", and also the way the community assess public libraries and other channels for information (Whipple and Nyce 2007, p. 694). Dissemination of information related to disaster preparedness and management that are heavily converged with wordy-context and technical jargons will in fact, decrease the effectiveness in public education. This is because individuals without in-depth knowledge on disaster preparedness and management will not understand the "fundamental ideas" of what needs to be done before, during and after disasters (Abdul Rashid et al. 2013, p. 103). Disaster managers have to ensure that presentation of information has to be clear and concise, preferably combining all the elements of learning styles, for example, visual, aural, verbal, physical and logical (Evans and Sadler-Smith 2006). This will allow individuals to evaluate their knowledge with their "significant experiences", and will retain what they have learnt more, strongly (Mumford 1995, p. 7).

Extended Use of Public Education Facilities: In holding disaster preparedness and management programs for the community, Goulding (2009) mentioned that partnerships with community centres, public libraries and schools will enhance relationship among individuals within the community. Whipple and Nyce (2007) emphasised the importance of public libraries in providing "services and resources to the community" is to provide "space for community activities", and involve volunteers to assist in various projects and also to contribute ideas for decision making (p. 694).

Interagency Co-ordination and Co-operation: Badpa et al. (2013) mentioned that strong co-ordination and co-operation between disaster managers and the community will mitigate any "significant impacts", and can save many lives (p. 786). The concept of *knowledge management* within public education allows disaster managers to work along with various NGOs and other governmental bodies and also reflects the analysis of the "ethnography" (Whipple and Nyce 2007, p. 694). In this way, disaster managers are able to lead the community to meet the requirements and various perspectives of DRM. In addition, by adopting local knowledge of the community, disaster managers would be able to acquire information of how a community has survived in a disaster (Singapore Civil Defence Force 2012a, b, c). As a result, both the community and disaster managers would propose solutions such as the implication of "Radio Frequency Identification" to boost the rate of response and recovery which will then prevent more vulnerable lives at risk (Badpa et al. 2013, p. 790).

Leadership and Accountability: According to Ismail et al. (2011), effective involvement of the disaster managers with the various activities of community would provide an opportunity for the disaster managers to identify their leadership skills and their competencies in "addressing issues and activities" (p. 828). In other words, disaster managers will comprehend whether their sharing skills of tacit knowledge in a form of "collaborative learning through research" is workable for a particular community (Bushell and Goto 2011, p. 1246). Engagement of youths in disaster preparedness and management programs will allow them to identify their strengths and weaknesses in their "transformational" pathway to gain better leadership skills (Bushell and Goto 2011, p. 1249). This will build up their confidence and communication skill. In other words, it is a personal grooming of their self-development. The continuous exchanging of information will allow individuals to respect the perspectives and ideas of one another (Turkay and Tirthali 2010). On the other hand, Hudson et al. (2013) added that mentors can also help to facilitate the development of leadership in individuals. Mentoring is another form of pedagogical method because it involves a two-way communication in gaining knowledge which therefore "reinforce positive culture" in community networking (Hudson et al. 2013, p. 1297). Hence, without doubt, it takes an individual with great leadership skills to maintain the practice of public education and to allow lifelong learning (Wisconsin Department of Public Instruction 2014).

12.2.3 Public Education and Disaster Risk Management in Various Countries

The International Federation of Red Cross and Red Crescent Societies (2011) have discussed the four main public education initiatives with regard to disaster management, namely educational campaigns, participatory learning, informal education, and formal school-based interventions. Different countries have adopted different initiatives to educate the public regarding pre- and post-disaster management. This section discusses some public education programs to respond to major disasters that took place in Indonesia, Iran, Japan, and Vietnam given the magnitude and the frequency of disasters in these countries. The aim is to provide readers to have an overview of what different countries have done to educate the public regarding disaster management.

12.2.3.1 Indonesia

Tsunami in Aceh 2004

In 2004, the 9.0 Richter scale earthquake followed by a tsunami has taken 170,000 lives and the impact was so strong that the waves have spread across the Indian Ocean to other parts of Asia - Sri Lanka, Myanmar and Thailand (Singapore Civil Defence Force 2012a).

Public Education Programs

Matsumaru et al. (2012) explained that there was no educational or mitigation programs in Indonesia. In fact, the government focused more on restoration, i.e. building of shelters to house the locals. But still, they have failed to consider safety factors in their course of building of the houses and roads. It is emphasised that Indonesia needs to address the issues on "better urban development" (Matsumaru et al. 2012, p. 19) and refer to other countries, for example, on how they set reconstruction and recovery plans. The Taipei Times (2014) reported that

Indonesia "struggles to fund basic infrastructure normally" although on contrary, Indonesia has spent more than Japan in disaster mitigation. Still, Indonesia failed to understand the risk that is increasing in its territory. IRIN (2014) added that the government needs to improve its leadership in disaster governance as a whole so that there will be greater integration with all the public and private sectors, and civil society. Hence, with greater collaboration, there will be more resilience in the country.

12.2.3.2 Japan

Tsunami in Japan, March 2011

According to The World Post (2014), Japan's 8.9 Richter earthquake was then followed by a tsunami. This tsunami affected Hawaii, California, Oregon, Washington and British Columbia ripped thousands of lives and damaged many properties (Sorkhabi 2011).

Public Education Programs

Selby and Kagawa (2012) explained that Japan implemented disaster education programs in schools. This initiative aims to enhance awareness among the students on how they should react and whom they should reach out for when there is a disaster. It is emphasised by the officials that community networks is essential. The program is planned in a way that students are to reflect the past disasters and to take note on how these disasters were managed back then. The government of Japan also welcomes any recommendations towards disaster preparedness and management in order to mitigate the impacts of similar disasters in the near future.

12.2.3.3 Iran

In the context of public education and disaster management, Parsizadeh and Mohsen (2010) explained that Iran's geographical location is prone to natural disasters such as earthquakes because it is a country that is considered to have "high seismic hazard regions of the world" (p. 32). The International Institute of Earthquake Engineering and Seismology (IIEES) was then set up in 1989 which involved in "research, knowledge development, capacity building, vulnerability reduction and public (especially children) awareness and preparedness" (Parsizadeh and Mohsen 2010, p. 32). Such studies have been conducted to overcome the impact of the earthquakes. Iran has promoted the awareness of disaster management through safety culture, and they have channelled the effectiveness of the knowledge through the concept of nurturing—beginning from school with children at a young age via the activities planned by their teachers. Over the time, the idea and the importance of earthquake disaster preparedness will be well developed in these children, and the dissemination of the knowledge will be made through family and friends. In fact, such programs are incorporated within the "curriculum framework of the three different educational levels in Iran: 1 year pre-school, 5 years of primary schools, 3 years of secondary schools, and 4 years of high schools" (Parsizadeh and Mohsen 2010, p. 37). It indicated that with much strategic planning made towards the implementation of the "earthquake safety education program" (p. 33), this will encourage the public to develop a willingness to save the lives of others in disasters.

Besides promoting public education on reducing earthquake risks in schools, Iran has introduced the following public education programs that educate the public at large. Firstly, media; television and radio programs provide the society with weekly information on earthquakes whether it was the Bam earthquake in 2003 or the past experiences of earthquakes. Such media reflections allow the society to understand the great economic losses earthquakes could contribute and therefore safety precautions programs such as "School Earthquake Safety Drill" educate the public on earthquake preparedness (Parsizadeh and Mohsen 2010, p. 34). The reason for using the media as a platform of public education is its cost-effectiveness. The government also wants to ensure that even the people living in rural areas are able to receive such information. With such constant provision of programs on earthquakes and ways to reduce its risks, the society is no longer living in fear. For those who are illiterate, they did not miss the chance of being educated via television and radio programs on how to react when such disaster is to occur again.

Secondly, printed media, such as books, brochures, newspapers, etc., have been designed to educate the various age groups. They have been tailored to ensure that the targeted audience are well informed on how they should be prepared for earthquakes. For example, if the targeted audience are children, then, the information is simplified with colourful pictures to accompany their level of understanding (Parsizadeh and Mohsen 2010). For adults, books (e.g. "Earthquake: Looking Deep Insider the Earth"), and brochures (e.g. "Educational Guidelines for Earthquake Reduction in Office Buildings") are mostly printed in simplified version to ensure readers can easily understand the content. Most of the times, the printed brochures are packaged with "posters and audio tapes", and these will also be disseminated to various organisations (Parsizadeh and Mohsen 2010, p. 34). Newspaper is still considered as the most popular way in providing more detailed information and current updates on earthquake risk reductions. Print advertisements, such as the poster "Make Your House Safer against Earthquakes", are also made available in simplified wordings and colourful pictures (Parsizadeh and Mohsen 2010, p. 34).

It is observed that the various integrated forms of public education programs made available in Iran, have proven its effectiveness in the dissemination of information on preparedness in the case of occurrences of earthquakes. Not only that the children are well-knowledged about risk reductions, the whole society is also well involved and informed.

12.2.3.4 Vietnam

According to Hoang et al. (2007), floods are common in Hanoi since there is an "overdevelopment of floodplains outside dyke" which has contributed to more occurrences of flooding (p. 246). To overcome the flooding issues, public education such as community-based interventions have been implemented (Central Project Management Office, Ministry of Agriculture & Rural Development Central Project Management Office, Ministry of Agriculture & Rural Development 2009). The government of Vietnam realised the importance to work together with the community groups by educating the community leaders and their community clusters first about floods and their impacts. Providing the correct information about floods and disasters has enhanced their understanding of such incidents. At the same time, community leaders would adopt an appropriate perception about the disasters, and thus could increase their level of preparedness. It is also recommended that communities should be involved in practical and hands-on community programs so as to enhance the knowledge of floods and disasters, and interaction skills with various stakeholders (Turnbull et al. 2013). When theories are put into practical forms, the public will know how to respond to similar situations in the future (Hoang et al. 2007). To ensure that public education is carried out effectively, trust and respect should be promoted since they will motivate participants to carry out their roles regarding "promotion and dissemination of community level flood preparedness measures" (Hoang et al. 2007, p. 256). It is also noted that all programs would require trainings for leaders so that they are equipped with updated knowledge and techniques with regard to information dissemination, mentoring and further promote stronger linkages between the public and community-local authorities (Hoang et al. 2007).

Overall, the above countries have introduced some educational initiatives to enhance the public awareness of disaster management. Such initiatives are limited and cover very basic information. Mechanisms to evaluate the effectiveness of such programs have not been adequately researched.

12.3 Theoretical Framework and Research Method

12.3.1 Theoretical Framework

The Wisconsin Model of Public education discussed in the previous section will be used to evaluate efforts in provision of public education to respond to disasters in Singapore. Public educational programs regarding disaster preparedness and response are assessed according to each of the five elements of the *Wisconsin Model of Public education* as illustrated in Fig. 12.1. This model is selected for this chapter since it includes various elements which are important in the planning, implementation and evaluation process of any public education program.

The four principles of public education pertinent to disaster management, namely consistency, legitimacy and credibility, scalability and sustainability,

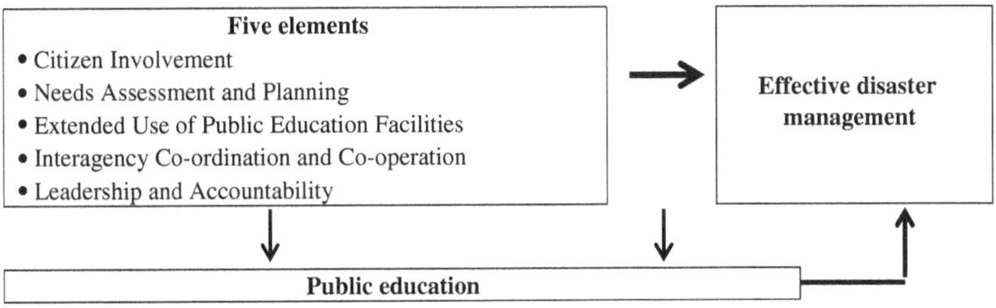

Fig. 12.1 Theoretical framework of public education and disaster management. *Source* by the authors

introduced by the International Federation of Red Cross and Red Crescent Societies (2011) have also been taken into consideration in the assessment process. Any messages sent to the public should be consistent in order to avoid confusion and misleading information. Any public education programs should be legitimised by relevant authorities and supported by different groups of stakeholders. They should be well designed and implemented in order to produce the desirable outcome. Such educational programs should be designed in a way that they can be multiplied and conducted in various locations and at various scales. Importantly, such educational programs should be sustainable.

A public education program which meets the assessment criteria would help providers and participants achieve their goals, and would contribute to make efforts in managing disasters effective.

12.3.2 Research Method

This is a conceptual chapter. Secondary data have been collected from existing academic and non-academic literature, such as country reports, reports from governments, NGOs (non-governmental organisations) and international organisations [i.e. the World Bank, the United Nations, UNESCO (United Nations Educational, Scientific and Cultural Organisation), ASEAN (Association of Southeast Asian Nations) secretarial, the APEC (Asia-Pacific Economic Cooperation), etc.]. The content analysis and policy evaluation method have been employed to analyse the collected data. Firstly, the authors have reviewed educational programs to enhance the public awareness of disaster management in other countries. Secondly, the authors have examined programs to educate the public on (i) how to prepare and respond to disasters in general, and (ii) how to manage particular post-disasters and prepare for similar disasters in Singapore. Thirdly, the authors have evaluated the extent to which such educational programs meet the five assessment criteria adopted by the *Wisconsin Model of Public education*. Finally, the authors have made some policy recommendations, drawn from the lessons learnt from the case of Singapore.

12.4 Findings and Discussion

12.4.1 Programs to Educate the Public on Disaster Management in Singapore

It is fortunate that Singapore has not experienced many natural and man-made disasters at a large scale as other countries have. This section discusses some programs to educate the public in general, and in responding to particulars incidents.

12.4.1.1 Severe Acute Respiratory Syndrome (SARS)—2003

The outbreak of the Severe Acute Respiratory Syndrome or commonly known as SARS in 2003, has caused many deaths, nationally and globally. Singapore's first patient of SARS was sent to Tan Tock Seng Hospital after returning from Hong Kong. Similar reports of the case increased which then activated the Ministry of Health (MOH) to issue control-guidelines for SARS (The Straits Times 2013).

Public Education Programs

- *SARS Containment Programs*: Singapore was able to keep the disease under control by educating its citizens on the ways the disease could be contracted (e.g. via disseminating documents such as the newspapers and health brochures, media coverage such as the television, internet, radio and other advertising mediums, and schools and offices), and also taking vital steps such as monitoring of the body temperature and avoiding crowded places in order to minimise the risk in the spread of the disease (Lai and Tan 2013).
- The Health Promotion Board (HPB) (2012a) issued the details of SARS on its website, which included information on how to determine the symptoms of the disease. It claimed that SARS is not curable, but it is best to prevent such disease from spreading by implementing controlling method. HPB encourages all networks of community to keep high level of personal hygiene by washing the hands thoroughly with soap and water after visiting the toilet, build up the immunity by eating healthily and keeping fit, also to keep the surroundings clean by not littering and spitting. HPB also advised the community to communicate diseases to doctors openly. The Ministry of Education (MOE) was also on guard and monitored students who went to overseas, especially to SARS affected countries during the June vacation in 2003 (Ministry of Education 2005). When the SARS situation worsened, schools had to close down temporarily. Temperature checks were carried out in airports, schools, hospitals (The Straits Times 2013).

- Mediacorp partnered with the HPB and came up with SAR-vivor Rap to participate in community service project by promoting an education campaign in combating SARS (Health Promotion Board 2012a). The HPB worked together with the Ministry of Health, MediaCorp, Singapore Press Holdings and Singapore Cable Vision. SARS channel was then set up to give viewers updates on SARS (Health Promotion Board 2012b). Nurses and doctors conducted medical talks to teachers at pre-school centres and they had to attend to public queries on SARS. They implemented "hospital-based surveillance system" to track down on people with highly suspected SARS cases (Health Promotion Board 2012c). Infection Control Procedures increased in hospitals and also in emergency departments in hospitals. Additional isolation facilities were located at the Communicable Diseases Centre (CDC) in Tan Tock Seng Hospital (Ministry of Health 2014).
- The Ministry of Health (MOH) (2014) and other governmental officials played a part by monitoring the SARS situation and advising travellers what they needed to do when they arrive in Singapore or when they depart to other countries (Ministry of Foreign Affairs 2004).
- *Post-SARS Management* It was noted that several groups of stakeholders have been involved in the SARS Preventive Programs. Players in the public sector are Ministry of Health, the Health Promotion Board, Ministry of Education, Ministry of Defence and other governmental officials, schools, and hospitals. The private sector and civil society have also contributed to this preventive activity. Some of them are Singapore Press Holdings, MediaCorp, Singapore Cable Vision (now known as Starhub), and various community groups.

12.4.1.2 Learning from Banda Aceh Tsunami 2004

Learning from experience, the Singapore government has also made efforts to provide education to the public when major incidents occurred in other countries, such as the Banda Aceh Tsunami 2004. This is a very proactive approach, instead of being reactive as in the past.

Public Education Programs

- Photo Exhibition: A photographic exhibition was launched at Atrium@Orchard, which showcased the efforts of the volunteers in providing relief for the tsunami victims (Ministry of Education 2005). Haines-Saah et al. (2013) described that a photo exhibition as both an informal and narrative way of expressing information. It is effective because, the public is free to interpret the meaning behind the photographs taken in Banda Aceh. Photographs are simply images without words, but they can create a strong influence the cognitive and emotional values of the public because somehow, it could reflect their relationship and the "collective lifestyles" or memories of the past events (Haines-Saah et al. 2013, p. 26).

- **Community Involvement in Relief:** Mercy Relief (2014) reported that relief efforts were sent to Indonesia involving various stakeholders. Stakeholder involvement was observed, included participants from all three sectors (the public sector, the private sector and civil society). As a way to educate students about post-disaster activities and the consequences of the disasters, students were involved in fund raisings by setting up bazaars and holding concerts. Thirty primary schools were involved in Mercy Relief's Play Packs Programme where they distribute gifts for the children, and write cards to other children to show their care (Mercy Relief 2014). Involvement of students in relief activities enabled students to learn about the impact of disasters.
- **Vodcasts:** The Singapore Civil Defence Force (SCDF), with the support of Home team units and schools, the private enterprises and industries, documented their disaster preparedness videos in YouTube. These videos are also linked to the SCDF website. There are videos on how they carried out their drills, such as Station With Immediate First Aid Treatment, Light Fire Attack Vehicle (Red Rhino), Watermist Gun, Dry Decontamination Vacuum Oven, Quick Coupling Device, Canine Camera, Hazmat Mobile Scrubber Vehicle, Fire Bike, Fast Response Paramedic, Emergency Ambulance, Aerial Ladder, Tracked Fire Fighting Vehicle, Combined Platform Ladder (Singapore Civil Defence Force 2014c).

12.4.1.3 Learning from the Tsunami in Japan, March 2011

Similar to the Tsunami in 2004, the Singapore government has again demonstrated her efforts to provide the public with relevant educational programs after the tsunami in Japan in 2011.

Public Education Programs

- **Exhibition:** The Science Centre partnered with Earth Observatory of Singapore to come up with an exhibition that showcased the occurrence of natural disasters to the public. There are various sections; Journey to the Centre of the Earth, Dynamic Atmosphere, The Earth Under Our Feet, Tsunamis, Volcanoes, Earthquakes and Science on the Sphere Show. In another section, the exhibition welcomes the public to send in their "eyewitness accounts of natural disasters" (Science Centre Singapore 2009). The stakeholders involved are Nanyang Technological University, Singapore Science Centre, scientists, engineers and academics.
- Wahab and Zuhardi (2013) explained that the exhibition involves "visual representations and interaction techniques" which allows the public to understand the events better because images tend to create a stronger impact in knowledge gain (p. 477). Exhibitions are suitable for school children and

adults. It is an effective approach because there are also hands-on activities or demonstrations in the "informal setting" of the exhibition environment that the public could freely participate in (Ahmad et al. 2013, p. 371). In this way, they are able to discover and uncover information.

- **Reaching the community**: To reach out easily to various communities and to continuously educate them, community centres provide free newspapers at reading corners, such as that in Tiong Bahru Community Centre. These free local newspapers are to update the community with information, especially for those without access to the internet via their hand phones or computers. In this way, the local newspapers are termed as a bridge that connects the community with the latest information will allow them to discuss about the current events. Not only that, the community will also be well informed of various disaster preparedness and management programs which are posted on the notice boards or banners. Residents living nearby the community centres and are interested in the programs, could sign up for it. In this way, disaster managers are indirectly fostering greater community ties within individuals living in the area.

12.4.2 Assessment of Public Education in Singapore

Given the limited information about public education in Singapore, and the instruments and mechanisms available to assess the effectiveness of public education programs with regard to disaster management, public education activities in Singapore are briefly assessed per the following five elements. Since no quantitative data is available, and thus the assessment was only carried at the minimal degree.

Citizen Involvement—Via the activities discussed in the previous sections, it is noted that the public sector (various government agencies and ministries, schools, hospital, statutory boards, etc.), the private sector (construction companies, communication organisations), and civil society (Mercy Relief, the media, professional centres, etc.) are involved. There has been no figure on the number of activities and the number of organisations participating in such activities. It is observed that public education initiatives are the same as those offered in other countries.
Needs Assessment and Planning—There is insufficient information about needs assessment. However, there has been a strategic plan for public education regarding disaster management. Yet, it is not clear how the training needs have been analysed, and how public education programs have been planned and implemented.
Extended Use of Public Education Facilities—Public education facilities have been provided widely at community centres, public libraries and online. For example, residents can obtain information from resources at community centres in Singapore. However, compared to Japan and Iran where disaster education has been incorporated in the school curriculum, Singapore may apply this approach, targeting pupils in primary schools and above.

Interagency Co-ordination and Co-operation—There has been some evidence of co-operation and co-ordination among government agencies, the private sector and civil society. In many cases, it is observed that government agencies are the key players in planning and executing public education programs, whereas other groups of stakeholders are the participants. Nevertheless, more evidence is required in order to assess the effectiveness of the co-ordination and co-operation among actors in different sectors.
Leadership and Accountability—Ministries and other government bodies have initiated a number of programs and activities to educate the public when major incidents have occurred either in Singapore or in other countries. Yet, it is not clear of whether mentorship programs have been implemented to guide young people.

12.5 Policy Implications and Recommendations

12.5.1 Policy Implication

Based on the limited information about consumer education and disaster management, the following points have been noted:

Firstly, most of the initiatives to educate the public of DRM are reactive. Government agencies and officials play the key role in managing educational programs since it is impossible for government to outsource its responsibility in disaster management. Empirical evidence has demonstrated that without clear government policies, responsibilities, and leaders' commitment and resources, any effort to manage natural disasters will be wasted as in the case of Hurricane Katrina in the USA (Col 2007).

Secondly, there is limited involvement of the private sector and civil society in consumer education of DRM in Singapore. Also, the participation of the private sector and civil society has been constrained via exhibitions, activities, forums, conferences, etc. It is difficult to obtain the statistics of (i) how many people have attended such programs, (ii) to what extent such programs can achieve the set objectives, and (iii) the financial aspect of such programs which affect their scalability and sustainability.

Finally, there has been lack of information and statistics about issues associated with the coordination among government agencies (Harkin 2006). Apparently, the efforts, legitimacy and resources to offer public education scatter among various government agencies. Sometimes, due to the lack of responsiveness and bad practices of some government departments, activities among different groups of stakeholders have been poorly coordinated which lead to waste of efforts and resources. Consequently, thousands of people, who might have been saved if better education had been executed, had perished in the case of Cyclone Nargis (Lai et al. 2009).

12.5.2 Recommendations

Some recommendations to address the above issues are made as follows:

Firstly, the government should promote stronger public-private partnership (PPP) in order to mobilise the available human and non-human resources of non-state actors to organise programs and activities to prevent and respond to disasters. Becker-Birck et al. (2013) explained that PPPs can improve cooperation and collaboration among sectors, i.e. creating opportunities for the public and private sectors to respond to disaster resilience building. Successful cases in disaster management and environmental protection have been found in many countries, such as the "Get Airports Ready for Disaster (GARD)" in Germany, and the "Lake Chilwa Basin Climate Change Adaptation Programme (LCBCCAP)" in Malawi (Becker-Birck et al. 2013, pp. 21, 27, 31).

Secondly, it is impossible to deny the role of the private sector in conversation, discussion and debates regarding instruments, mechanisms and processes to enhance the public awareness of disaster management. According to the International Fund for Agricultural Development (2007), businesses do have the ability, capabilities and resources which they can share with the public sector in order to increase the effectiveness and efficiency of disaster management. Thus, government should actively engage the private sector in all educational programs and all the stages of disaster management.

Finally, it is impossible to deny the role of Civil Society Organisations (CSOs) in implementing public education and projects regarding disaster preparedness, disaster resilience building, and disaster relief (Becker-Birck et al. 2013; Forsyth 2005). Thus, governments should also engage CSOs in the design, delivery and evaluation of public education programs with regard to disaster management.

12.6 Conclusion

In conclusion, this chapter has discussed the importance of public education in disaster management, the *Wisconsin Model of Public Education* and different educational programs offered in different countries. It has also discussed various public education programs conducted in Singapore in order to enhance the public awareness of disaster risk reduction and disaster preparedness. Nevertheless, it is not clear how such programs and initiatives have been planned, designed, coordinated and evaluated. Thus, an integrated approach to public education, including government, the private sector and civil society, should be adopted in disaster management since (i) all groups of stakeholders will be affected by any plans, programs or decisions made by government, and (ii) the impact of natural disasters are profound and long-lasting, and does not spare any individuals and any sectors.

This suggests that further research on public education regarding disaster management should focus on stakeholder engagement and participation, and assessment mechanisms of the effectiveness of any education efforts to enhance the public awareness of various issues pertinent to disaster management. However, limitations are unavoidable due to lack of information.

References

Abdul Rashid N, Mohd Nasir T, Sahrim L, Norizam S, Zunairah M, Ros Shilawani SAK, Samsudin K (2013) Summative EEG-based assessment of the relations between learning styles and personality traits of openness. Procedia—Soc Behav Sci 97:98–104

Abdul Wahab MH, Ahmad Zuhardi AF (2013) Human visual quality: art gallery exhibition. Procedia—Soc Behav Sci 101:476–487

Ahmadi A, Laei S (2012) Public education: compulsory and free? a paradox. Procedia—Soc Behav Sci 47:868–873

Ahmad S, Abbas MY, Yusof WZM, Taib MZM (2013) Museum learning: using research as best practice in creating future museum exhibition. Procedia—Soc Behav Sci 105:370–382

Badpa A, Shakiba M, Singh MJ (2013) Effects of knowledge management system in disaster management through RFID technology realization. Procedia Technol 11:785–793

Becker-Birck C, Crowe J, Lee J, Jackson S (2013) Resilience in action: lessons from public-private collaborations around the world. Meister Consultants Group Inc. Boston

Buccholtz AK, Carroll AB (2012) Business and society: ethics and stakeholder management, 8th edn. South-Western, Cengage Learning, Melbourne

Bushell B, Goto M (2011) Leadership for sustainable society: a transformative learning approach. Procedia—Soc Behav Sci 29:1244–1250

Central Project Management Office, Ministry of Agriculture & Rural Development (2009) Community based disaster risk management. Central Project Management Office Ministry of Agriculture & Rural Development, Hanoi

Col JM (2007) Managing disasters: the role of local government. Public Adm Rev 67(s1):114–124

Dobos A, Jenei A (2013) Citizen engagement as a learning experience. Procedia—Soc Behav Sci 93:1085–1089

Evans C, Sadler-Smith E (2006) Learning styles in education and training: problems, politicisation and potential. Educ + Training 48(2/3):77–83

Fakhri LS, Talebzadeh F (2011) A framework for Professional citizenship education based on knowledge management principles. Procedia—Soc Behav Sci 29:1133–1142

Forsyth T (2005) Building deliberative public–private partnerships for waste management in Asia. Geoforum 36(4):429–439

Goldstein B (1990) The problem with the margin of safety: toward the concept of protection. Risk Anal 10(1):7–10

Goulding A (2009) Engaging with community engagement: public libraries and citizen involvement. New Libr World 110(1/2):37–51

Ha H (2013) Climate change governance—Singapore case. In: Ha H, Dhakal TN (eds) Governance approaches for mitigation and adaptation of climate change in Asia. Palgrave McMillan, London, pp 182–199

Haines-Saah RJ, Oliffe JL, White CF, Bottorff JL (2013) "It is just not part of the culture here": young adults' photo-narratives about smoking, quitting, and healthy lifestyles in Vancouver. Health Place 22:19–28

Harkin C (2006) Coordination of international humanitarian assistance in tsunami-affected countries: the 2004 tsunami: civil military aspects of the international response. Tsunami Evaluation Coalition, London

Health Promotion Board (2012a) Health promotion programmes: SARS containment activities, Retrieved 17th Mar 2014 from, http://www.hpb.gov.sg/rpt/2003/programmes_5.htm

Health Promotion Board (2012b) Preventing and controlling public health threats, Retrieved 17th Mar 2014 from, https://www.hpb.gov.sg/HOPPortal/health-article/476
Health Promotion Board (2012c) Severe acute respiratory syndrome (SARS), Retrieved 17th Mar 2014 from, http://www.hpb.gov.sg/HOPPortal/article?id=578
Hoang VH, Shaw R, Masami K (2007) Flood risk management for the RUA of Hanoi: importance of community perception of catastrophic flood risk in disaster risk planning. Disaster Prev Manage 16(2):245–258
Hudson P, Hudson S, Gray B, Bloxham R (2013) Learning about being effective mentors: professional learning communities and mentoring. Procedia—Soc Behav Sci 93:1291–1300
International Federation of Red Cross and Red Crescent Societies (2011) Public awareness and public education for disaster risk reduction: a guide. International Federation of Red Cross and Red Crescent Societies, Geneva
International Fund for Agricultural Development (2007) Private sector development and partnership strategy. International Fund for Agricultural Development, Rome
IRIN (2014) Indonesia: report card on disaster preparedness, Retrieved 23rd Mar 2014 from, http://www.irinnews.org/fr/report/95602/indonesia-report-card-on-disaster-preparedness
Ismail IA, Silong AD, Asmiran S, Hassan Z (2011) Development of educational leadership in research university through community of practices among professors. Procedia—Soc Behav Sci 15:828–832
Lai AY, He JE, Tan TB, Phua KH (2009) A proposed ASEAN disaster response, training and logistic centre enhancing regional governance in disaster management. Trans Stud Rev 16(2), 299–315
Lai AYH, Tan SL (2013) Impact of disasters and disaster risk management in singapore: a case study of Singapore's experience in fighting the SARS epidemic. ERIA Discussion Paper Series. Jakarta, Indonesia: Economic Research Institute for ASEAN and East Asia
Matsumaru R, Nagami K, Takeya K (2012) Reconstruction of the Aceh Region following the 2004 Indian Ocean tsunami disaster: a transportation perspective. IATSS Res 36(1):11–19
Mercy Relief (2014) Boxing day tsunami 2004, Indonesia, Retrieved 17th Mar 2014 from http://mercyrelief.org/web/Contents/Contents.aspx?ContId=16
Ministry of Education (2005) Speech by Mr. Tharman Shanmugaratnam, Minister for Education, At the Mercy Relief's Glimpses of Light Exhibition, On Saturday, 19 Mar 2005, At 3.00 Pm, At Atrium@Orchard, Ministry of Education, Singapore
Ministry of Foreign Affairs (2004) Press statements and speeches: how Singapore is coping with SARS. Ministry of Foreign Affairs, Singapore
Ministry of Health (2014) Measures taken to control the SARS outbreak in Singapore. Ministry of Health, Singapore
Mumford A (1995) Learning styles and mentoring. Ind Commercial Training 27(8):4–7
Nielsen S, Lidstone J (1998) Public education and disaster management: is there any guiding theory? Aust J Emerg Manage 13(3):14–19
Parsizadeh F, Mohsen G-A (2010) Iran public education and awareness program and its achievements. Disaster Prev Manage 19(1):32–47
Pitoska E, Lazarides T (2013) Environmental education centres and local communities: a case study. Procedia Technol 8:215–221
Salehi M, Gholtash A (2011) The relationship between job satisfaction, job burnout and organizational commitment with the organizational citizenship behaviour among members of faculty in the Islamic Azad University–first district branches, in order to provide the appropriate model. Procedia—Soc Behav Sci 15:306–310
Science Centre Singapore (2009) Earth: our untamed planet. Science Centre Singapore, Singapore
Selby D, Kagawa F (2012) Disaster risk reduction in school curricula: case studies from thirty countries. UNESCO and UNICEF, Paris, France and Geneva
Singapore Civil Defence Force (2012a) Asian tsunami disaster, aceh, Indonesia (2004). Singapore Civil Defence Force, Singapore

Singapore Civil Defence Force (2012b) Collapse of hotel new world. Singapore Civil Defence Force, Singapore

Singapore Civil Defence Force (2012c) Community and volunteers: social studies syllabus. Singapore Civil Defence Force, Singapore

Sorkhabi R (2011) Japan's megaquake and killer tsunami: how did this happen? Available from http://www.earthmagazine.org/article/japans-megaquake-and-killer-tsunami-how-did-happen. Accessed 20th Mar 2014

The Straits Times (2013) Sars in Singapore: timeline. Available from http://www.straitstimes.com/the-big-story/sars-10th-anniversary/story/sars-singapore-timeline-20130316. Accessed 17th Mar 2014

The Taipei Times (2014) Feature: Indonesia continues to lag in tsunami preparedness. Available from http://www.taipeitimes.com/News/world/archives/2011/04/06/2003500059. Accessed 23rd Mar 2014

The World Post (2014) Japan earthquake 2011: 8.9 magnitude earthquake hits, 30-foot tsunami triggered. Available from http://www.huffingtonpost.com/2011/03/11/japan-earthquake-tsunami_n_834380.html. Accessed 20 Mar 2013

Turnbull M, Sterrett CL, Hilleboe A (2013) Toward resilience a guide to disaster risk reduction and climate change adaptation. Practical Action Publishing Ltd and the Schumacher Centre, Warwickshire

Turkay S, Tirthali D (2010) Youth leadership development in virtual worlds: a case study. Procedia—Soc Behav Sci 2(2):3175–3179

Whipple M, Nyce JM (2007) Community analysis needs ethnography: an example from Romania. Libr Rev 56(8):694–706

Wisconsin Department of Public Instruction (2014) Wisconsin's components of public education. Wisconsin Department of Public Instruction, Madison

Chapter 13
Participatory Environmental Governance and Climate Change Adaptation: Mainstreaming of Tidal River Management in South-west Bangladesh

Kazi Nurmohammad Hossainul Haque, Faiz Ahmed Chowdhury and Kazi Rabeya Khatun

13.1 Introduction

Climate change is not an unprecedented phenomenon in human history. However, the distinction of the climate change of our generation with the previous spells of climate change is the former's primarily anthropogenic origin. In other words, the ongoing climate change is overpoweringly caused by human behaviour rather than natural shifts in environment. This is evident from just one of the major climate change processes—the rising global temperature. The global average surface temperature has increased by nearly 0.74 °C over the past 100 years between 1906 and 2005. The warmest years in a century or so are 1998 and 2005. In fact, 11 of the 12 warmest years have occurred in the 12 years from 1995 to 2006 (IUCN 2011). These times especially the twentieth century unlike the centuries beforehand is marked by growth and globalisation of the unsustainable development model—the carbon economy.

Bangladesh is one of the most vulnerable countries to climate change according to Inter-governmental panel on climate change (IPCC). It observed that up to one third of the country may inundate in case of full impact of sea level rise (IPCC 2007). But climate change is about broader ecological changes and so the rising sea level is not the only climate challenge that the country is faced with. Projections by Agarwala et al. (2006) stipulated that between 2030 and 2100, the

K.N.H. Haque (✉) · F.A. Chowdhury · K.R. Khatun
BRAC Institute of Governance and Development (BIGD), BRAC University, Dhaka, Bangladesh
e-mail: kazi@igs-bracu.ac.bd

K.N.H. Haque · F.A. Chowdhury
Save the Children, Oslo, Norway

K.R. Khatun
Independent Researcher based in Trondheim, Oslo, Norway

H. Ha (ed.), *Land and Disaster Management Strategies in Asia*,
DOI: 10.1007/978-81-322-1976-7_13

country's surface average temperature may increase by 1–2.4 °C, annual mean rainfall may increase by 4–10 %, monsoon may be 5–12 % wetter and winter may be drier. The impacts of such climatic variations would be much wider.

While these are slow-onset processes that are taking their toll gradually, the sudden-onset processes are taking their toll by surprise. Once every few years, there is a severe flood that covers greater parts of the country with more significant damages to lives and livelihoods. Between 1988 and 2008, the country was visited by seven major floods causing thousands of deaths, millions displaced and economic costs worth billions of dollars. Every 3 years, on average, the country is hit by a severe tropical cyclone. Its numbers and frequency are likely to increase due to climate change (IOM 2010). While the whole country is vulnerable to climate change and its consequences, south-western part is in the frontline. Like most other parts of the country, the region is characterised by population density, flat topography and low elevation. As a monsoon-influenced river delta, south-west has its share of floods, river erosion, waterlogging and drought that are becoming more and more intense. Their effects are exacerbated by cyclonic storm surges, tidal surges, costal erosions and salinity intrusions.

Since the humans are the primary agents of current climate change, addressing it is as much political, economic and social affairs as scientific and technological affairs. So, environmental governance with wider and deeper participation of civil society is instrumental in mounting comprehensive response to climate change. In soliciting widest possible human efforts in responding to climate change, the civil society is globally recognised as a key actor. That is amply manifested in global regimes of climate change, i.e. Agenda 21, the United Nations framework convention on climate change (UNFCCC), the Kyoto Protocol and the Bali Action Plan as well as the national regimes like Bangladesh climate change strategy and action plan (BCCSAP). Specially, with respect to empowerment of weaker sections of population, expansion of popular participation in programmes and enhancement of current actions—civil society actors are deemed large part of the answers. As this is generally the case in many parts of the world, this is more so in Bangladesh where civil society especially non-governmental organisations (NGOs) have been a major actor in development and governance besides government and private sector. However, with respect to environmental governance including climate adaptation, civil society's participation is still to be consolidated. There is greater participation of national NGOs and think tanks in environmental decision-making at national level. On the contrary, CSOs and local NGOs are still found to be waging movements to secure their participation at local environmental governance.

This research aims to investigate how civil society especially CSOs secure their participation in local environmental governance through mobilisations and movement, and how in this process they eventually contribute in climate change adaptation. The answers to the research question are illustrated through a case study of Paani Committee's (PC) movement for mainstreaming of tidal river management (TRM) in environmental governance of water resources and flood control in the south-west Bangladesh. The study is divided into six sections: introduction, literature review, conceptual framework, research methodology, case study and conclusion.

13.2 Literature Review

The root of the term *govern* derives from the Latin and Greek terms for *steering* or *piloting* a ship. Hence, the linguistic connotation of *governance* has been traditionally *the act or process of governing* primarily associated with governmental steering through regulation and sanctions (Lafferty 2004, pp. 4, 5). Such essentialisation of governance as how government exercises power and deliver responsibilities started to change in late twentieth century. During the 1980s, political scientists increasingly started to draw distinction between *government* and *governance* (Kjaer 2008, p. 3). This happened due to *enormous changes* that led to diffusion of decision-making power and authority from government to numerous other actors. The context of the shift from government to governance is succinctly summarised by Volger and Jordan (2003, p. 143) as they draw on Rosenau (1992) and Rhodes (1997). Rosenau (1992) observed that authority in the modern world is simultaneously moving up and down, respectively, towards supranational and subnational actors. Rhodes (1997) found modern nation states to be more differentiated and hollowed-out by the centrifugal forces created by globalisation in one hand that is pushing power upwards from the state and greater local autonomy on the other hand that is pushing power downwards to local service delivery agencies.

Although the term *environmental governance* started appearing much later, interestingly enough, environment is one of the public spheres where the government to governance shift first occurred. If governance is understood as diffusion of power and authority from government to other non-governmental actors, i.e. businesses, civil society, NGOs and community then such phenomenon was already evident in some countries specially the developed ones. For example, civic environmentalism in the United States, whose other variations include community-based environmental protection, collaborative natural resources management and ecosystem management. Before environment became a wider global concern, civic environmentalism led to a rising tide of watershed councils, river keepers, land trusts, voluntary monitoring programmes and sustainable community plans. The approach was geared for tailoring responses to local conditions, overcoming bureaucratic fragmentation and making environmental decision-making democratic (Durant 2004, p. 180). Also in the developing world even in the most traditional societies, the ones without any modern government, some pedigree of environmental governance can be found. Ostrom (1990) showed that many small-scale traditional societies were smart enough to develop governance arrangements for appropriation of common-pool natural resources including water bodies and forest and pastoral lands even before creating a government.

With the human dimension of environment and the government to governance shift in the background, World Bank (WB) and World Resources Institute (WRI) (2003, p. 2) seems to be right when it claims that *environmental governance (is) how we as humans exercise our authority over natural resources and natural systems*. It emphasises that how we make decisions about the environment, who participates in the decision-making and what decisions are finally made can make huge differences, i.e. environmental improvement or harm, effective or ineffective environmental policy and, success or failure in managing ecosystems and natural resources. With the scope of the current research in mind, the most important observation made by WB and WRI (2003, p. 2) is that the most direct routes to better environmental governance is to provide easy access to environmental information and encourage broad participation in environmental decisions. It is rightly argued that when those affected by a decision can participate in the process, the result is likely to be fairer, more environmentally sound and more broadly accepted.

Discussion of participation in governance including environmental governance cannot be effective without a prior discussion of civil society. It is long-recognised concept in social sciences particularly in political science and sociology. However, the root of civil society can be traced as back as classical Western political philosophy where Aristotle discussed about *politike koinona* or political community and Cicero talked about *Societas Civilis* or civil society (Kaldor 2002, p. 3; Davis and McGregor 2000, p. 49). Fast forwarding to modern liberal tradition, civil society is celebrated as the source of social capital (Putnam 1993). In the context of global south especially with donor support for NGOs, there has been *NGOisation of … civil society* in many cases. NGOs tend to be considered synonymous with civil society and are seen as vehicles for democratisation and citizenship (Connolly 2007, p. 8).

Over last two decades, civil society is also regarded as a manifestation of deliberative democracy that is not hierarchical, bureaucratic and party bound (Newman et al. 2004, p. 204). The idea is claimed to be different from mainstream democracy on at least two accounts: emphasis on reasoned argument and public reflection instead of exercising ballot as core of democracy and identification of public interest to generate from deliberation rather than summing of pre-existing preferences. Democratic deliberation is deemed to constitute reasoned debate, public justification and political equality. Collective choices would be made through reasoned debate not through blind acceptance of the views of established authorities or deals among vested interests. There should be public justification of policy proposals and outcomes. All constituents should enjoy equal access to the decision-making process (Meadowcroft 2004a).

Meadowcroft (2004b) gives a detailed overview of the participation discourse in relation to environmental decision-making aka governance since the 1970s. He classifies it into three strands: citizenship, community-centred and stakeholder. The criticism of traditional environmental governance for being over-centralised, exclusionary and unable to recognise important values and interests are common to all the three strands. However, they differ about benefits, representatives and mechanisms of participation.

From citizenship perspective, the main benefit of participation is the *direct* involvement of *ordinary* citizens in the environmental decision-making process leading to breaking-up of hitherto policy monopoly of technical, business and political elites. The participatory mechanisms preferred by citizenship perspective are typically formal consultation and public enquiries. Some of the latest or deliberative mechanisms are citizen advisory panel, citizen jury and referendum. From the community-centred perspective, enhanced participation within and by *local communities* generates functional benefits. Its participatory mechanisms are often common to those of citizenship and stakeholder perspectives with the exception that they operate at community level. From the stakeholder perspective, functional benefits are gained through inclusion of *group representatives.* In addition to the participatory mechanisms used by citizenship perspective, stakeholder perspective further prefer some innovative mechanism, namely environmental mediation, negotiated regulation, environmental covenants and environmental co-management (Meadowcroft 2004b, pp. 170, 172–176).

The literature review has given a glimpse of participatory environmental governance specially its actors, mechanisms and benefits. However, the experience of securing participation in the first place remains almost untouched. It is not clear whether the participatory mechanisms were spontaneously allowed by government or results of public demand. There is still lack of taking participatory environmental governance lenses on climate adaptation issues. Most empirical evidence of participatory environmental governance is drawn from developed northern contexts while their bigger potential is in the global south countries like Bangladesh which are also in the receiving end of climate change. The outcomes of civil society participation such as policy or governance innovation are not highlighted as such. These are the areas where the current research will make some contributions.

13.3 Participatory Environmental Governance for Climate Change Adaptation: The Conceptual Framework

One of the things reflected in the literature review above is that *participation* is at the heart of the government to governance shift. The downward and upward diffusion of power and authority that generally characterises the government to governance shift is largely materialised through broader scope of non-governmental, inter-governmental and external actors' participation in *governing.* The tasks of governing that were traditionally recognised to be in the realms of individual governments began to be shared with various actors outside government: the non-governmental actors including businesses, civil society, NGOs and community, the inter-governmental actors like International Organisations (IOs) in general and those in the United Nations (UN) system in particular, and, the external actors specially donor countries and agencies.

The literature review also shows that environment is one of the public spheres where the government to governance shift through participation of the abovementioned actors in governing first took place. In fact, *environmental governance* as the public policy phenomenon preceded the coining of the term. Environmental governance eventually needed to be *more* participatory at all its levels—global, regional, transnational, national and local—as human impact on environment came to the fore with sustainability debate through 1970s–1980s and more so with climate change phenomenon since the 1990s onwards. The widening and deepening of participation in environmental governance is also the main reason for this study to use the prefix *participatory* before environmental governance despite the fact that participation is at the heart of the government to governance shift.

In order to conclude on a basic understanding of environmental governance, this research would like to keep in mind the two conceptual developments mentioned earlier: the recognition of human impact on environment as the primary source of challenges to sustainability including climate change and the shift from government to governance in politics and public policy. Keeping these conceptual developments in mind, we would like to understand environmental governance in terms of how decisions are made about environment, who can participate in that decision-making and what are the outcomes of the decisions made. Accordingly, participatory environmental governance is understood as environmental governance as defined above with added emphasis on its course towards widening and deepening participation of various actors in addition to government.

Since the current climate change is primarily driven by human actions and environmental governance is mainly about decision-making on environment, participatory environmental governance can lead to necessary institutional changes in addressing climate change. Because, greater the scope for wider and deeper participation of all concerned actors in- and out-side government in decision-making on climate change, greater the chance for the resulting decisions to be representative of diverse interests. Otherwise, the decision(s) to address climate change will have neither legitimacy nor effectiveness especially with the shift from government to governance. There are also chances for them to be technically flawed even if they are informed by expert knowledge, a fact that is amply reflected in the case study, the mainstreaming of TRM in south-west Bangladesh.

Based on the literature review, we mentioned earlier three kinds of environmental governance actors from outside government. However, this research involves focused discussion of civil society especially CSOs and associated citizen mobilisation in favour of securing people's participation in environmental governance arrangements of their concern in the context of climate change adaptation. Because, civil society in its various denominations including community, CSOs and NGOs are recognised in global and national regimes of climate change as facilitators of wider public participation, specially the weaker sections of society, in policies and programmes. Moreover, this research follows a case study-based methodology as elaborated in the following section that constrains from dealing with examples of a handful of actors at the same time. Civil society, for the purpose of this research, is understood as an organised section of citizens who are working for positive changes in society. Such sections of citizens may or may not be formally organised under certain institutional arrangements such as name, leadership structure, defined membership, regular activities and organisational rituals. But when they are formally organised as such, they are considered as CSOs whether they are registered as NGOs or not.

Meadowcroft's (2004b, pp. 169–170) overview of (civil society) participation in environmental governance amply demonstrated its actors, mechanisms and benefits as informed by different participation approaches. We would see in our case study that if participation mechanisms are not spontaneously created by government then civil society may deploy strategies to be heard and secure participatory access. Once participation is allowed, some of the strategies may turn into mechanisms as well. Meadowcroft (2004b, pp. 169–170) also showed that there can be overlapping of mechanisms irrespective of actors and approaches. We would like to argue from our case study that even same actor can adopt different strategies/mechanisms in different stages of its participation trajectory: from initial mobilisation and movement to securing participation and then up to its consolidation.

13.4 Research Methodology

In order to get answer to our research question, how civil society especially CSOs secures their participation in local environmental governance through mobilisations and movement, and how in this process they eventually contribute in climate change adaptation, the study adopts case study as the core method. As part of undertaking the case study, several other methods were also used: literature review, key informant interview (KII) and focused group discussion (FGD). The fields for KII and FGD were Dhaka, Khulna and Jessore. A total of 20 KIIs were conducted with central and *upazilla* committee officials of Paani Committee, civil society leaders, NGO leaders, and environmental and water experts. The 10 participants each of the two FGDs were *upazilla* and basin committee members of Paani Committee, community leaders and local representatives.

Initially, the literature review was mainly used to identify an appropriate case study that best illustrates the research question. The selection criteria were as follows:

- Significance of CSO/movement concerned as an actor/phenomenon that demands in-depth investigation.
- Any original or indigenous contribution of the CSO/movement in terms of policy or governance innovation vis-à-vis environmental governance and climate change adaptation.
- Substance of the CSO/movement in terms of strategies, activities and outcomes.

Following the above criteria, the following case study was selected: PC movement for introduction and mainstreaming of TRM in flood and water resource management in south-western Bangladesh. *Paani* is the word for water in *Bangla*, the main language of Bangladesh. After case selection, the literature review was used for both secondary and primary data collection. The secondary data were available literature about environmental governance, climate change adaptation, Paani Committee and TRM. A small portion of the primary data, mainly policy documents, was collected from literature review.

With the research question in mind, the key points of discussion in KII and FGD were as follows:

- Genesis of Paani Committee and TRM movement.
- TRM as an environmental governance innovation and climate adaptation approach.
- Evolution of civil society and associated movements in the south-west.
- Institutional architecture of Paani Committee.
- Strategies, activities and outcomes of TRM movement.

13.5 Case Study: Paani Committee's Movement for Mainstreaming of TRM

The case study findings are discussed and analysed in the following four subsections: an introduction of TRM, emergence of participatory environmental governance in the south-west through PC and its TRM movement, institutional architecture of PC and, strategies, activities and outcomes of PC as part of TRM movement.

13.5.1 TRM: An Indigenous Approach for Water Management and Climate Adaptation

For millennia, the Bengal delta that constitutes Bangladesh and India's West Bengal has been formed by sediments carried downstream by *Ganges*, *Meghna* and *Brahmaputra* river systems. The south-west region of Bangladesh is a tidal basin of the Bengal delta and large parts of it is still going through active deltaic formations. Geographically, the region is bounded in the north by the Ganges floodplain and in the south by the Sundarban mangrove tidal forest (UNDP 2011, p. 4). Traditionally, river tides would inundate vast tracts of lowland twice a day. Along tide came sediments from upstream that were rich with micronutrients. Much of these lowlands are *beels*, a natural depression that serve as tidal basins. There is distinct ecosystem of these areas that is rich in biodiversity. During high tides and wet seasons, the beels were home to various marine and amphibious life forms including fishes. During dry seasons, they produced many varieties of rice and other vegetation.

TRM is the scientific term given to the age-old practice of flood and water management in the south-west Bangladesh that complies with the region's floodplain ecology. It capitalises on natural movement of tidal water facilitated by low-intensity structures which are seasonal, temporary and flexible. Islam and Kibria (2006, p. 23) nicely articulated the TRM.

The basic idea of TRM is to allow tidal flow into wetland basin, known as *jowar-bhata khelano* (free play of tidal flow) in local vocabulary, and releasing the tidal flow back to the river. As a result of this process, sediments carried by tidal flow deposits on the wetland basin instead of riverbed. The process is continued for several years (usually three years, the duration depends on the size of the wetland basin). It gradually raises the land on the wetland basin with formation of alluvial soil from silt .... (It) prevents sediment deposition on the riverbed and ensures the drainage and smooth navigation in the river channels.

TRM therefore specially suits the floodplain ecology of the south-west. The estuarine rivers of the coastal region have two tidal cycles a day. Their high tides bring in muddy water flow with thick concentration of sediments. At that time, had local people cut embankment at appropriate point, the natural high tide would gush into floodplain and deposit large chunk of sediments. Thus, the low lands of the south-west would be nourished and elevated while the rivers and water bodies would be free of congestion (Kibria 2011, p. 1). Floods and waterlogging would be effectively controlled without affecting the region's ecological balance. Soil elevation as well as increased depth and navigability of the water channels specially the rivers would address challenges from sea-level rise. These are not just assumptions about TRM but empirically substantiated by successful outcomes of people's TRM in *Beel Dakatia* and *Beel Bhaina.* They are testaments of TRM's effectiveness in flood and water management as well as climate change adaptation.

The effectiveness of TRM is also acknowledged by credible technical authorities on water resources and flood control. Traditionally, water experts of the country showed least interest in TRM for its humble and "unscientific" origin. Later on, Centre for Environmental and Geographic Information System (CEGIS) and Institute of Water Modelling (IWM)—two of the country's premier scientific authorities on water resources—conducted detailed studies on TRM. The CEGIS (1998) was about economic and social impacts. The IWM (2008), on the other hand, built sophisticated mathematical modelling for planning and designing of TRM. Both studies, undertaken a decade apart, gave resounding scientific evidence in favour of TRM.

13.5.2 Paani Committee and TRM Movement: Emergence of Participatory Environmental Governance in the South-west

The PC and its TRM movement are rooted in the historical evolution of flood and water management in south-west Bangladesh that eventually gave rise to participatory environmental governance that region.

Prior to the twentieth century, there was less population pressure and lower level of agricultural production in the region. Modern water engineering technologies were yet to arrive and there was no water development intervention. The agricultural and water management practices of that era were compatible with local ecological characteristics (Interview of Hashem Ali Fakir, General Secretary, Paani Committee, 28 May, 2013). For example, in the seventeenth and eighteenth centuries, the local landlords or *Zaminders* used to build low earthen dykes around the tidal flats to prevent tidal intrusion and wooden sluice gates to drain off surplus rainwater. The tidal flats were then cultivated with indigenous varieties of flood and salinity tolerant rice. The dykes and sluice gates were seasonal installations. They were dismantled after each harvest and, people grazed cattle and fished in the tidal floodplains (Tutu 2005, pp. 118–119).

Such agro-ecological balance of south-west Bangladesh first started to disrupt with expansion of railway in the then East Bengal (today's Bangladesh) during the late nineteenth century. The British Indian colonial government did not necessarily keep up with floodplain ecology of the East Bengal in laying the railway tracks. In the process, many smaller rivers vanished and many bigger rivers were choked. In case of the former, the narrow river channels were earth filled to lay the railway tracks over them. In case of the latter, culverts were built on them to lay the railway tracks over that narrowed their channels. Some of the bigger rivers which were narrowed in the process of railway expansion as such are Kumar, Nabaganga, Chitra, Bhairab and Ichamoti (Hashem Ali Fakir, op. cit.).

Major interruption in the agro-ecological balance of south-west Bangladesh got underway during the Pakistan era. With increasing population pressure, there was ever growing demand for food production. So, there was greater eagerness for flood control in order to bring more land under cultivation all-round the year. It was also the era of green revolution. The availability of modern flood control and irrigation technologies was greater than ever especially with foreign aid. In the 1960s and early 1970s, flood control and drainage provisions were two key interventions in the Government's strategy for increasing monsoon season rice crops, particularly wet season (*aman*) rice. The East Pakistan Water and Power Development Authority (EPWAPDA) was created as the concerned government agency (Kibria 2011, pp. 13–14).

Coastal Embankment Project (CEP) introduced in the 1960s and initially financed by United States Agency for International Development (USAID) was the pioneering flood and water management initiative. CEP involved two sets of interventions: embankments and drainage structures. High earthen embankments were built across the southern coastal areas as a bulwark against tidal inundation, tidal surges, extreme monsoon rainfall and flash floods. Such embankment better known as polder is a classic flood control solution that has been successfully used in the Netherlands for several centuries. The polders were supplemented by drainage structures like sluice gates set-up at different river sections (Kibria 2011, op. cit.).

The initial outcomes of the polders were quite rewarding both economically and socially. Daily tidal intrusion became a thing of the past and floods were effectively brought under control. For the following two decades, agricultural output multiplied significantly. With two seasons of bumper rice production annually, the majority landowner and farming populations of the south-west region gained more wealth than ever before. There was more disposable income and investible surpluses at their disposal. As farming became less difficult and less time-consuming, there was influx of social and cultural activities through community initiatives (Hashem Ali Fakir, op. cit.).

Gradually, the polders and gated structures, i.e. regulators restricted tidal flows and de-linked the rivers from the catchments. The obstruction by the polder system led to accelerated silt deposition and sediment accumulation in region's rivers and channels. River flows during monsoon season brought vast volumes of water with high sediment loads washed down from the Himalayas. The polder system also impeded these sediment-laden flood flows. Then, floodwaters accumulated upstream of the polders leading to widespread flooding and worsening of sedimentation in the riverbeds and channels. Since the 1980s, vast tracts of land went under water semi-permanently, i.e. for 6 months or more in a year (Kibria 2011, p. 15). People of the south-west were initially clueless that the polders were causing siltation and waterlogging. They initially accepted the official explanation of sluice gate or regulator malfunctioning from Bangladesh Water Development Board (BWDB), the successor of EPWAPDA (Hashem Ali Fakir, op. cit.). But the actual causes started to be revealed; partly from people's day-to-day experience and partly from civil society's awareness raising.

The south-west already had a century-old tradition of citizen activism especially in relation to water and environmental governance. The tradition of building earthen dykes and wooden sluice gates seasonally for flood management was not driven by local landlords alone as suggested by Tutu (2005, op. cit.). Some community leaders and senior citizens argue instead that such indigenous flood management derived from cross section of local peoples' collective action. The landlords used to patronise what was already initiated by local farming population (FGD with members of Paani Committee, 29 May, 2013, *Dumuria*, *Khulna*). Their claims also echo in the term *Dosher Badh* (People's Embankment) used in local language to refer to the earthen dykes. The term is suggestive of people's collective efforts. In the late 1930s, *Badhbondi Andolon or* Embankment Movement against the local landlords spread over large parts of the south-west region. It was people's reaction when the landlord class was not keeping up their traditional patronage for indigenous flood management. Through over a decade, the embankment movement gradually evolved from a collective demand for embankment to their collective development. People started cooperating more frequently and widely in setting up people's embankment and wooden boxes, their maintenance and conducting TRM. To these ends, they mobilised resources, gave voluntary labour and guarded the embankments from any breach in its defences (Hashem Ali Fakir, op. cit.). *Krishak Shangram Shamity* (Farmers' Movement Association) was at the forefront of embankment movement. They also pioneered the TRM movement. Besides, separate issue-specific movements started in different communities across the south-west, total number of which reached up to 30. When negative impacts of a polder or regulator continued unabated, the affected community started a movement in response (FGD with members of Paani Committee, op. cit.).

While some of those movements could cause more stir than others, their overall success was limited. They suffered of the following limitations: smaller membership and resource bases, lack cooperation with each other and limited outreach of their campaigns and lobbying. Consequently, the separate community movements could cause little revision in the environmentally disruptive water interventions. The PC was formed in order to develop a broad-based, well-organised and coordinated movement. Its formation was led by a nucleus of *Krishak Shangram Shamity* together with their like-minded colleagues in local civil society. PC brought the small separate community movements spread throughout south-west under one umbrella and took up TRM as its main agenda (Hashem Ali Fakir, op. cit.).

13.5.3 Institutional Architecture of Paani Committee

PC was officially established in 4 April, 1990. While a citizen-driven initiative, its formation was facilitated by Uttaran, a national NGO based in Tala, Satkhira. PC is well organised throughout much of polder-affected areas of south-western Bangladesh. It is organised in three tiers: central committee, *upazilla* or sub-district committees and *nodi abobahika* or river basin committees. The 51-member central committee is based in Tala, Satkhira. There exist twelve *upazilla* committees in three polder-affected districts: Jessore, Satkhira and Khulna (Hashem Ali Fakir, op. cit.).

Upazilla and river basin committees consist of 31 members each. Each *upazilla* committee is stipulated to have representation of all river basins and unions concerned and to have minimum one-third women members. The central committee is constituted of representatives from *upazilla* committees. Each *upazilla* committee send three members to central committee—president and general secretary as ex officio members and another one as elected member. President and general secretary of central committee are elected by basin committee members.

All the committees—central, upazilla and basin—are elected for 2 years' duration. Upazilla and basin committees convene on bi-monthly basis while the central committee convene quarterly. Besides, all the three types of committees hold annual general meetings (AGMs) (Hashem Ali Fakir, op. cit.).

In general, the memberships of PC units are drawn from among various conscious sections of local population. They are development workers, local NGOs, members of *Union Parishads* and other local governments, local leaders and activists of political parties, teachers, social workers, so on (Paani Committee 2001, p. 1). Many of them have been part of the embankment movement who later initiated many of the anti-polder movements. So, the committee members are from diverse professional backgrounds. However, there are three dominant characteristics in the membership which are also overlapping at times: leftist political orientation, NGO linkage and Uttaran connection.

South-western Bangladesh is known for leftist political activism both civil and armed. Prior to the embankment movement and its successor TRM movement, the region has long been a hotbed of agrarian movements.[1] The main driving forces of these agrarian movements were often leftist political parties. The latter had large following among agricultural labourers and small farmers. However, due to political suppression of leftist political parties in the new state of Pakistan, the former often used front or affiliated organisations to facilitate agrarian movements. *Krishak Shangram Shamity* that pioneered the embankment movement was the farmers' front organisation of Biplobe Communist Party. There was also *Khetmajur Shamity* or Farm Labourers' Association that was affiliated with Workers' Party. When leftist party workers as such were instrumental in embankment and TRM movements, they were not necessarily advancing respective party agenda. However, the ideological solidarity and the broad-based network that came along facilitated the movements' expansion and consolidation.

Another dominant characteristic of PC's membership is NGO linkage. Many of the members are either in the staff of local and national NGOs or involved with them. This is not so surprising since there was convergence among the agenda of PC and those of the many NGOs working in south-west. The growing environmental issues linked to polders were development challenges as well. So many of the NGOs were implementing advocacy projects on waterlogging, salinity, flood and so on. While the NGOs found a powerful civic forum in the form of PC that reinforced their advocacy campaigns, the latter was enriched by the former's expertise and outreach. Besides, many of the NGO staffer also had leftist backgrounds and thus their association with many of the PC veterans predated its formation.

Uttaran connection is the other dominant characteristic of PC's membership. Despite being an NGO, Uttaran's involvement with PC should be seen in separate

[1] One significant agrarian movement was *Tebhaga Andolon* that took place in pre- and post-participation Bengals for several decades. South-west was one of the regions where this movement was prominent. Many of the activists of this movement were later instrumental in other agrarian and civic movements.

light than from other NGOs. Uttaran facilitated its formation and continues to provide financial and secretarial assistance. There are apparently three main reasons for Uttaran's greater involvement with PC.

First, unlike most other national NGOs, Uttaran is based in a south-western *upazilla* instead of a major urban centre like Dhaka or Khulna. Its base Tala is at the heart of polder-affected river basins of the south-west. So Uttaran combined first-hand on-site knowledge and action bases typical of local NGO with broader programmatic approach and advance capacity of national NGO. Uttaran witnessed unfolding of both environmental degradation from polders and resulting civil society activism. Riding on the momentum, Uttaran facilitated separate anti-polder civil society actions graduate into a broad-based united forum in the form of PC.

Second, a number of founding members of Uttaran including its founding head Shahidul Islam were part of leftist political movements beforehand. Uttaran's leadership thus consisted of some veterans of embankment and TRM movements. Uttaran was therefore, in a way, an outcome of those movements and shared their mission most among the NGOs operating in the region.

Third, close personal relationship between leaderships of PC and Uttaran was nurtured through but was not limited to their long years of collaboration in leftist political activism as well as embankment and TRM movements. For example, Uttaran's founding head Shahidul Islam was also long-time colleague of both President and General Secretary of Paani Committee at Chuknagar College in Dumuria, Khulna.

13.5.4 TRM Movement: Strategies, Activities and Outcomes

PC's movement for TRM is a rich tapestry of strategies, activities and outcomes. Strategy in this case is simply understood as goal, objective and guidelines that set the movement's general parameters. Activities are deemed as individual action or set of actions to achieve goal and objectives set-up-front while results of those activities are considered as outcomes.

13.5.4.1 Strategies

PC articulated the development-induced environmental challenges from polders and sluice gates holistically without missing inter-linkages among their community-specific manifestations. Accordingly, the goal set for the ensuing citizen movement was drawn upon integrated approach. While at basin and upazilla levels, PC emphasised on environmental issues relevant to the respective areas, overall they were approached comprehensively as different manifestations of environment–development nexus produced by water management interventions. This is the first strategic formulation of TRM movement.

The second strategic formulation—in line with the goal discussed above—is overarching design of the movement's structure. When PC was formed, there were already a number of environmental movements going on in the region. In a smart move, the founders of PC did not go about for crowding in the region's civic space with yet another movement. They rather went for a collectivisation of existing movements. Accordingly, PC designed the TRM movement as a broad-based one that combined the on-going movements utilising their points of convergence towards achieving a greater goal.

A conscious policy advocacy orientation is the third strategic formulation of TRM movement. The other pre-existing movements also targeted environmental consequences of water development interventions and demanded policy remedies. However, PC took up policy advocacy as one of its core objectives. Since the water interventions in question were part of national programmes, PC did not limit its campaigns and lobbying with concerned local administrations, local governments and local units of BWDB. They also covered concerned government departments and ministries and BWDB head office in Dhaka in their advocacy campaigns. TRM movement was the first among the environmental movements of south-west to effectively identify the complexity of governance system associated with water interventions.

13.5.4.2 Activities and Outcomes

With such comprehensive strategies, the activities of TRM movement were diverse both in terms of themes and typology. Some of the movement's major activities are information gathering and dissemination, awareness raising, publishing, mobilisation, advocacy, campaigning, lobbying, training, participatory planning, contribute in designing TRM interventions and facilitate their execution. As we discuss the outcomes of TRM movement next, there are elaborate manifestations of the above activities.

Amid worsening waterlogging situation and other environmental fallouts, popular resentments and oppositions to polders were also on the rise. In an apparent move to address waterlogging specially in the *Beel Dakatia* area, the Khulna Coastal Embankment Rehabilitation Project (KCERP) was introduced in late 1980s with support from Asian Development Bank (ADB). Again, there was no consultation with concerned population and there was hardly any reflection of their needs in the project design. As of consequence, waterlogging of *Beel Dakatia* further deteriorated leading to widespread public protest. The project was ultimately suspended in 1990 as local people led by some PC officials initiated TRM by cutting the embankments (Fakir 2008, p. 2).

After the widespread success of *Beel Dakatia* TRM, another successful citizen-driven TRM was in *Beel Bhaina*. In 1997, water upstream of *Bhabadaha* flooded the beel. In the absence of any useful government action to address the situation, local people came together to initiate TRM. During October and November, they cut one embankment at the *beel* itself and another at 1 km upstream. The TRM continued for the next 4 year.

With *Beel Dakatia* and *Beel Bhaina* together, there was a decade of citizen-driven TRM in the south-west. PC's advocacy at local, national and international levels also continued simultaneously. However, for a long period of time, instead of accepting the ecological approach of TRM, government and multilateral donors continued structural water interventions. Following suspension of KCERP, Khulna Jessore Drainage Rehabilitation Project (KJDRP) was formulated in 1993 with support from ADB. Its core objective was rehabilitating the existing drainage infrastructure to reduce congestion and enhance protection from tidal and seasonal flooding. The total project area was approximately 100,000 ha of low-lying flat lands with *beels* supporting a population of 800,000 (ADB 2007, p. v).

The renewed emphasis on structural water interventions and continuous de-emphasis on natural drainage processes made local population apprehensive of KJDRP. There was widespread opposition, protest and advocacy over implementation of KJDRP in its proposed form. The stand-off between local citizens and KJDRP was also the one between two approaches: TRM and structural water interventions. While this is written large throughout overall KJDRP phenomenon, there are stark manifestations in cases of the death of *Hamkura* River and the resistance to *Tiabunia* and *Kashimpur* regulators.

The death of *Hamkura* River was a tragedy of somewhat epic proportions. It was not something unique since many rivers died in that region both before and after *Hamkura*. What is so tragic about this river is that it was almost rescued by people through TRM from the damages done to it by polders. The *Beel Dakatia* TRM restored the navigability of the river by increasing both its depth and width within a short span of time. But then, the river largely died when the cut points made in embankments under *Beel Dakatia* TRM were filled up (Paani Committee 2001, p. 3).

The *Hamkura* River was an important drainage channel for a handful of *beels* that cut across Dumuria upazilla of Khulna: Dakatia, Madhugram and Singa beels on the north-west of the river and Madhabkati, Khajura and Khalshi beels on the south and the east. It was also connected with *Lower Bhadra* River 15 km downstream. Previously, tidal water would flow all the way from *Hamkura* through *Lower Bhadra*. Now, it flows up to 14 km from *Dighulia* towards north while the rest of the river is completely silted up. The dry river bed is covered by paddy fields, fisheries' enclosures and homesteads (Islam and Kibria 2006, p. 13). KJDRP made the demise of the river permanent. Subsequently, the downstream *Lower Bhadra* River is also fast dying. There was repeated demand from local population to rescue *Hamkura* by resumption of TRM in the relevant beels specially *Dakatia*. KJDRP instead decided to let the river die and use alternative rivers as drainage channels. That option was not viable—already suggested by indigenous knowledge—as other neighbouring river basins were in higher plains (Paani Committee 2001, p. 3).

Although the local citizens could not persuade KJDRP to revive *Hamkura* River, they were able to prevent construction of *Tiabunia* and *Kashimpur* regulators. The same plan that let *Hamkura* die also proposed *Tiabunia* regulator to facilitate use of

Salta River as the alternative drainage channel. But the regulator could not be constructed in the face of massive public protest led by Paani Committee. This made any work towards regulator construction even land acquisition impossible. Similarly, KJDRP planned *Kashimpur* regulator taking death of *Upper Bhadra* River for granted. Local population resisted the regulator's construction and initiated TRM in *Beel Buruli Pathra* as counter-measure. There was day-and-night vigil by people in *Kashimpur* point. Even a conference was organised in Khulna city where all stakeholders were invited. In the end, KJDRP had to abandon the construction of *Kashimpur* regulator (Islam and Kibria 2006, pp. 12, 15).

In addition to public mobilisations and protests in field, PC stepped up policy advocacy with government, donors and civil society. For example, a broad-based public consultation was organised in Khulna city on KJDRP in July 2001. With mounting public pressure and relentless advocacy of PC, ADB commissioned environmental impact assessment (EIA) and social impact assessment (SIA) of KJDRP to CEGIS. The study findings gave scientific and engineering evidence in favour of TRM as the most suitable design option for dealing with waterlogging and drainage congestion in the project areas (Kibria 2011, p. 17).

Following the CEGIS studies, TRM was accepted and incorporated in the project design of KJDRP in the last phase of its implementation. PC advocacy also succeeded in persuading operations evaluation division (OED), the independent evaluation arm of ADB, in undertaking a project performance evaluation report (PPER). The report rightly observed that lack of appreciation for indigenous knowledge systems and resistance to adopting non-structural solutions in favour of structural solutions were the main factors contributing to the rift between the local people and BWDB, ultimately, bogging down KJDRP. It also acknowledged TRM as technically feasible, economically viable, and socially acceptable (ADB 2007, p. vi).

The original position of PC on polders and associated environmental problems as part of a human-induced environmental degradation instrumented by water development interventions was further reinforced with experiences of KJDRP. In the words of PC's general secretary, "It's a long time when polders and waterlogging were main targets of TRM movement; we are now more of an environmental movement for rescuing the south-west from ongoing ecological disaster and climate change" (Hashem Ali Fakir, op. cit.). Therefore, TRM movement has been giving lot of attention lately to participatory planning and long-term strategies for overall water–environment nexus in the south-west Bangladesh. Two major works in that direction are peoples' plan for integrated river and water management in the south-west and peoples' plan for *Kopotakkhyo* river system.

13.6 Conclusion

In the backdrop of rising environmental challenges often due to climate change, the environmental governance of south-western Bangladesh has been based on water development interventions. The government water agency BWDB with advice and finance from donors—first USAID and Dutch government, and then ADB—introduced structural solutions in water management. This was done with total disregard to the region's ecological characteristics. Initially, the polders and regulators could bring relief from tidal inundations and frequent floods. However, within decades, the decline in environment was worse off. Both scale and intensity of environmental challenges worsened than pre-polder days.

The TRM movement led by Paani Committee—a civil society organisation—emerged in this juncture to secure people's participation in the region's environmental governance and mainstream indigenous knowledge-based TRM as the primary flood and water management approach. PC made at least three strategic strides: a holistic integrated approach to the region's environmental challenges keeping in view the environment–development nexus and the climate change, collecitivisation of existing environmental movements and a conscious policy advocacy orientation targeting wider policy audience ranging from local to regional to national. The activities or participation mechanisms deployed by PC under TRM movement have been diverse: information gathering and dissemination, awareness raising, publishing, mobilisation, advocacy, campaigning, lobbying, training, participatory planning, contribute in designing TRM interventions and facilitate their execution. It is interesting and to some extent unique that with passing time, the movement adopted participation mechanisms of different participation strands.

Through TRM movement, PC could certainly secure considerable people's participation in south-west region's main environmental governance arrangement, the flood and water management system. Despite continuous opposition from the concerned government agency and the donors, PC successfully conducted TRMs one after another with people's support and participation. The successful cases of TRM complimented with people's mobilisation and policy advocacy drew attention of scientific and policy communities to the indigenous knowledge system. Thus, a handful of scientific studies were undertaken that produced scientific and engineering evidence vindicating the effectiveness of TRM as flood and water management approach suitable for south-west region's floodplain ecology.

While primarily a flood and water management approach, TRM is also a climate adaptation approach concerning some major climatic challenges to Bangladesh, i.e. flood and sea-level rise. Being one of world's most active deltas, the country's ecology is characterised by regular floods. Flood flows are instrumental for availability of water, navigability of rivers and waterways, agriculture, food production, so on and so forth. Flood and water flows of the south-west region have been affected by structural water development interventions. They negatively altered local ecological conditions resulting into waterlogging and reduced availability of fresh water. Due to climate change, floods have increased in numbers, become irregular and more disastrous. There is increasing toll on flood and water management. TRM appears as an effective adaptation approach to climatic challenges pertaining to flood and water flows. TRM also has potential to adapt with sea-level rise since it increases land elevation and river depth.

References

Agarwala S, Ota T, Ahmed AU, Smith J, van Aalst M (2006) Development and climate change in Bangladesh: focus on coastal flooding and the Sundarbans. OECD, Paris

Asian Development Bank (2007) Project performance evaluation report on loan no. 1289-BAN(SF): Khulna-Jessore drainage rehabilitation project and TA no. 2012-BAN: Khulna-Jessore drainage rehabilitation proposed evaluation approach. ADB-OED, Manila

Centre for Environmental and Geographic Information Services (1998) Environmental and social impact assessment of Khulna-Jessore drainage rehabilitation project. CEGIS, Dhaka

Connolly E (2007) The role of civil society in poverty alleviation and development. Working Paper 1/2007 from research project on engagement with civil society for poverty reduction, The advisory board for Irish aid, Dublin

Davis P, McGregor J (2000) Civil society, international donors and poverty in Bangladesh. Commonwealth Comp Polit 38(1):47–64

Durant RF (2004) Reconnecting with stakeholders. In: Durant RF, Fiorino DJ, O'Leary R (eds) Environmental governance reconsidered: challenges, choices and opportunities, 1st edn. The MIT Press, Cambridge, pp 177–182

Fakir HA (2008) Tidal river management (TRM): Khulna-Jessore drainage rehabilitation project (KJDRP) area. Position paper, Uttaran and Paani Committee, Tala, Satkhira

Institute of Water Modelling (2008) Mathematical modelling for planning and design of Beel Kapalia tidal river management (TRM) and sustainable drainage management. IWM, Dhaka

IPCC (2007) Climate change 2007: The physical science basis. contribution of working group 1 to the fourth assessment report of the intergovernmental panel on climate change. Cambridge University Press, Cambridge and New York

International Organization for Migration (2010) Assessing the evidence: environment, climate change and migration in Bangladesh. IOM, Dhaka

International Union for the Conservation of Nature (2011) Climate change issues and climate change negotiations: an overview from Bangladesh perspective. Bangladesh Country Office, IUCN, Dhaka

Islam S, Kibria Z (2006) Unraveling KJDRP: ADB financed project of mass destruction in southwest coastal region of Bangladesh. Uttaran, Tala, Satkhira

Kaldor M (2002) Civil society and accountability. occasional paper 2002/6, background paper for HDR 2002. Human Development Report Office, UNDP, New York

Kibria Z (2011) Tidal river management (TRM): climate change adaptation and community based river basin management in southwest coastal region of Bangladesh. Uttaran, Dhaka

Kjaer AM (2008) Introduction: the meanings of governance. In: Kjaer AM (ed) Governance, 5th edn. Polity, Cambridge, pp 1–18

Lafferty WM (2004) Introduction: form and function in governance for sustainable development. In: Lafferty WM (ed) Governance for sustainable development: the challenge of adapting form to function, 1st edn. Edward Elgar, Cheltenham, pp 1–31

Meadowcroft J (2004a) Deliberative democracy. In: Durant RF, Fiorino DJ, O'Leary R (eds) Environmental governance reconsidered: challenges, choices and opportunities, 1st edn. The MIT Press, Cambridge, pp 183–217

Meadowcroft J (2004b) Participation and sustainable development: modes of citizen, community and organisational involvement. In: Lafferty WM (ed) Governance for sustainable development: the challenge of adapting form to function, 1st edn. Edward Elgar, Cheltenham, pp 162–190

Newman J, Barnes M, Sullivan H, Knops A (2004) Public participation and collaborative governance. J Soc Policy 33(2):203–223

Ostrom E (1990) Governing the commons: the evolution of institutions for collective action. Cambridge University Press, Cambridge

Paani Committee (2001) Position paper presented at the public consultation on ensuring experience, management and peoples' participation in KJDRP for addressing water logging in iKhulna-Jessore Region. ADAB, Khulna, 8 Jul 2001

Putnam R (1993) Making democracy work: civic traditions in modern Italy. Princeton University Press, Princeton.

Rhodes R (1997) Understanding governance. Open University Press, Milton Keynes

Rosenau JN (1992) Citizenship in a changing global order. In: Rosenau JN, Cziempel EO (eds) Governance without government: order and change in world politics, 1st edn. Cambridge University Press, Cambridge

Tutu AA (2005) River management in Bangladesh: a people's initiative to solve water-logging. Participatory learning and action (PLA) 51: civil society and poverty reduction Apr 2005, pp 117–123

United Nations Development Program (2011) Water logging in Satkhira district: an analysis of gaps between needs and response. Early warning facility, UNDP Bangladesh, Dhaka

Volger J, Jordan A (2003) Governance and the environment. In: Berkhout F, Leach M, Scoones I (eds) Netotiating environmental change: new perspectives from social science, 1st edn. Edward Elgar, Cheltenham, pp 137–158

World Bank, World Resources Institute (2003) World resources 2002–2004: decisions for the earth—balance, voice, and power. WB, WRI, Washington DC

Chapter 14
Ecology and Culture: Innovative Approaches to the Educational System of Indigenous People of Altai

Irina Zhernosenko and Huong Ha

14.1 Introduction

14.1.1 The Altai Republic

The Altai Republic (Altai) is located in the centre of Asia, and at the crossroad of several countries, natural zones and cultures. It is mountainous with a very charming and scenic landscape. China, Mongolia and Kazakhstan are its neighbouring countries. The Altaians account for 31 % of the total population of the Altai Republic, whereas the Russian population and the Kazakhs and other races population constitute of 63 and 5.6 % of her population, respectively (Department of the Government Communication and Information of the Altai Republic 2013a).

The indigenous population, the Oirot-Altaians, reside mainly on the mountains and foothills of the geographical Altai, and other areas, such as Ulagansky, Ust-Kansky and Ongudaisky regions. There are two ethnographic groups of the indigenous population, namely the northern Altaians (the Ural type) and the southern Altaians (Central Asian or south Siberian types) (Department of the Government Communication and Information of the Altai Republic 2013b).

Given the current environmental conditions in many countries where there has been an increase in globally ecological crises, such as climate change, disappearance of many biological species, noise, air and water pollution and uncontrollable consumption of natural resources, the value of "untouched landscape"

I. Zhernosenko (✉)
Department of Culture and Communication Technologies, I.I. Polzunov Altai State Technical University, Altai, Russia
e-mail: iaj2002@mail.ru

H. Ha
UON Singapore (University of Newcastle, Singapore Campus), Singapore, Singapore
e-mail: artp6075singapore@yahoo.com

H. Ha (ed.), *Land and Disaster Management Strategies in Asia*,
DOI: 10.1007/978-81-322-1976-7_14

and the photogenic sceneries and cleanliness of Altai has also increased (Almashev 2008).

Mountain Altai is considered as one of the centres of ecological and biological varieties in central Asia. More than 2,200 species of plants, more than 80 kinds of mammals and over 300 kinds of birds can be found in the country (WWF 2001). However, many of these species constituting the rich ecological systems in the country have disappeared during the "reform" period in the 1990s. Economic stagnancy causing a sharp fall of the standard of living of the population has led to an increase in illegal hunting and smuggling of the scarce resources.

Altai can attract attention of the whole world because of her rich natural, historical and cultural resources. Although the territory of Mountain Altai is small, it has all diversified landscape zones, from high-mountainous glaciers to steppes, from tundra to desert, the mountain taiga alternates with the alpine meadows and a light forest of larch (Forestry Department, Food and Agriculture Organization of the United Nations 2001). However, the people in Altai have felt strong pressure on their tradition and culture, which has been shown in the growth of globalisation and the steadfast interest of transnational and multinational corporations to her resources, especially in wildlife tourism. The modern technocratic modernisation and globalisation have become less compatible to the processes of natural development in Altai. The Mother Nature has openly begun to revenge the mankind via numerous catastrophes and disasters, such as floods, tsunami, earthquakes, cyclones and others. People in Altai, and elsewhere, convulsively start to search for ways to adapt to new requirements of the nature.

To make things worse, the young people of Altai head to the cities, looking for blessings of modernisation, without realising the uniqueness and the eternal value of their native land and hometown. Therefore, it is necessary to introduce new approaches to provide ecological and cultural education to the natives in respond to the threats of modernisation and globalisation (Mamyev 2012).

14.1.2 Research Objective

The objective of this chapter was to share the experience of the development of the textbooks to educate students on the importance of environmental protection and cultural preservation in Altai. In the current conditions, modern techniques and traditional practice as well as knowledge of the indigenous peoples are important in the development of educational programs. This has been highlighted by Emery (2000) that.

> Traditional knowledge systems of indigenous peoples, while highly variable in their content and style, nonetheless all have a great deal to offer in sustaining life on the planet. Most traditional knowledge systems assume that people are part of the land, not that they own the land, so they consider themselves as true guardians. The wisdom derived from this philosophy can be used to advantage when planning for sustainability (p. 3).

14.2 Case Studies of Ecological and Cultural Education in Altai

A number of initiatives have been carried out in Altai in order to raise the awareness of the young generation about Altai's culture, ecological and biological systems as well as how to preserve them.

Since 2002, the first author has been the supervisor of the educational project, at the federal level, titled "*Designing and realisation of the cultural models at schools*" in the territory of Altai. This project aimed to enhance the awareness of children living in the Altai region regarding ecological consciousness and cultural values (Zhernosenko 2008a, b).

Since 2006, educational experiment has been carried out for the Altaians in the Karakolsky valley, a protected natural territory with the natural park Uch Enmek. This area has rich historical and cultural roots, including five thousand-year-old cliff drawings, debris of the ancient civilisations of the Karakolians and Scythians, the mediaeval sites of ancient settlement and fortresses of northern branch of the Silk way and other modern ethnocultural and religious spectacles. Scientists have found about 5,000 archaeological objects in Uch Enmek Park (Zhernosenko and Mamyev 2007).

In 2008, another project under the United Nations Development Program and Global Ecological Fund (UNDP/GEF) framework was carried out to equip the residents with knowledge of various aspects of the Karakolsky valley. The project aimed to engage local residents in tourism business, which could partially address the unemployment problem, and also to raise their interest in preservation of the wildlife and the cultural heritage of Altai. This, in turn, would help to attract more tourists, but at the same time to protect the local environment and culture.

The project was carried out via conducting 500-h courses by the experts. Many experts were involved in the development of the teaching material for this course, and the first author was the scientific editor of the teaching material. The curriculum included different themes, namely geography, physical geography and geomorphology, bio resources, culture and history of Altai. A great attention was also given to philosophical, spiritual, noosphere (or sphere of mind/human thought) and ecological aspects so that participants could learn about the uniqueness of Altai.[1]

The course also included a compulsory practical component, i.e. an internship or employment opportunity on excursion business, ecotourism and marketing of excursion services. Participants also learnt about safety practice and precautions during a tour, and knowledge of basic first aid. Upon completion of the course, all participants received a certificate, which certified that they are qualified tour guides. This project has continued to operate independently since then (Zhernosenko 2008a, b).

---

[1] The noosphere—"Sphere of the Mind"—the term of V.I. Vernadsky.

In 2009, the first author was the head of a project, which developed a set of three textbooks on the reservation of Altai's land, culture and environment, including multimedia applications and two instruction manuals. The first author was also the Editor-in-Chief of the textbooks. This project was initiated and financed by the UNDP and the GEF. The developers employed innovative approaches, incorporating the principles of modern natural science, humanity knowledge, spiritual culture and traditional knowledge of Altaians, to design the content of the textbooks. The textbooks also include explanation about the structure of the universe and the use of land in Altai. This was the first time that the Altai Republic attempted to produce a full set of textbooks and manuals, which aimed to educate the young generations on the country's culture, traditions and the unique and diverse landscapes.

The first two textbooks on "Nature Reservation in Altai" and "Cultural Reservation of Altai" were developed for Year 8 and Year 9 students, respectively. The first textbook was rich in practical examples, including cultural realities. It also described how wildlife has been traditionally managed by the indigenous population of Altai. The second textbook focused on the interconnection between historical events and cultural development. It also discussed archaeological artefacts, traditional beliefs and ceremonies, which demonstrate the ecological reciprocity between human and nature, i.e. if people took care of the Mother Nature, the later would do the same. The third textbook on "Ecology and Culture" was developed for the senior classes. It integrated various aspects of the natural science and cultural approaches, including spiritually ecological and philosophical concepts, to explain the nature environment in a creative and interesting manner.

The multimedia material contained a rich collection of power point slides and video clips. Instruction manuals were also developed and attached to the textbooks. Students, taking the courses, also took part in campaigns, excursions and scientific expeditions (including overseas trips), which were organised by the park administrators. They learnt various topics on archaeology, botany, geomorphology, biophysics, ethnography, etc.

In November, 2009, the Russian Academy of Education approved the use of the textbooks. The academy also commended the textbook for its innovative approach. In 2010, the textbooks were awarded the "Best educational Book of the year" at the All-Russia publisher competition (Zhernosenko 2010). Other research activities include (i) "Let's restore springs of our village!" project, aiming to restore and protect the water resource of the sacred springs located in the valley, (ii) "Kolagash" project, studying the influence of the thrown mercury mine on the health of local residents and (iii) "ProViktimis" project, studying the influence of flight of rockets on adjacent to the commodore areas.

It was noted that the new educational approach could only be fully realised when training was combined with practical activities. Also, partnership between different groups of stakeholders would improve the effectiveness of educational programs. For example, all five schools in the Karakolsky valley constantly work with the administrators and the research assistants of Uch Enmek Park to conduct similar educational activities.

14.3 Conclusion

This chapter has discussed a number of educational projects carried out in Altai to educate the young generation about ecological and cultural preservation. The projects were launched with the hope that the young generation in Altai would wisely learn how to love for the Mother Nature and their native place, be filial and loyal to the traditions and their ancestors. At the same time, they should learn to open to advanced achievements of science and technology in the new modern world. Government and other stakeholders should invest in similar educational programs, which teach the young generations how to preserve the natural resources and spiritual treasures of Altai.

References

Almashev CD (ed) (2008) Specially protected places of Altai. Excursions and visiting rules of protected and sacred places of Altai (Особо охраняемые места Алтая. Экскурсии и правила посещения охраняемых и священных мест Республики Алтай). Foundation for Sustainable Development of Altai, Gono-Altaisk, pp 3–7

Department of the Government Communication and Information of the Altai Republic (2013a) Territory and geography of the Altai Republic. Department of the Government Communication and Information of the Altai Republic

Department of the Government Communication and Information of the Altai Republic (2013b) Ethnic history of the Altai Republic. Department of the Government Communication and Information of the Altai Republic

Emery AR (2000) Integrating indigenous knowledge in project planning and implementation. Canadian International Development Agency, Quebec

Forestry Department, Food and Agriculture Organisation of the United Nations (2001) Global ecological zoning for the global forest resources assessment 2000. Forestry Department, Food and Agriculture Organisation of the United Nations, Rome

Mamyev DI (ed) (2012) Altay cosmizm as a basis of the ecologically focused ability to live (Алтайский Космизм Какоснова Эколого-Ориентированной Жизнедеятельности). Eurasianism: Philosophy. Culture. Environment (Евразийство: Философия. Культура. Экология), 4:1–174

WWF (2001) Ecoregional climate change and biodiversity decline. AltaiSayan Ecoregion 1. WWF, Moscow

Zhernosenko IA, Mamyev DI (2007) The sacred centres of Altai as a historical and cultural and geophysical resource of the formation noosphere type of consciousness. Natural conditions, history and culture of western Mongolia and adjacent regions. In: Proceedings of VIII international conference, Gorno-Altaisk, 19–23 Sept 2007, 1:46–51. Rio Gasu, Gorno-Altaisk

Zhernosenko IA (2008a) The organization of the innovation activity of teachers in terms of conservation and sustainable development of traditional culture in the Altai. The cultural-creative school. Innovative model of general education and its staffing. In: Proceedings of the seminar of the association of Russian cultural-creative schools, St. Petersburg, 27–28 Mar 2008, RSPU A.I. Gertsena, St. Petersburg, pp 118–125

Zhernosenko IA (2008b) The experience of realization of the project of guides training for protected areas of the Altai Republic. Continuous environmental education: problems, experiences, perspectives. Paper read at materials II interregional scientific and practical conference, Tomsk, 6–7 Nov 2008, pp 184–185

Zhernosenko IA (2010) Culture: the basis of environmental consciousness. New textbooks at the Altai. Conserv Area 1(76):6–7

Chapter 15
Peacebuilding Model in Diverse Conflict Lines Southern Philippines

Charlita Andales-Escano

15.1 Introduction

The twentieth century is controlled by the legacy of distressing worldwide wars, colonial struggles, and ideological conflicts, including efforts to establish international systems that would foster global cooperation, peace and prosperity. In Southern Philippines, there are areas in which peace and security condition is very unstable due to the presence of conflict drivers and varied conflict lines. Peacebuilding is the never ending effort to create a peaceful environment that challenges the actors and sectors at all levels from grassroots to national and transnational levels. The immediate implication is how peacebuilding variously and discursively practiced within the country or region, specifically its attribution to the attainment of the millennium development goals (MDGs) targets in 2015 and conclusion of armed conflict in Southern Philippines.

Peacebuilding is often context-bound and can vary among actors such as the government, non-government organizations (NGOs), communities, policy and decision-makers, politicians, business sector, donor agencies, among others. Different actors pursue different practices and approaches to peacebuilding as revealed in various literature, which frequently do not agree on the path to follow that produce further confusion about the concept. There have been significant numbers of peace and development frameworks and paradigms, both Philippine government and donor agencies working, in Southern Philippines. However, despite the experiences in using different models and approaches, it is still directed to major gaps and challenges. Apparently, the different frameworks and paradigms of donor agencies and government extending assistance to conflict-affected and

C. Andales-Escano (✉)
Mindanao Development Authority, University of Southeastern Philippines, 4F SSS Bldg., J.P. Laurel Avenue, 8000 Davao, Philippines
e-mail: charlie.escano@minda.gov.ph; charlie_escano@yahoo.com

H. Ha (ed.), *Land and Disaster Management Strategies in Asia*,
DOI: 10.1007/978-81-322-1976-7_15

vulnerable communities have strengths and weaknesses, which may affect the just conclusion of the peace problem, and in addressing the prevailing socio-economic challenges.

Effective peacebuilding initiative necessitates addressing both the root causes and drivers of conflict to attain peace and sustainable development. In the formulation of recommended activities or sub-projects that would address the needs and priorities of conflict affected areas, the joint needs assessment (JNA), a project of the World Bank, employed the conflict sensitive approach and peace and development framework. The framework has four (4) pillars of peace and development. Good governance provides the foundation in achieving the desired goals of each pillar, namely: human security, basic services, economic opportunities and social cohesion. It emphasises that assistance will be primarily provided to help reduce tension and conflict, to help build and restore trust and confidence within, between and among people, communities and state in the conflict affected areas. Extensively, peacebuilding refers to the initiatives which foster and support sustainable structures and processes that strengthen the prospects for peaceful coexistence and decrease the likelihood of outbreak, recurrence or continuation of violent conflict. Another framework adopted from Colletta and Cullen (2000) emphasised the interaction of forces and nature and the degree of their dysfunctional state that is manifested through the intertwining dynamics of poverty, inequality and social exclusion that are aggravated by corrupt and inefficient and ineffective governance platform which resulted in accumulated neglect in the provision of basic services especially to the vulnerable and conflict-ridden communities.

The Philippine development plan (PDP) 2011–2016 has included a peace agenda that stresses the achievement of a stable national security environment, as a major sectoral outcome that prioritizes development and security needs of areas affected or have remained vulnerable to armed conflicts. To achieve this, the plan calls for reforms and initiatives that effectively address the causes of all armed conflicts and other issues that affect the peace process as well as the satisfactory conclusion of ongoing and past negotiated political settlements with rebel groups. This strategy underpins the *Payapa at Masaganang Pamayanan* (Peaceful and Progressive Communities) which serves as the current peace and development framework of the government (NEDA 2011). Relatedly, the action for conflict transformation (ACT) for Peace Programme in Southern Philippines was the only donor funded initiative that was classified as a peacebuilding initiative that employs theory of change in measuring its outcomes.

Given the commonalities and gaps of the frameworks and paradigms adopted by various actors, it has been observed that peacebuilding strategies are done in packets, sporadic, not comprehensive, not coordinated, and lack of scale. This condition makes the peace and development scenario in Southern Philippines more fragile posing greater challenges for a more comprehensive and integrated approach in peacebuilding and conflict transformation efforts. Thus, it is imperative to outline the strands of peacebuilding framework for Southern Philippines, notwithstanding the recent development on the peace process through the signing of the comprehensive agreement on the Bangsamoro (CAB) between the Government and the Bangsamoro, which is a notable impetus towards apt implementation of programs and projects in the coverage area.

The basic intention of this study is to develop a peacebuilding model for Southern Philippines that would attribute to the achievement of peace and security. Specifically, it aims to assess the outcomes of the strands of peacebuilding and conflict transformation initiatives in Southern Philippines by comparing the ACT for Peace Programme outcomes with other local programs and projects. It also aimed to determine the contribution of peacebuilding model of the ACT for Peace Programme in addressing peacebuilding challenges in Southern Philippines. Various studies were conducted on instability in Southern Philippines and Mindanao in particular, yet, there is still a need for another study about peace, conflict, and development in Southern Philippines in the context of peacebuilding and conflict transformation. To date, program donors and government engagements have often been disparate and are still uncoordinated, in part, because many studies were narrowly focused. Hence, this study identified and described the critical strands of peacebuilding that may help improve programmatic responses by national government, local government, donor agencies, NGOs and other key stakeholders on peacebuilding and conflict transformation. Generally, this study may serve as a reference for peacebuilders and development practitioners in pursuit of their agenda. Building on the case studies and rich lessons from the ACT for Peace Programme, it may provide a perspective on the attributes of the programme as a peacebuilding and conflict transformation initiative in conflict affected areas and vulnerable communities.

15.2 Literature Review

15.2.1 Demography and Challenges

In this study, Southern Philippines is composed of Mindanao and Palawan. Mindanao has a total land area of 133,656 km^2 approximately 39 % of the total land area of the Philippines, the second largest island in the country that consists of six (6) regions, 26 provinces, 33 cities, 422 municipalities, and 10,082 barangays, and Palawan has a land area of 14,650 km^2 (NSCB 2010). Based on national statistical coordination board (NSCB) Philippines census (2010), Mindanao population had reached 21.968 million or an annual growth rate of 1.79 %, meanwhile Palawan (excluding Puerto Princesa) posted a population of 771,677 which posted an annual growth rate of 2.66 % during the period 2000–2010. Muslims, Lumads and Christians comprise the population in Mindanao and Palawan. The Muslims have 13 ethno-linguistic groups with a vast majority in the provinces of Maguindanao, Lanao del Sur, Tawi-Tawi, Sulu, Basilan and in the city of Marawi and a minority in Palawan. These areas (excluding Palawan) comprise the autonomous region of

Muslim Mindanao (ARMM). The Lumads have 18 ethnic tribes with ancestral domains covering 17 provinces and 14 cities in Mindanao. This current composition has resulted in a multi-faith and multi-ethnic population with a shared identity. Yet, its distinct history of migration, colonisation and secessionist clashes as well as its unstable power structures and feudal relations have contributed to its complex and fragmented environment of conflict.

Southern Philippines is strategically located in the East association of Southeast Asian Nations (ASEAN) region, almost equidistant to the eastern sections of Indonesia, Malaysia, and Brunei Darussalam, that underscores its potential to be a major transhipment point and center of trade in the region (ADB 2002). Cultural diversity and the island's natural beauty make Mindanao one of Asia's favored tourist destinations with its white sand beaches, scenic volcanoes, vast orchid gardens, and various ethnic festivals are only some of its unique tourist attractions (DOT 2011). Yet, Mindanao in particular has not shown significant impact in reducing poverty incidence because growth has not been sustained. NSCB (2011) showed high levels of poverty, income inequality and wide disparities across regions in Mindanao.

Studies revealed that various types and levels of conflict characterise Mindanao's regions. These include: protracted conflicts led by non-state armed actors such as Moro National liberation front (MNLF), Moro Islamic liberation front (MILF), Communist party of the Philippines-new peoples army-new democratic front (CPP-NPA-NDF), violent clan-related conflicts; armed violence from lawlessness and banditry, and conflicts over control of resources (MEDCo 2006). The armed conflict scenario in the country that has predominantly attributed to Southern Philippines was confirmed in the recent results of failed state index that ranked the Philippines at 59 of the 178 countries with a score of 82.8 under the very high warming category, along with Mozambique (FP 2013). Meanwhile, the country ranked 129 out of 162 countries in the global peace index (GPI) with a score of 2.374 under medium state of peace category in 2013 (IEP 2013). Major factors considered were: militarisation (2/5), society and security (3.1/5) and domestic and internal conflict (3.2/5), in which one is very low and five is very high.

Current peace and development considerations in Southern Philippines require a comprehensive approach that addresses vulnerabilities, builds human security, and enables long-term peace and development across the whole spectrum from fragile to relatively stable settings in the region (MinDA 2011b). It is deemed important to pursue accountable and conflict-sensitive local governance as a critical element in ensuring stability and sustainable development in both conflict-affected and non-conflict areas (MinDA 2011a). The recent development of the peace process between the Philippine government and MILF through the signing of a comprehensive agreement on the Bangsamoro (CAB) (Philippine Official Gazette 2014) would require a very comprehensive and integrated approach in the implementation of its annexes. The CAB is a milestone that concludes the 17-year peace negotiations between the government and the MILF.

15.2.2 Peacebuilding

The term peacebuilding can be used to describe a varied set of activities or programmes, the manner that these programmes are implemented as well as their potential outcomes and impacts. It embraces efforts undertaken at different levels of the intervention and implemented by different actors (Lewer 1999). Galtung (1975) defined peacebuilding as encompassing the practical aspects of implementing peaceful social change through socio-economic reconstruction and development. In his latest contribution, peacebuilding is re-defined as part of 'third generation' of peace approaches evident after the Cold War, when a reaction against simplistic approaches to building peace finally recognised the deep-rooted nature of conflict and its links to development. Galtung (2002) argues that issues of culture, human needs, and 'fault-lines of the human condition such as gender' have now entered the peace debate and are recognised as crucial. The intent of peacebuilding is to create peace as stressed by Galtung (1996), where he suggested two (2) different concepts of peace: negative peace and positive peace. Peacebuilding promotes positive peace that had become a starting point of analyses in addressing the root causes of conflict (Carbonier 1998; OECD 1997; UN 1995). Boutros-Ghali (1995) defined peacebuilding as a wide range of activities associated with capacity building, reconciliation and societal transformation. It is more than just a post-accord reconstruction, but should be understood as a comprehensive concept that encompasses, generates, and sustains the full array of processes, approaches, and stages needed to transform conflict towards a more sustainable and peaceful relationships and governance modes and structures (Lederach 1997; Morris 2000). It is often context-specific and can vary among actors such as the NGOs, communities at large, policy-makers, politicians and donors (Hamber and Kelly 2005). There seem to have a consensus that peacebuilding requires a long-term commitment to addressing the underlying causes of conflict through structural and relational transformation (Lederach 1997).

15.2.3 Conflict Transformation

It is now the leading approach to peacebuilding that combines short-term conflict management with long-term relationship building, and transformation of the root causes of conflict (Rupesinghe 1995). It is context-responsive, taking into account the culture, history, and ethos of conflicted groups or societies (Lederach 1997). The goal of conflict transformation is not about resolving any particular conflict but transforming the way people deal with their conflict (Miall 2001). It focuses on relationships and transactions between conflicting parties while addresses wider social, economic, and political sources of conflict and transform negative energy and hostilities into a positive social change (Lederach 1997). Miall lays out five (5) types of "transformers" of conflicts—context, structure, actor, issue and personal/elite

(Article 1, 2001: 12). An example of a contextual transformer is the change to the rules of diamond trade which may have an impact on conflicts in Sierra Leone and Angola. As a structural transformation, he points to the Black Consciousness Movement in the conscientisation of people in township areas in South Africa during the Apartheid regime. In terms of actor transformation he notes the decision of leaders to initiate a peace process (2001:11–14). Conflict transformation is considered as an 'open-ended' process (Reimann 2001, p 17; Rupesinghe 1995, p 76).

15.2.4 Social Cohesion

It is described as affective bonds between citizens (Chipkin and Ngqulunga 2008), local patterns of cooperation (Fearon et al. 2008) and the bond that stick society together, promoting harmony, a sense of community, and a degree of commitment to promoting common good (Colletta et al. 2001). Discourses on social cohesion arise in analysing the causes and consequences of social upheaval, violence, misallocation of financial assistance, deep-rooted poverty, slow or negative economic growth, and failure to realise welfare gains from market-oriented economic reforms (Colletta and Cullen 2000; Easterly et al. 2000, 2006; Winters 2008). Several theories and frameworks provided support to the ideas that human and social dimensions are important factors in economic growth and development, including sustainable livelihoods (World Bank 2005), capabilities approach (Sen 1999), and theories emphasizing the role of institutions (Chang 2002).

15.2.5 Basic Social Services

It includes access to healthcare, clean water and sanitation, and education. Improved service delivery for the constituencies, particularly the poor and disadvantaged can facilitate strengthening civic engagement, rebuilding public trust and confidence in government institutions, contributing to state legitimacy, and reducing the probabilities of recurrence of conflicts (Collier 2007). Governance reforms deemed critical in promoting longer-term social and political transformation have more chances of success if connected to reforms in basic service delivery, which generate concrete results benefiting the constituencies (Berry et al. 2004). However, providing basic services in fragile and conflict affected states is certainly difficult and challenging because this would imply rewarding existing poor performing and dysfunctional local structures (Pavanello and Darcy 2008). Yet, insufficient and/or inequitable access to education for example was a factor in the decision of adolescents and youth to join armed groups in Sierra Leone (Ashby 2002). Apparently, education potentially provides a sense of normality, and shared values and identity to children and youth in conflict areas, thus offering them hope for the future (Baird 2011).

15.2.6 Community Economic Development

Community Economic Development refers to livelihood and income generating initiatives provided to conflicted communities. This assistance is effective as an incentive if it can provide a credible counterbalance to political ambitions, support pro-peace constituencies, and affect the social and economic environment key to conflict (Griffiths and Barnes 2008 in Wennmann 2010; Uvin 1999). Notably, there is an increasing interest in employing economic propaganda to address other root causes of conflict, such as ethnic tensions, and elite capture of natural resources (Banfield et al. 2006). The capacity of former combatants to reintegrate is dependent on many factors, including what motivated them to fight and what role they played during the conflict and what alternatives are available to them (Specht 2010).

15.2.7 Conflict Management

It is usually associated with conflict containment (Hamad 2005). Theorists see violent conflict as an unavoidable result of differences of values and interests within and between communities which they regard resolving such conflicts as unrealistic. The best that can be done is to manage and contain them, and sometimes to reach a compromise in which violence may be set aside and normal politics is resumed (Miall 2004). Conflict takes place when two (2) or more parties find their interests incompatible; while peace is more than just the absence of war and is not a static condition but a dynamic process that involves a network of legitimised social relationships which include conflict but are able to manage it in a constructive manner (International Alert 1996).

15.2.8 Conflict Over Natural Resources

Many studies were devoted to the relationship between environmental degradation and violent conflicts and not to environmental cooperation and peacebuilding. However, since Kofi Annan's speech at the World Day for Water 2002 this notion has changed. He emphasized that:

> […] the water problems of our world need not be only a cause of tension; they can also be a catalyst for cooperation. […]. If we work together, a secure and sustainable water future can be ours.

The 2004 report of the UN Secretary-General's High-Level panel on threats, challenges and change highlighted the fundamental relationship between the environment, security, and social and economic development in the pursuit of global peace in the twenty first century. Meanwhile, a historic debate at the UN

Security Council in June 2007 concluded that poor management of high-value resources constituted a threat to peace (UNSC 2007). Recently, UN Secretary-General Ban Ki-moon confirmed that the basic building blocks of peace and security for all peoples are economic and social security, anchored in sustainable development, because they allow us to address all the great issues—poverty, climate, environment and political stability as parts of a whole (UNEP 2009). The struggle about access to and control over important resources (as water, oil, gold, diamonds, productive land etc.) is one specific cause of conflict. Conceivably, it is intuitive that natural resources could become conflict issues, but less obvious is the role that resources may have in specific instances of a given conflict. Inequities in the distribution, use, needs, desires, and consequences of resources management have been sources of tension and international and intrastate disputes (e.g. Philippine claim over Sabah State, etc.)

Some resource conflict researchers (Ehrlich, Gleick and Conca 2000) stressed four (4) important conditions that influence the likelihood that resources will be the object of military or political action: (1) the degree of scarcity; (2) the extent to which the supply is shared by two (2) or more groups/ states; (3) the relative power of those groups; and (4) the ease of access to alternative sources. Studies revealed the most popular approach in the resource conflict literature which is a "resource scarcity" as a central conflict contributor. This approach that links resources considers resource scarcity (e.g. supply induced, demand induced or absolute scarcity), and environmental degradation as key conflict issues (Homer-Dixon 1999). Homer-Dixon and Percival (1997), highlighting the causal pathways between conflicts and resources in some developing countries, argue that under certain conditions, the scarcity of renewable resources such as cropland, forests and water generate social effects (e.g. poverty, migration, and weak institutions) and produce tensions and conflicts. Accordingly, dealing with conflict proves to be a continuing process, with conflict prevention as its most effective measure. Yet, it is generally accepted that this form of conflict dealing ranks among the most complex and most expensive ventures, for which conflict-vulnerable regions are mostly not in a position and are rarely convinced from the fugal "peace islands" (Weller and Kirschner 2005, p 24).

15.2.9 Land Administration

It involves the process which has to be based on sound policy and manageable procedures. Land is often a significant factor in a widespread violence and is also a critical element in peacebuilding and economic reconstruction in post-conflict situations (USAID 2004; Batson 2007). Land continues to be a cause of social, ethnic, cultural and religious conflict if not secured because land rights is germane to vulnerable groups such as the poor, women, orphans, displaced persons and ethnic minority groups (Bell 2007). The land and land related issues are increasingly recognised by international community as important element where all conflict activities take place before, during and after hostilities, during the period of peacebuilding (Todorovski 2011). Unsolved land tenure problems and ineffective land administration can result in economic and political failure (Pienaar 2009; Ocheje 2007). Moreover, even where land is not necessarily the root of conflict, tenure disputes often emerge in the course of conflict and serve to effect insecurity and instability (Van der Zwan 2010). As pointed out, a highly skewed distribution of land ownership and patterns of land access can provoke social conflict and violence (Deininger 2006). The likelihood of violent conflict increases when gross inequities characterised by land-holding patterns, particularly when a large landless or land-poor population has limited livelihood opportunities (USAID 2004). Attention to land policy and land governance needs to be a part of any development or peacebuilding efforts in conflict environment (FAO 2009).

15.2.10 Good Governance

It refers to quality and effectiveness of institutions, the interactions between different levels of government within a country and the interactions between members of public, including business and NGOs and government. Good governance is an essential component of sustained economic performance, particularly in transition economies (UNECE 2003); an economic policy-making and implementation, service delivery and accountable use of public resources and of regulatory power (World Bank 2000); the ability of government to develop an efficient, effective, and accountable public management processes that are open to citisen participation and that strengthen rather than weaken a democratic system of government (USAID 2009); how institutions, rules and systems of state—executive, legislature, judiciary, and military—operate at central and local levels and how the state relates to individual citisens, civil society and the private sector (DFID 2010); and the exercise of economic, political and administrative authority to manage a country's affairs at all levels (UNDP 1997). In essence, it concerns norms of behaviour that help ensure that governments deliver to their citisens the services they deserve.

15.2.11 Peacebuilding as a Change Theory

Peacebuilding is about change—knowing when it happens, understanding how it happens and working on how to sustain and scale up such change. The process of peace building and conflict transformation looks deep into the four (4) dimensions of change as brought about by social conflicts—personal, relational, structural and cultural. The four (4) dimensions of change are described as: (1) Personal—this involves changes in attitude (the way people think and approach a given topic, situation or relationship) and behaviour (the way people actually act, respond, express themselves and interact with others); (2) Relational—this involves changes in communication patterns, cooperation, decision-making and mechanisms in handling conflict situations; (3) Structural—this involves changes in social conditions (i.e. disparity, inequity, racial/religious/ethnic disadvantage), procedural patterns (lack of transparency, equality, access, participation, fairness) and institutional patterns (i.e., lack of access, historical patterns); and, (4) Cultural—this involves changes in cultural patterns (culture defined as the process of how meaning is constructed and shared by a group) although such change is often very slow and difficult to monitor since culture is embedded in all three of the other dimensions (Lederach et al. 2007).

15.2.12 Theory of Change

A theory of change is a set of beliefs about how change happens. Behind each peacebuilding and development initiative, there is at least one theory of change. While such theory broadly explains the logic behind the program's approaches, it also looks for specific changes that can easily be monitored and evaluated. The program looks at significant changes in three (3) of the four (4) dimensions of conflict transformation—personal, relational and structural. These are reflected in the indicators which describe specific changes in knowledge, skills, attitudes and behaviours of partners or clients, as well as changes in communication patterns and working relationships between and among claim holders and duty bearers, among others. The concept of theory of change is gaining greater attention within the conflict studies community (Shapiro 2002, 2005, 2006; Church and Shouldice 2002, 2003; Lederach et al. 2007; OECD-DAC 2008). The concept originated in the theory-based program evaluation literature of the 1970s (Weiss 1972) and has reverberated throughout the social sciences. Schon (1983) focused on the need to name problems in order to bring technical expertise to address them.

15.2.13 Conflict Lines in Mindanao

Informed of the conflict analyses of the USAID (2003) and the AusAID (2008), Mindanao stakeholders involved in the conflict analysis and peace needs workshops in 2009 (MEDCo 2009) agreed to focus on the following conflict lines in Southern Philippines and Mindanao in particular:

1. Armed conflict between Moro forces seeking greater autonomy or a separate state and the GPH defending state sovereignty and territorial integrity;
2. A long-standing communist insurgency (splintered into different factions) which is national in scope;

3. An assortment of typically localised conflicts caused by elite political and economic rivalries;
4. Local, often community level conflicts, over access, control and use of land and natural resources; and,
5. Conflicts related to criminal activities including drug trafficking, arms trading, narco-politics, extortion, kidnapping for ransom, illegal logging, smuggling, human trafficking and illegal gambling.

A narrow viewpoint on conflict in the study area in terms of religion, culture or ethnicity, that illustrate the approach of many donor agencies and other development institutions, national government and some scholars, ignores the class divisions and dynamics within and between Christian and Muslim communities, and presents only very limited understanding of the roots and character of the current violent conflict (Vellema et al. 2011).

15.2.14 ACT for Peace Programme

Building on the gains of the three (3) phases of the government of the Philippines United Nations Multi-donor Programme (GOP UN-MDP) initiatives in Mindanao, the fourth phase known as the ACT for Peace Programme was designed as a peacebuilding and development intervention to respond to the changing peace and development context in Southern Philippines and to address the continuing vulnerability of many communities. Evolving from being a humanitarian and socio-economic response to the GOP-MNLF final peace agreement in its earlier phases, the Programme embarked on a purposive and integrated program for the promotion of human security and the culture of peace in conflict-affected and conflict-vulnerable communities. These were inclusive of MNLF or MILF-affiliated communities and other vulnerable or marginalised areas due to conflicts related to presence of other insurgent and armed threat groups (UNDP 2005). The ACT for Peace Programme operated in 278 PDCs which were distributed in 151 municipalities and 12 cities in Mindanao and Palawan that comprises Southern Philippines in this study.

The core of the ACT for Peace Programme had been the peace and development communities (PDCs). Initially established during the third phase of the GOP-UN MDP the PDCs evolved from conflict-affected/vulnerable and fragmented communities to stable and resilient communities that can undertake local peacebuilding and conflict transformation initiatives and participate in and benefit from development. Beyond delivering services and projects to communities, the PDCs represented not only a transformative strategy at the grassroots level but also the significance and benefits of transformation. The progress of PDCs was measured and tracked against several stages of transformation and development (MinDA 2010).

15.3 Theoretical Framework

15.3.1 Peacebuilding Frameworks

Lederach (1997) integrates the peacebuilding concepts to the field of development planning, programming, and implementation. This is complemented with the Peacebuilding Levels (Lederach et al. 2007) that describes the critical roles of key actors, from top level, midlevel and lower level, where community partners and civil society have critical roles in achieving peacebuilding and conflict transformation goals. Lederach (1997) proposes an intervention structure as a pyramid with the apex representing the top military, political and/or religious leadership; the middle level representing a mid-range leadership including sector, professional, ethnic or international non-government organisation (INGO) leaders; and the grassroots leadership placed at the pyramid's base. Each of the three 'spaces' contains activities to advance the peacebuilding process. For instance at the top, high-level negotiations, cease-fire agreements and highly visible mediation would be used. The middle level problem-solving workshops, conflict resolution training, and peace commissions would be better suited. Thus, grassroots interventions may include local peace commissions, grassroots training, prejudice reduction activities, and psychosocial work in post-war trauma recovery (ibid, p 39).

The peacebuilding pallet (Smith 2004) indicates that physical security is equally important as establishing good governance and as socio-economic foundation of a long-term peace. The peacebuilding wheel (Hart 2008), reveals the interfacing of tangible and intangible elements of peacebuilding. The interplay of various forces, the nature and degree of their dysfunctional state is clearly described through the interweaving dynamics of poverty, inequality and social exclusion that are aggravated by poor governance (Colletta and Cullen 2000). The conflict sensitive approach (WB 2005) on peace and development efforts display the four (4) pillars in which good governance serves as the critical platform in attaining the goals of each pillar, such as human security, basic services, economic opportunities and social cohesion. In addition, the framework on building resiliency against violence (Barron 2010) showed the risks of violence would increase when external and internal stresses are coupled with weak institutions, which is basically anchored on good governance. In this framework, the need to restore confidence between and among people requires transforming institutions that provide security, justice and jobs to the citisenry. Meanwhile, the people's agenda in Mindanao (MPW 2010) focuses on genuine peace and sustainable development. Further, the integrated peace and development framework (MEDCo 2008) provides guidance for understanding Mindanao context through conflict analysis and other appropriate methods; integrating peacebuilding and conflict-sensitive concepts and principles to programs and projects; and, formulating and developing the Mindanao 2020 Peace and Development Framework Plan. At the community level, the PDC provides a venue to actively engage the marginalised conflict-affected community as the framework under the ACT for Peace Programme.

15.3.2 Analytical Framework

Based on various concepts, theories, studies, and operating context in Southern Philippines, this study considered four (4) key strands of peacebuilding, namely: social cohesion, basic social services, community economic development, and conflict management. The four (4) strands are intertwined through the nexus of good governance as the anchor of peacebuilding. Each peacebuilding strand is examined through the lens of the four (4) dimensions of change, which are personal, relational, structural and cultural, to determine the levels of change and transformation that are in place after the programme intervention. These strands have specific theory of change that all geared towards achieving the overarching goal of peacebuilding and conflict transformation in Southern Philippines. The theory of change in each strand is interlinked with the theory of change on good governance as an essential precondition of peacebuilding and conflict transformation. The practice of good governance is not confined to the government sector as traditionally viewed by many. Sound management practices in the private sector are equally crucial to a nation's development particularly in achieving effectiveness, efficiency, accountability, transparency, relevance and impact of the private sector operations as the engine of growth. Good governance, therefore, is a shared responsibility of government, market and civil society dealing with societal concerns.

Social cohesion seeks to contribute to sustainability in the transformation of PDCs and other conflict-affected areas and harness community efforts to develop and advance their own initiatives for peace. Sustainability of peacebuilding and conflict transformation efforts require strong peace constituency and support from various actors at the local, regional, national and international levels. Through these networks and partners, LGUs link communities to peace actors and macro policy makers to generate support in sustaining local peace initiatives. For peacebuilding to be effective, it is deemed vital to have a more active engagement and participation of multiple stakeholders. In particular, the initiative acknowledges that LGUs play strategic roles in conflict transformation as front-liners in the countryside. Deliberate efforts to identify and meaningfully involve and engage multiple stakeholders are regarded as important to the development undertaking. This suggests the desirability of engaging as many stakeholders as possible and the purposive inclusion of marginalised groups while building consensus among the engaged groups and actors. Moreover, re-building trust or confidence, social capital and social cohesion at the community level, while important have their limits because they are constrained by the discourses and agreements of leaders of the parties in conflict.

Obviously, human security encapsulates basic social services and community economic development as most development practitioners posit that security has become the watchword since the early part of the decade as the country is becoming more vulnerable to terrorism, armed conflict and insurgency. In strict sense, what matters most is not the abstract security of state but security of people.

The Philippine Human Development Report (1997) identified common threats to human security that may fall to any of the seven (7) categories, namely: economic security; food security; health security; environmental security; personal security; community security; and, political security. In this study, land is classified as an element of economic development, being a primary and germane resource of the people in the conflict and vulnerable areas.

During conflict situation, the poor and disadvantaged groups suffer more where their vulnerability, lack of opportunities and recognised helplessness come to fore. The effects of armed conflict are most apparent in certain areas of Southern Philippines that affects the country's GDP as well as in achieving the MDGs targets in 2015. Conflict management is seen as one of the durable solutions to this difficult problem to widen prevention and settlement efforts. The presence of local conflict resolution mechanisms or local structures that facilitate the resolution of conflicts or prevent their recurrence is one of the critical elements of the peace infrastructure in Southern Philippines. These forms of structures include legislated, traditional or customary, religious, and contemporary. In support of this, the culture of peace was developed to educate and empower people, institutions, organisations and governments to cultivate values, attitudes and to promote relationships, actions and practices to move from a culture of violence to a culture of peace.

Good governance, however, as an anchor and enabler in building a strong foundation of peace infrastructure, establishes stable and reliable social, economic and political institutions that are responsive to the needs of the constituency and foster equity, justice, confidence, inclusiveness and fair allocation of resources, accountability and transparency. There is a clear relationship between poverty and poor governance, while the recurrence of conflict is considered as the ultimate confirmation of the failure of governance. Conflicts flourish in weak states and fragile institutions. Ultimately, building or rebuilding governance systems is the responsibility of citisens and leaders in post-conflict societies. The interventions of international community cannot, by themselves, fix a country's governance structure, though they can support reconstruction and reforms. Nonetheless, the greatest challenge for developing a governance reconstruction roadmap is to develop processes, systems and tools for bringing altogether local and external actors in manners that gainfully contribute to improving legitimacy, security and effectiveness. Peacebuilding and development thru conflict-transformation processes require important factors including confidence building initiatives and political support to sustain programs after completion; thus, impacts of programs continue to benefit conflict-affected communities.

15.4 Research Methodology

This study employed data mining for qualitative methods of research. It used metasynthesis, multiple case narratives and vignettes. This study assessed the condition of the areas covered by the ACT for Peace Programme during its implementation period, employing context analysis and metasynthesis to examine the four (4) parameters, namely; social, economic, political, and conflict environments. Metasynthesis, multiple case narratives, and vignettes were also used to determine the results of the implementation of the ACT for Peace Programme in terms of social cohesion, social services, economic security, conflict management and good governance. Multiple case narratives were used to capture and analyse the various programs and projects implemented in, on and around conflict areas. Vignettes were used to visually present the good practices and success stories of peacebuilding programs in the study area. Focus group discussions and key informant interviews served as venues to further validate the findings from the earlier cited methods.

In this study, the theory of change applied for peacebuilding and conflict transformation considers that: "If programs and policy reforms that enable the poor to have equitable access to and control of productive resources, inclusive mechanisms for participation of disadvantaged sectors in the bureaucracy, and effective basic services delivery for marginalised communities to address their conditions, then community and government structures are strengthened to be resilient and their capacities are enhanced to manage differences thus prevent the situation from escalating into armed conflict".

15.5 Findings and Discussion

The ACT for Peace Programme employed a peacebuilding and conflict transformation model in Southern Philippines. In this model, it sought to respond to the complex situation of conflict in this part of the country, which involved a range of on-going, potential and post-conflict situations in the target areas, in addition to responding to physical and social reconstruction of conflict affected areas. It necessitated interventions to support and sustain peace as well as build human security in these conflict scenarios, such as but not limited to: (a) building local capacities for conflict prevention and peacebuilding in areas of potential conflict, as well as in post-conflict situations; (b) basic service delivery and capacity-building for conflict prevention in conflict and post-conflict areas, and, (c) relief and rehabilitation of areas affected by on-going armed conflict, and also in post conflict areas.

Context assessment revealed that the areas covered by the ACT for Peace Programme have been the object of a long unending journey for survival on peace and security. Most LGUs in these areas have poor governance capacity and the institutions were notably fragile. It was evident in the lack or even absence of basic social services which have been the primary needs of the conflict affected and vulnerable communities especially the internally displaced peoples (IDPs). Most of these communities have failing marks in terms of human development index and poverty incidence including meeting the MDG targets in 2015. Due to the unstable peace and security in these areas, it had been difficult for them to entice private sector investments, thus, the economy continue to lag behind the rest of Southern Philippines and the country in general. The needs of these communities are very huge that the local government units (LGUs) can hardly supply. The political condition also aggravated the situation. Evidently, "rido" (clan war) and armed skirmishes erupted because of the weak governance and fragile institutions in the communities including the lowly capacities of the people and groups. Specifically, five (5) challenging conflict lines are present in Southern Philippines, namely: (1) armed conflict between Moro forces; (2) communist insurgency; (3) assortment of typically localized conflicts caused by elite political and economic rivalries; (4) conflicts related to land and natural resources; and, (5) conflicts related to criminal activities.

The strands of peacebuilding and conflict transformation model of the ACT for Peace Programme are composed of: (1) Strand 1—social cohesion; (2) Strand 2—basic social services; (3) Strand 3—community economic development; and, (4) Strand 4—conflict management. The four (4) strands are intertwined through a nexus of Good Governance. These strands were evaluated in terms of four (4) dimensions of change such as personal, relational, structural and cultural. The theory of change on social cohesion, stated that: "If isolation, division, and discrimination between individuals and groups are broken down through established and functional institutional mechanisms and processes while addressing common interests and differences to promote equitable benefits from economic activities, social and economic gaps then constructive, non-violent, mutually reinforcing relationships, mechanisms and practices can be established and strengthened".

Basic social services that refer to the capacities in accessing services especially primary health care and potable water, and managing and sustaining these services through the collective efforts of the communities were examined to determine the links and effects on the capacities of the individuals, groups and LGUs particularly in governance. The theory of change on basic social services, stipulated that: "Increased access to basic social services such as education, health and social welfare would increase equity in participating in public and private institutions, systems and processes to nurture a more self-governing and ultimately more stable peace and development."

The community economic development initiatives were venues that provided capacities in utilising local resources and the establishment of market links toward environment-friendly, industry-level production. These were reviewed to find out its contribution to the attainment of good governance outcomes. It basically included the capacities in generating local employment, increase in household income and overall economic growth from the interventions, e.g. community livelihood, enterprise, infrastructure, including the capacities of the communities to utilise their local resources and establish market-connected industry-level production that generate local employment and peace-promotive and equitable economic growth. The theory of change on community economic development, posited that: "If reliable institutional mechanisms, systems and processes are established to promote equitable benefits from economic opportunities and activities, social and economic disparities will be reduced thus peace and development will be sustained".

Conflict management refers to the capacities of the communities in transforming and preventing conflicts from getting violent through conflict analysis, mediation and resolution using traditional, religious, legal and other appropriate systems and mechanisms. The theory of change in conflict management states: "If various actors and sectors in the society respect the viable indigenous, customary or community-based dispute resolution mechanisms where duty-bearers and claim-holders sustain the promotion of culture of peace, then greater intra-ethnic, inter-cultural and intergenerational understanding and peaceful coexistence of community people is warranted".

Good governance as a nexus that intertwined the strands of peacebuilding refers to the capacities of local leaders and stakeholders in promoting and broadening community participation in local governance processes, resource mobilization, implementation and management of community development initiatives that promote peace. In effect, the theory of change on good governance, states that: "Improving conflict management and good governance capacities of LGUs and other local stakeholders would promote effectiveness of conflict transformation and peacebuilding initiatives, responsiveness to concerns of the conflict affected and vulnerable communities, and effectively inform macro policy and institutional response".

The results of the metasynthesis, multiple case narratives and vignettes showed the vital role and contribution of each peacebuilding strands in achieving good governance outcome. Specifically, it was found that social cohesion is deemed necessary as a bridging, binding and bonding element for peacebuilding particularly in improving governance capacities of the local leaders (especially LGUs) and stakeholders. The programme strengthened and mobilised the people's organization (POs) and local social formations (LSFs) to support social capital formation as well as capacitated them to support peacebuilding and human security initiatives. To achieve this, community organising was conducted and capacities were improved through intensive training workshops that resulted in increased trust and confidence that led to a significant increase in the level of appreciation and participation of individuals and groups in peace and development activities and processes in the community.

Theorists supported the findings that working together towards a common goal builds trust and affective bonds between community members (Heyneman and Todoric-Bebic 2000; Heyneman 2003; Levy-Paluck 2007). Rebuilding of interpersonal and intergroup networks, trust and reciprocity, is crucial for sustainable peace (Colletta et al. 2001; Woolcock 2000) that bring great benefits to the people (Putnam 2000). Likewise, the creation of community mechanisms such as the peace and development communities (PDCs) or zones of peace was found to be the formula for people's participation and enabled the communities (UNDP 2004) to formulate their peace plans and barangay development plans as well as a venue for regular interfaith activities (Kessler 2010). Community-based capacity building activities that supported peacebuilding and human security initiatives had empowered the actors especially former MNLF combatants, thru improved efficiency, confidence and involvement in group work and activities aimed at bridging, bonding and binding them. The transformation of the actors at the personal and community level is critical in pursuit of achieving its common goals. The process of transforming individual and groups' ideologies and mindset would take longer; however, it is essential in search for self-determination and in improving good governance capacities of actors. Building the confidence of former MNLF combatants through capacity building is integral to their transformation process into peace and development advocates.

Providing basic social services to the communities served as an avenue for local leaders to improve their governance capacity including resource mobilisation for sustainability of such services. The Programme provided core shelters, water and sanitation facilities, health care and health facilities, education and children's playground along with organising and capacitating the individuals, groups and mechanisms to manage the facilities as well as in responding to challenges. In pursuit of attaining such targets, through the rigorous support of the communities and LGUs, it was able to achieve its targets as set in the logical framework of the Programme. With the improved capacities, there was a significant improvement in the access to and utilisation of basic services in PDCs. Improved health conditions was translated to additional income and time to improve the quality of life. The formation of community-based mechanism as an offshoot of the improved capacities acquired after the training is deemed imperative in sustaining prompt response to community needs and in promoting ownership and shared responsibilities in maintaining the facilities such as water system and health stations.

Moreover, community economic development (CED) initiatives facilitated a more cohesive community and enhanced capacities of actors on good governance particularly in the provision of support services and enabling policy. Reports revealed that improved access and capacities on CED were noted in successful livelihood and enterprise projects. Thus, capacities of communities to identify, implement and manage these projects that essentially resulted in self-actualisation are critical in ensuring the success of such initiative even after the end of the programme intervention. A novel project of the Programme involves former MNLF combatants which forged a final peace agreement with the Philippine government on September 2, 1996. The coco sugar production venture is managed by the Mindanawan Cocosugar Corp. (MCC), composed of the coalition of Peace and Development Advocates Leagues from South Cotabato, Sarangani, Davao del Sur, Sultan Kudarat mainland, General Santos City and Cotabato City. The transformation—from being former combatants to entrepreneurs—of these PDAs is a great testament that the efforts of the Programme in cooperation with other partners have succeeded in its aim of transforming lives. Legitimate economic opportunities that bring warring factions together can offer opportunities to re-establish relationships and trust (Gerstle and Meissner 2010). Some practitioners expressed interests on economic programming in addressing other root causes of conflict, such as ethnic tensions and elite capture of natural resources (Banfield et al. 2006).

Meanwhile, conflict management capacities of actors mainly the LGUs and traditional leaders were found critical in ensuring a sustainable peace and security conditions through the promotion of culture of peace including the Islamic leadership in Governance (ILG), and conflict sensitive processes in addressing conflicts and/or preventing from protracting to violence. Results showed the improved relationship among peace actors resulted in the forging of multi-sectoral partnership for peace constituency building. The breakthroughs are anchored on culture of peace through sustained interfaith activities. Moro-settlers atrocities in a certain PDC rooted from a land conflict was resolved through series of traditional conflict management sessions of community elders as triggered by the multi-stakeholders peace dialogue involving the Philippine National Police, Armed Forces of the Philippines, LGUs (i.e. provincial, municipal, barangay and sitio), religious leaders, citisen armed force geographical units (CAFGUs), and MNLF. Other atrocity of Moro-settlers in another PDC triggered by the Memorandum of Agreement on Ancestral Domain caused the revival of the ILAGA, a Christian militia. However, these were resolved through multi-stakeholders peace dialogues with the media's monitoring that produced traditional council of elders' support. It was disclosed that strengthened capacities on individuals, groups and institutions on conflict management and community transformation processes enabled them to be resilient during recurrence of armed conflicts and natural calamities. The enhanced abilities for and adherence to the culture of peace among former MNLF combatants were shown through active involvement in their communities as peace champions or peace advocates. Relatedly, the POs and LSFs such as the PDALs are now representing local structures and are implementing projects as well.

Results showed that through the interventions of the ACT for Peace Programme, peace platforms and mechanisms to manage local conflicts were created. Schools of peace were strengthened as among the venues to help promote the culture of peace and peace-based concepts through peace-based exemplars and lesson plans. Strengthened local mechanisms, systems and procedures are therefore necessary in managing local conflicts to prevent from protracting to structural violence, and a means in nurturing peace gains through the schools of peace. In conflict-affected situations, education is more than service delivery but a means of socialisation and identity development through the transmission of knowledge, skills, values and attitudes across generations (UN Report 2009). Ultimately, the PDC became a model that can systematise the transformation of conflict actors in the context of peace agreement. Peace education was also institutionalised through issuance of an executive order that mandated the Department of Education to integrate peace education in the curriculum of basic education. The LGU support in mainstreaming peace and development gains was shown through the issuances of local ordinances and resolution, adopting peace initiatives such as membership of youth in local peace and order councils.

Culture of peace is an effective tool in establishing trust and confidence among peoples in the community to broaden peace constituency across cultures. Traditional way of conflict management by the council of elders (CoE), however, still remains an effective way of resolving local conflicts. The ILG mainstreaming, and the pursuit of more intra and interfaith dialogues and consultations proved to be a predisposing and enabling factor in harnessing the spirituality of community residents and Muslim barangay leaders for peacebuilding. It also worked the same way in the mobilisation of community elders for conflict mediation and transformation.

Good governance in conflict context as an element of peacebuilding was revealed as the encompassing nexus of social cohesion, basic social services, community economic development and conflict management, that intertwined all the elements in achieving the overarching goal of the peacebuilding change theory. The purposive enhancement of the capacities of actors and partners particularly the former MNLF combatants and LGUs from barangay, municipal to provincial levels enabled them to effectively participate in the local structures and conferences. For example, the barangay officials were empowered to represent in conferences where they shared experiences, lessons and practices on enhancing their knowledge, skills and attitudes towards governing their respective areas. Also, the capacity of the LGUs to practice transparency and accountability in its planning, programming and resource mobilization indicated an improved governance capacity. Good governance practices in the private (corporate) and public (state) sectors, especially concerning policy, planning, decision-making, management and administration is critical in any endeavours (Pienaar 2009; Ferreira-Snyman and Ferreira 2006; Botha 2008). LGUs established the Provincial Peace Resource Centres as venues for knowledge and skills enhancement on peacebuilding. Strengthened and functional mechanisms and institutions established at the local and national levels are vital as a display of political support in sustaining the gains on peacebuilding and conflict transformation initiatives. Peacebuilding requires creating stronger state institutions, encouraging broader political participation, undertaking land reform, deepening civil society and respecting ethnic identities (Doyle and Sambanis 2000). Significant breakthrough of the ACT for Peace Programme included the good practices and mechanisms on peacebuilding initiatives that were mainstreamed and replicated outside the coverage area.

Findings in each peacebuilding strand were extracted from the results of assessment employing the four (4) dimensions of change. Anent to the findings from metasynthesis, were the key findings extracted from multiple case narratives and vignettes in each dimension of change. These findings exemplified the following: Strand 1. Social cohesion is deemed critical as bridging, binding and bonding agent for peacebuilding particularly in sustaining a cohesive environment and improved governance capacities of the local leaders and LGUs in conflict affected and vulnerable communities; Strand 2. Basic social services are critical instruments that provide a massive link for peacebuilding and good governance capacity of the LGUs and the local population; Strand 3. Community economic development serves as a vehicle that propels peacebuilding efforts towards a more sustainable environment of peace and security, and good governance; and, Strand 4. Conflict management capacities of state and non-state actors are critical in sustaining peace and security mechanisms that contribute to the improved governance capacities of local leaders and LGUs. Meanwhile, good governance in conflict context, as the fifth encompassing element of peacebuilding, serves as the nexus of the four strands that provides the basic foundation and avenues for the achievement of peacebuilding change theory. Noteworthy, findings from other programmes and projects implemented in, on and around conflict complemented the findings of the ACT for Peace Program as a peacebuilding model.

15.6 Conclusion and Recommendations

The ACT for Peace Programme was implemented, during the period 2005–2010, within complex environment characterised by unstable peace and security situation, imbalanced development and largely low human security conditions. The coverage area has varied levels of development requiring the adoption of context-specific interventions to ensure its relevance and responsiveness. The four (4) strands of peacebuilding, namely: social cohesion, basic social services, community economic development and conflict management which are interlinked altogether through the nexus of good governance are deemed critical in pursuit of stable peace and security conditions, and improved governance capacity of the LGUs. Seemingly, Strand 1—social cohesion is deemed critical as an agent in bridging, binding and bonding individuals, groups and institutions for peacebuilding and improved governance capacity. Strand 2—basic social services as a critical formula draws and strengthens people and institutions alike in pursuit of peacebuilding thru transformative process that contributes to the enhancement of governance capacity. Meanwhile, Strand 3—community economic development initiatives serve as a vehicle for peacebuilding and conflict transformation that contributes to the improvement of the local governance capacity of the local structures. Moreover, Strand 4—conflict management capacities of actors particularly the LGUs are very vital in contributing to the re-shaping of and sustaining the governance landscape in the conflict affected and vulnerable communities. Results showed that the culture of peace is an important factor that strengthened the foundation of conflict management capacities of actors.

Finally, findings revealed that the Peacebuilding Model for Southern Philippines consists of Strand 1 social cohesion, Strand 2 basic social services, Strand 3 community economic development, and Strand 4 conflict management is anchored on Good Governance as its ultimate strategy for success. Thus, any combinations of the peacebuilding strands in conflict context could work in fragile and conflict areas when intertwined to Good Governance. These results may help concerned authorities and peace and development players in discharging their duties and responsibilities with respect to peacebuilding and MDG efforts. Each strand consists of several sets of projects and activities as results of community consultations and dialogues. Creative strategies and innovative practices have to be constantly explored and shared to effect changes and strengthen weak and fragile institutions. Essentially, a policy framework from the national government that guides in responding to the needs and concerns of the communities in conflict affected and vulnerable communities must be anchored on strong institutions and policy support with dependable management capacities in particular good governance to effectively and efficiently deliver the needed services to the people.

Based on the foregoing results, the following imperatives which are categorised according to policy direction, program agenda and community level agenda may help the concerned authorities in discharging their duties and responsibilities with respect to peacebuilding and conflict transformation efforts in Southern Philippines:

15.6.1 Policy Direction

1. Revisit the country level (national) thrusts on peacebuilding and conflict transformation efforts for Southern Philippines. This will include the review of the Philippine Development Plan roadmap including the approaches and strategies of concerned agencies such as the office of the presidential adviser on the peace process (OPAPP), National Security Council, Cabinet Cluster on Peace and Security, among others.
2. Harmonize all donors, government agencies, LGUs, I/NGOs and private sectors initiatives on peacebuilding in Southern Philippines through appropriate agencies to ensure complementation of efforts.
3. Review the Philippine laws and other relevant issuances on donor assistance to address the issues on inclusivity and external influence on peacebuilding initiatives.
4. Capitalise public-private partnership (PPP) particularly the corporate social responsibility (CSR) as a source of funds from corporations and companies operating in Southern Philippines especially for basic social services and community economic development projects.
5. Maximise the role of local NGOs in promoting the conflict management systems, processes and practices as trainers or resource persons. In this view, there is a need to strengthen conglomeration of local resource partners in Southern Philippines that can ably handle training on conflict management, culture of peace, among others.
6. Help resolve the roots of conflict, specifically those pertaining to economic issues. Among the steps to be taken is for the national government, in partnership with the international community, to assist the Bangsamoro build a capable bureaucracy with streamlined and transparent procedures.

7. LGUs to institutionalise Council of Elders and scale up through capacitation as a formal mechanism in conflict management that will work in tandem with the Lupong Tagapamayapa (Barangay Justice System).
8. Allocate sufficient funds for Southern Philippines considering its very unique characterisation that needs sufficient fund to meet the MDGs in 2015 thus pushing up the areas which are still way below targets.

15.6.2 Program Agenda

1. Institutionalise the convergence of project management offices (PMOs) operating in Southern Philippines as a venue for complementation and harmonisation at the program level.
2. In terms of economic development projects, enhance economic development in communities where economic recovery has been successful and assist economic recovery in those communities where economic recovery has not yet been successful. Enhancing economic development in promising communities is a rapid way to increase Mindanao's regional gross domestic product, helping both government and business. Meanwhile, assisting economic recovery in poorer areas, especially the conflict affected and vulnerable communities, will reduce the uneven development that may threaten the stability of Southern Philippines in the medium to long term.
3. Set-up mechanism at the LGU level (e.g. province, municipal, barangay) for the continued capacitation of the peace and development advocates leagues (PDALs), peace and development advocates (PDAs) and barangay officials on conflict management, culture of peace and conflict sensitive and peace-promotive processes.
4. Implement and Institutionalise the Program Monitoring and Evaluation System to ensure complementation and partnership of initiatives between state and non-state actors in Southern Philippines.
5. Implement strategic communication and outreach plan to communicate the results of all programs, projects and activities implemented in, on, and around conflicts in Southern Philippines. This will facilitate addressing the image problem, thus attracting more investments and tourists.

15.6.3 Community-Level Agenda

1. Institute reforms and provide capacities to LGUs particularly from barangay to municipal levels as frontlines of the communities.
2. Capacitate the local actors on the culture of peace and conflict sensitive and peace-promotive (CSPP) processes to reach and involve all sectors of society, not only individuals but also groups, particularly families, teachers, civil

servants, young people, political leaders, policemen and soldiers, non-governmental and community organization.
3. Roll-out the CSPP monitoring and evaluation system at the community level to enhance transparency and improve governance.
4. Promote gender mainstreaming and environmental management and governance in all policies, programs, projects and activities in the communities as cross-cutting themes.
5. Support the establishment of peace centres or link with existing structures (e.g. but not limited to social action centres, and academic institutions).

15.6.4 Other Researchable Areas

1. Conduct a comprehensive study on the five (5) conflict lines in Southern Philippines in order to come up with explicit and appropriate strategies and approaches in addressing the more than 40-year persistent armed conflicts.
2. Conduct an impact evaluation on the contribution of community economic development initiatives to peacebuilding in Southern Philippines particularly those implemented in, on and, around conflict.
3. Conduct an in-depth study on the issue on inclusivity and elite capture of donor programs and projects in Southern Philippines.
4. Conduct a social research on natural resource conflicts in Mindanao particularly in Caraga and Central Mindanao to come up with a more scientific and data-based information on the complexity of the conflict in these areas.

References

Ashby P (2002) Child combatants: a soldier's perspective. Lancet 360:s11–s12
Asian Development Bank (2002) Guidelines for the financial governance and management of investment projects financed by the asian development bank. ADB, Manila
AusAID (2008) Southern Philippines peace, conflict and development analysis. Australian agency for international development
Baird M (2011) Service delivery in fragile and conflict-affected states, background paper, world development report 2011. World Bank, Washington, DC
Banfield J, Gündüz C, Killick N (eds) (2006) Local business, local peace: the peacebuilding potential of the domestic private sector. International Alert, London
Barron P (2010) Community driven development in post conflict and conflict affected areas: experiences from east Asia. University of Oxford, Oxford
Batson ED (2007) Registering the human terrain: a valuation of cadastre. National Defense Intelligence College Press, Washington, DC

Bell KC (2007) Good governance in land administration, plenary session III—responding to the global agenda—policies and technical aspects. FIG Working Week, Hong Kong, China SAR. The World Bank, Washington, DC, 13–17 May 2007

Berry C, Forder A, Sultan S, Torres MM (2004) Approaches to improving the delivery of social services in difficult environments. PRDE (DFID) working paper. Policy Division, UK Department for International Development, 3 Oct 2004

Botha MM (2008) JCRDL Botha MM can whistle-blowing be an effective good governance tool?. Journal of Contemporary Roman-Dutch Law

Boutros-Ghali B (1995) Agenda for peace, 2nd edn. United Nations Publications, New York

Carbonier G (1998) Conflict, postwar rebuilding and the economy: a critical review of literature. United Nations Research Institute for Social Development (UNRISD), Geneva, Feb 1998

Chang H-J (2002) Kicking away the ladder: development strategy in historical perspective. Anthem Press, London

Chipkin I, Ngqulunga B (2008) Friends and family: social cohesion in South Africa. J S Afr Stud 34(1). Human Sciences Research Council Library: shelf number 540

Church C, Shouldice J (2002) The evaluation of conflict resolution interventions, parts i: framing the state of play. INCORE, Derry/Londonderry

Church C, Shouldice J (2003) The evaluation of conflict resolution interventions, part ii: emerging practice and theory. INCORE, Derry/Londonderry

Colletta NJ, Cullen ML (2000) Violent conflict and the transformation of social capital: lessons from cambodia, Rwanda, Guatemala, and Somalia. The World Bank, Washington, DC

Colletta NJ, Lim TG, Kelles-Viitanen A (2001) Social cohesion and conflict prevention in Asia: managing diversity through development. The World Bank, Washington, DC

Collier P (2007) The bottom billion: why the poorest countries are failing and what can be done about it. Oxford University Press, Oxford

Deininger K (2006) Land policy reforms. In: Coudouel A, Paternostro S (eds) Analyzing the distributional impact of reforms, vol 1. The World Bank, Washington, DC

Department for International Development (DFID) (2010) The politics of poverty: elites, citizens and states. A synthesis paper findings from 10 years of DFID-funded. Research on Governance and Fragile States 2001–2010. Available at http://www.research4development.info/politicsofpoverty.asp

Department of Tourism (2011) Profile of Philippine tourist destinations, Manila, Philippines. Year-end report. Available at www.tourism.gov.ph/downloadablefiles//2011/DOT

Doyle MW, Sambanis N (2000) International peacebuilding: a theoretical and quantitative analysis. Am Polit Sci Rev 94(4):2000

Easterly W, Ritzen J, Woolock M (2006) Social cohesion, institutions, and growth. Economics Politics 18(2). Available at http://www.cgdev.org/content/publications/detail/9136

Ehrlich AH, Gleick P, Conca K (2000) Resources and environmental degradation as sources of conflict. Paper presented at 50th Pugwash conference on science and world affairs, Cambridge. http://www.studentpugwash.org/uk/WG5.pdf

Fearon JD, Humphreys M, Weinstein J (2008) The behavioural impacts of community driven reconstruction: evidence from an irc community-driven reconstruction project in liberia. Stanford University and Columbia University, Typescript

Ferreira-Snyman MP, Ferreira GM (2006) Global good governance and good global governance. South African Yearbook of International Law, pp 55–94

Fund for Peace (2013) Failed states index in 2013. Fund for peace, foreign policy. Available at www.failedstatesindex.org

Galtung J (1975) Three approaches to peace: peacekeeping, peacemaking, and peacebuilding. In: Galtung J (ed) Peace, war and defence—essays in peace research, vol 2. Christian Ejlers,

Copenhagen, pp 282–304 (in Miall et al (1999): Contemporary Conflict Resolution; Cambridge: Polity Press)
Galtung J (1996) Peace by peaceful means: peace and conflict, development and civilization. Sage, London
Galtung J (2002) Peace by Peaceful Means: peace and conflict, development and civilization. Sage Publication, London
Gerstle T, Meissner L (2010) Peacebuilding essentials for economic development practitioners. Practice note 1: market development in conflict-affected contexts. International Alert, London
Griffiths A, Barnes C (2008) Incentives and sanctions in peace processes. In: Griffiths A, Barnes C (eds) Powers of persuasion: incentives, sanctions and conditionality in peacemaking. Accord issue 19. Conciliation Resources, London
Hamad AA (2005) The reconceptualization of conflict management, peace, conflict and development. Interdisciplinary J 7, July 2005. Available at http://www.peacestudiesjournal.org.uk
Hamber B, Kelly G (2005) A place for reconciliation? conflict and locality in Northern Ireland no. 18. Democratic Dialogue, Belfast
Hart B (2008) Peacebuilding in Traumatized Societies. University Press of America, Lanham MD
Heyneman SP (2003) Education, social cohesion and the future role of international organizations. Peabody J Educ 78(3)
Heyneman SP, Todoric-Bebic S (2000) A renewed sense for the purposes of schooling: challenges of education and social cohesion in Asia, Africa, Latin America, Europe and Central Asia. UNESCO Prospects XXX(2):145–166, June 2000
Homer-Dixon T (1999) Environment, scarcity, and violence. Princeton Princeton University Press, Princeton
Homer-Dixon TF, Percival V (1997) Environmental scarcities, state capacity and civil violence: the case of Bihar, India, Toronto
IEP (2013) Global peace index 2013. Measuring the state of global peace. Institute for economics and peace. Available at http://www.visionofhumanity.org/#page/indexes/global-peace-index/2013
International Alert (1996) Conflict sensitive approaches to development humanitarian assistance and peacebuilding. International Alert, London
Kessler E (2010) Building bridges of hope: lessons from british faith-based, Humanitarian Aid Organisations, Woolf Institute, Cambridge. Building bridges of hope, Pontifical Gregorian University, Oct 2010
Lederach JP (1997) Building peace: sustainable reconciliation in divided societies. United States Institute of Peace Press, Washington, DC
Lederach JP, Neufeldt R, Culbertson H (2007) reflective peacebuilding: a planning, monitoring, and learning toolkit. Edited by University of Notre Dame and Catholic Relief Services Southeast Asia Regional Office. The Joan B. Krock Institute for International Peace Studies. The Joan B. Krock Institute for International Peace Studies, University of Notre Dame, Mindanao
Lewer N (1999) International non-governmental organisations and peacebuilding—perspectives from peace studies and conflict resolution. Working Paper Series, Centre for conflict resolution, Department of peace studies, University of Bradford, Working Paper 3
Miall H (2001) Conflict transformation: a multi-dimensional task. Berghof handbook for conflict transformation
Miall H (2004) Conflict transformation: a multi-dimensional task. In: Austin A, Fischer M, Ropers N (eds) Transforming ethnopolitical conflict: the berghof handbook. Berghof Research Center for Constructive Conflict Management, Berlin, p 2004
Mindanao Development Authority (2010) The act for peace school of peace (sop) assessment. MinDA, Philippines

Mindanao Development Authority (2011a) The action for conflict transformation (act) for peace programme terminal report. MinDA-UNDP, Philippines
Mindanao Development Authority (2011b) Mindanao 2020 Peace and Development Framework Plan. MinDA, Davao
Mindanao Economic Development Council (2006) A report on the multi-stakeholders' conflict analysis workshop of the act for peace programme. MEDCo, Davao
Mindanao Economic Development Council (2008) Integrated Peace and Development Framework. MEDCo, Davao
Mindanao Economic Development Council (2009) Conflict analysis and peace needs workshops of the mindanao working group. MEDCo, Philippines
Mindanao Peaceweavers (2010) Towards a Mindanao peoples' peace agenda. Mindanao Peaceweavers, Initiatives for international dialogue. MPW, Matina
Morris C (2000) What is peacebuilding: one definition. Available at http://www.peacemakers.ca/publications/peacebuildingdefinition.html
National Economic and Development Authority (2011) The Philippine development plan, 2011–2016. Official Gazette, Republic of the Philippines May 2011
National Statistical Coordinating Board (2010) Philippine national statistical yearbook, National statistical information centre (NSIC). NSCB, Makati
National Statistical Coordinating Board (2011) Philippine national statistical yearbook, national statistical information centre (NSIC). NSCB, Makati
Ocheje P (2007) In the public interest: forced evictions, land rights and human development. J Afr Law 51(2) Oct 2007. School of oriental and african studies, Cambridge University Press
OECD (1997) The OECD report on regulatory reform: synthesis. Organisation for Economic Cooperation and Development, Paris
OECD-DAC (2008) Evaluating conflict prevention and peacebuilding activities factsheet 2008. Organization for Economic Cooperation and Development, Paris
Paluck EL (2007) Reducing intergroup prejudice and conflict with the mass media: a field experiment in Rwanda. Households in conflict network working paper. Harvard University, Weatherhead Center for International Affairs, Paluck. Available at http://www.hicn.org/wordpress/wp-content/uploads
Pavanello S, Darcy J (2008) Improving the provision of basic services for the poor in fragile environments, international literature review, synthesis paper. ODI Humanitarian Policy Group. Prepared for AusAID/ODE, Dec 2008
Philippine Official Gazette (2014) The comprehensive agreement on the bangsamoro. Official Gazette, Republic of the Philippines Mar 2014
Pienaar G (2009) Aspects of land administration in the context of good governance. 2009/2 Potchefstroom Electronic Law J (PER), pp 15–56.
Putnam R (2000) Bowling alone: the collapse and revival of american community. Simon & Schuster, New York
Reimann C (2001) Towards conflict transformation: assessing the state-of-the-art in conflict management-reflections from a theoretical perspective in Berghof handbook for conflict transformation at http://www.berghof-center.org/handbook/index.html
Ritzen J, Easterly W, Woolcock M (2000) On good politicians and bad policies—social cohesion, institutions, and growth. World Bank policy research working paper series, no. 2448. World Bank, Washington, DC
Rupesinghe K (1995) Conflict transformation. St Martin's Press, New York, p 270
Schon D (1983) The reflective practitioner. Basic Books, New York
Sen A (1999) Development as freedom. Oxford University Press, Oxford
Shapiro I (2002) Mapping theories of practice and change: a comparative analysis of intervention programs addressing racial and ethnic tensions in U.S. communities. George Mason University, Fairfax

Shapiro I (2005) Theories of change. beyond intractability. Guy Burgess G, Burgess H (eds) Conflict research consortium. University of Colorado, Boulder
Shapiro I (2006) Extending the framework of inquiry: theories of change in conflict interventions. Berghof handbook dialogue Nr. 5. Berghof Center for Constructive Conflict Management, Berlin
Specht I (2010) Peacebuilding essentials for economic development practitioners. Practice note 4: socio-economic reintegration of ex-combatants. International Alert, London. ISBN 978-1-906677-55-8
Smith D (2004) Towards a strategic framework for peacebuilding: getting their act together. Overview report of the joint utstein study of peacebuilding. The Royal Norwegian Ministry of Foreign Affairs. Norway International Press Centre, Norway
Todorovski D (2011). Characteristics of post-conflict land administration with focus on the status of land records in such environment, facing the challenges—building the capacity. Marrakesh, Morocco, Republic of Macedonia, 18–22 May 2011
UNEP (2009). From conflict to Peacebuilding. The role of natural resources and the environment. United Nations Environemnt Programme P.O. Box 30552, Nairobi, Kenya, 2009. Available at http://unep.org
United Nations (1995) Copenhagen declaration on social development, World summit for social development. Copenhagen, Denmark, United Nations, 6–12 Mar 1995
United Nations (2009) General assembly, report of the secretary-general on peacebuilding in the immediate aftermath of conflict. United Nations, New York, 11 June 2009
United Nations Development Programme (1997) Human development report, UNDP. Oxford University Press, Oxford
United Nations Development Programme (2004) An evaluation of the GOP-UN multi-donor programme: assessment of performance and results. GoP-UNDP, Philippines
United Nations Development Programme (2005) ACT for peace programme document. UNDP, Philippines, May 2005
United Nations Economic Commission for Europe (UNECE) (2003) The economic survey of Europe, vol 1, Chap. 5. Economic Commission for Europe, Geneva, United Nations
USAID (2003) Philippines conflict assessment. United States of America for International Development
USAID (2004) Land and conflict—a toolkit for intervention. USAID—Office of Conflict Management and Mitigation, Washington, DC
USAID (2009) Governance: decentralization and democratic local governance programming handbook. USAID, Washington, DC
Uvin P (1999) The influence of aid in situations of violent conflict. Organization for Economic Cooperation and Development, Paris
Van der Zwan J (2010) The need for conflict-sensitive land policy and land governance in Africa. Initiative for Peacebuilding. International Alert, London, Dec 2010
Vellema S, Borras SM Jr, Lara FJr (2011) The agrarian roots of contemporary violent conflict in Mindanao, Southern philippines. J Agrarian Change 11(3):298–320, July 2011
Wennmann A (2010) Peacebuilding essentials for economic development practitioners. Practice note 5: supporting the economic dimensions of peace processes. International Alert, London. ISBN 978-1-906677-56-5
Weiss CH (1972) Evaluation research: methods of assessing program effectiveness. Prentice-Hall, Englewood Cliffs
Weller C, Kirschner A (2005) Zivile Konfliktbearbeitung—Allheilmittel oder Leerformel? Möglichkeiten und Grenzen eines viel versprechenden Konzepts, in Internationale Politik und Gesellschaft, 4, pp 10–29, (www.fesportal.fes.de/pls/portal30/docs/FOLDER/IPG/IPG4_2005/03_WELLER.PDF)
Winters MS (2008) The impact of government accountability on foreign aid effectiveness. Working Paper, Niehaus Center for Globalization and Governance, Princeton University
Woolcock M, Narayan D (2000) Social capital: implications for development theory, research, and policy. World Bank Research Observer 15(2), Aug 2000

World Bank (2000a) Beyond economic growth: meeting the challenges of global development, glossary. World Bank, Washington, DC

World Bank (2005) Joint needs assessment for reconstruction and development of conflict affected areas in mindanao. World Bank, Washington, DC

Zakout W, Wehrmann B, Torhonen MP, FAO (2009) Good government in land administration: principles and good practices. Available at http://www.fao.org/docrep/011/i0830e/i0830e00.htm

Chapter 16
Community Governance for Climate Change Adaptation in Nepal

Tek Nath Dhakal and Ishtiaq Jamil

16.1 Background

Conceptually governance is taken as a process of accelerating public affairs in the interest of people and also for intra-/inter-institutional cooperations and institutional arrangements in policymaking, development, and service processes to pursue collective interest. Governance discourse also seeks more about coordination, coherence, and collaborative networks among the wide variety of actors such as public–sector organisations, corporations, private organisations, civil society organisations, and international organisations (Pierre and Peters 2000). Thus, it has been realised that the solutions should be sought at various levels among different institutions and the sectors. Community forest user groups (CFUGs) as one of the actors of community governance can play effective roles especially to initiate people's participation. CFUG as the civil society organisations could have both 'conflicting' and 'interdependence' and 'partnership' relations with government (Salamon & Anheier 1998). In case of sharing expertise, experiences, and resources, there could be a situation of complementing one another. In such situations, CFUG could not be alternatives to the state in the forest governance, but rather the state and the CFUG could grow in parallel and even in cooperation with each other for expansion of public goods and work hand to hand.

Up to the half of the twentieth century, the forests resources in Nepal used to be considered "*hariyo ban Nepal ko dhan*" (green forest is Nepal's property). The main

T.N. Dhakal (✉)
Central Department of Public Administration, Tribhuvan University, Kathmandu, Nepal
e-mail: dhakaltn@gmail.com

I. Jamil
Department of Administration and Organisation Theory, University of Bergen, Bergen, Norway
e-mail: Ishtiaq.Jamil@aorg.uib.no

H. Ha (ed.), *Land and Disaster Management Strategies in Asia*,
DOI: 10.1007/978-81-322-1976-7_16

reason behind this is that forest resources were the main sources of livelihood of the people who depend on forest resources and also one of the important sources of revenue of the government. In those days, almost 90 % of the total population relied on forest for fulfilling their household fuel need, fodder of animals, and construction materials. As an essential good of the daily lives, the local communities used to involve in the management of forest resources for both consumption of the forest product and also protection of the forests around their locality. More specifically, people's participation could also be found for governing the forest in Nepal.

In the early 1960s, the Government of Nepal had adopted some major policy shift to govern forest by nationalising the private and the community-managed forests and adopted some legal measures to giving absolute powers to the forest related government institutions. The fast decline of forests in Nepal started after nationalisation of forest in early 1960s, and particularly the 1970s deforestation reduced the total forest area to just 29 %. In addition, it had also affected the ecosystem and became the cause of climate change in Nepal. As a result, there has been a hue and cry for demanding people's role in the management of forest in Nepal.

The present share of forest coverage is nearly 40 % of the total land in Nepal. Despite government's key role in managing the forests, around 28 % of the total forest area is managed by 'CFUGs' at the local level (MoFSC in www.mof.gov.org). This indicates that communities had also high concerns for the protection and conservation of forest of the country. Such efforts not only support their livelihoods at community level but also contribute to the national economy and also can support for managing the climate change adaptation initiatives.

In this context, this paper discusses to what extent the CFUGs are contributing to govern the forest in Nepal. It also explores the strengths and weakness of CFUG governance to adapt climate change in Nepal. For the purpose of study, 30 CFUGs were selected from two village development committees (VDCs)—9 CFUGs from Hekuli VDC in Dang district of the Mid-Western Development Region and 21 CFUGs from Markhu VDC of Makwanpur district in Central Development Region of Nepal. These two districts—Dang and Makwanpur—have 440 CFUGs for managing the community forest. Total 102 general member of the community were also asked regarding the benefits of community forest from these two VDCs.

16.2 Forest Governance in Nepal

For managing the forest in Nepal, the Department of Forests (DoFs) was established as a state agency, for the first time in 1942. Until 1950s, a number of forest lands were given to the private individuals comprising the near and the dear ones of the then ruling class. The Department was established in order to protect the forest resource of the country. In fact, the department did not have the mandate to accommodate communities or the local people in forest protection. This initiative obviously made the foresters more powerful. From the 1950s, the Government initiated a number of legal instruments and policies to govern the forest in Nepal (see Table 16.1). It was only in the mid-1980s, the government had initiated the partnership approach to govern forest in Nepal (Acharya 2002). Before that, i.e. up to the end of the 1970s, the government's approach was coercive and centralised for managing the forests in Nepal.

As stated by Kanel (n.d.)

> To strengthen state control over the forest management process the private forests were nationalised with the promulgation of *Private Forest Nationalisation Act* in 1957. This act legally centralised the authority of forest management in the Terai (lowlands) and Hills under state control, although local communities were informally managing certain plots of forests adjoining to their settlements. Furthermore, special rights to issue permits even to harvest trees for household purposes were assigned to forest officials with the promulgation of the *Forest Act* of 1961. This Act further strengthened the authority of forest officers to take legal actions. Moreover, *Forest Protection (special arrangements) Act* was promulgated in 1967 to provide additional power to state foresters for taking legal action to the offenders. It also established one-person special court run by the Divisional Forest Officer (p. 4).

This legal provision did not accept the role of local people in the management of forest up until the late 1970s (Kanel 1993). Looking into the deteriorating situation of forest, especially in the 1970s, the Government again slightly back-tracked in its policy initiatives and brought 'National Forest Plan' (NFP) in 1976 which opened windows for the people's participation in the management of forest in Nepal (Kanel et al. 2005; Kanel and Niraula 2004). Incorporating the spirit of

Table 16.1 Policy and legal measures adopted for forest management in Nepal

Year	Policy and legal instruments	Remarks
1957	Private forest nationalisation act	Transfer of private management forest to the Government
1961	Forest act	Forest categorisation, forest official empower
1967	Forest protection act, special provision	Entrusted judicial power to forest officials
1976	National forestry plan	Concept of PF and PPF initiated recognising people's participation
1977	Amendment in forest act	Provision of PF and PPF in the Forest Act
1978	PF and PPF rules	Handover of some forest to PF and PPF
1982	Decentralisation act	Promotion of Users' Committee
1987	Revision of PF (panchayat forest) and PPF (panchayat protected forest) rules	Provision of User's committees for forest protection
1989	Master plan for the forestry sector	Incorporated the concept of CFUG
1993	Forest act	More rights given to CFUGs
1995	Forest rules	Specifies more rights and duties of CFUGs, forest officials' role to be facilitator
1999	Revision of forest act	25 % to be spent in forest activities
2000	Forest policy 2000	CFUGs in Terai to give 40 % of income from sale of surplus timber to government

Source HMGN (1956, 1961, 1967, 1973, 1978, 1996, 1993, 2000), NPC (2002); also see Kanel (n.d.)

NFP, the government amended the existing 'Forest Act 1961' in 1977 which also made provisions to hand over part of government forests to a local "Panchayat", a local politico-administrative unit. The government also had brought panchayat forest (PF) and panchayat protected forest (PPF) Rules 1978 which initiated the implementation of the community forestry programme in Nepal. The PF and PPF rules clearly specified provisions for the transfer of responsibility for forest management from the government to the local Panchayat as PF and PPF (Joshi 1993; Bartlett 1992; Kanel n.d.). Again,

> For better forest management and implementation of these rules, the government launched several community forestry projects with international support. One of the main early projects was the World Bank funded Hill Community Forestry Project, which operated in 38 hill districts. Other implemented projects were the Nepal Australia Forestry Project (in two hill districts), and the Integrated Rural Development Projects with forestry components in the Koshi Hills, Rapti, Rasuwa and Nuwakot Districts in the central and eastern Hills (Kanel n.d., pp. 4–5).

The government again revised the PF and PPF in 1987 specifying the role of users' committee for forest protection. This, in fact, initiated the process forming CFUGs in Nepal.

Looking into the long-term perspectives for managing the forestry sector, the government brought 'Forest Master Plan 1989' in Nepal (HMGN 1989). One of the important provisions specified in this Master Plan was about the role of CFUGs for protection, conservation, and utilisation of forest resources at community level. To implement the Forest Master Plan, the Government enacted 'Forest Act 1993' which specifies the authorities and the responsibilities to be given to the CFUGs for managing forest at community level. Under this Act, the Government brought 'Forest Rules 1995' which specifies more rights and duties of the CFUGs and the role of forest officials have been taken as facilitator of the CFUGs. In 1999, the Government revised the existing 'Forest Act 1993'. The new revised Forest Act added new provisions with regard to CFUGs, especially on spending up to 25 % of the total revenue generated from the forest products and the activities of CFUGs. Finally, the government brought 'Forest Policy 2000' which also specifies about the importance of CFUGs in the protection of forest in Nepal. The policy also made a provision of paying revenues to the Government up to 40 % of the total income earned by CFUGs. Such provision mainly become mandatory particularly in Terai-based CFUGs.

16.3 Institutional Involvement for Managing Community Forestry

There are many institutions involved in Nepal to support the community forestry programme. Some of them are governmental organisations; others are civil society organisations and private institutions. The Community Forestry Federations (CFF) is the key umbrella organisation of CFUGs for mobilising, facilitating, and activating the CFUGs. There are other networks of development partners or the donors who are also involved directly or indirectly in this initiative.

16.3.1 Role of Public Sector Organisations

The Ministry of Forests and Soil Conservation (MoFSC) is the focal ministry of the Government of Nepal for formulation of policies, legislation, and supervision and monitoring of community forestry activities in the country (Kanel n.d.). Under the MoFSC, "there are five Regional Directorates of Forests which are responsible for planning, coordinating and evaluating forestry activities including planning and monitoring of community forests. The regional directorates also participate in the project steering committee of various community forestry projects" (Kanel n.d., p. 6). The DoF supports the community forestry through its Community Forest Division to implement and coordinate community forestry plans and programmes, help to maintain uniformity in implementations, and to provide technical support for the better management of community forests. There are 74 District Forest Offices (DFOs) under the DoFs which are the key actors to facilitate and monitor the CFUGs. The major activities of the DFO related to community forestry include "handing over of community forests to user groups, support user groups by providing technical skills in field level activities, prepare forest inventory and operational plan, and monitor activities of the user groups" (Kanel n.d., p. 7). In addition to the abovementioned governmental offices, other offices such as the Department of Forest Research and Survey, the Department of Plant Resources, and some of the academic and training institutions support the community forestry sector in Nepal.

16.3.2 Role of Civil Society Organisations

There are many civil society organisations involved in community forestry in Nepal which directly or indirectly supports to advocate, train, disseminate information, and create awareness including pro-poor campaign, etc. These are New Era, Nepal Forum for Environmental Journalists (NEFEJ), Nepal Foresters' Association (NFA), the WATCH, etc. In addition, "the Federation of Community Forestry Users, Nepal (FECOFN) works as an umbrella organisation of CFUGs represented in all 74 districts. FECOFUN has been active in advocating policy formulation and amendments in favour of community forestry. It is also equally advocating in providing rights to the users in decision making and benefit sharing" (Kanel n.d., p. 10).

In addition some of the private organisations are also involved in community forestry-related strategic plan preparation, operational plan preparation, community forestry research and studies, and work as service providers. "The MoFSC has also developed a guideline for acquiring services from such private sectors" (Kanel n.d., p. 9).

16.3.3 Role of Donors and the Development Partners

A number of international development partners such as Danida, Department of International Development, UK (DFIF), SNV Netherlands Development Organisation, AusAID, Swiss Agency for Development and Cooperation (SDC), CARE Nepal, International Union for Conservation of Nature (IUCN) and WWF are involved for supporting the community forestry. In addition, some of the international institutions such as Regional Community Forestry Training Centre (RECOFTC), International Centre for Integrated Mountain Development (ICIMOD), National Trust for Nature Conservation (NTNC), and Centre for International Forestry Research (CIFOR) are also found for training, research, and awareness building at global and national level regarding community forestry.

16.4 Community Forest User Groups: Key Actors for Community Forestry Management

In Nepal, the size of land coverage by forest is 5.83 million hector (39.6 % of the total land of the country), and, of this, dense forest occupies 4.27 million hector and the shrubs area cover 1.56 million hector (see Table 16.2).

As of June 2013, there are 17,810 CFUGs established across the country. These CFUGs manage 1.66 million hectors of forests, which is 28.47 % of the total forest (MoFSC 2013). For this, a huge number of people and households are involved for managing the community forest. Altogether 2.21 million households are directly involved in the CFUGs which accounts to more than 40 % of the total population. This shows that Community Forest Users' Groups are ever increasing grassroots institutions in Nepal and give a notion of a people oriented programme.

16.4.1 CFUGs' Group Formation Process

The Forest Act (1993), Forest Rules (1995), and the Community Forestry Guideline (2002) give clear direction to follow the process for creating the community forestry users' groups in Nepal. The interested villagers should complete the following five steps to form such groups as proposed by Kanel (n.d.) below:

(1) "Identification of users and formation of user group;
(2) Formalisation of CFUGs constitution;
(3) Preparation of operational plan;
(4) Implementation of operational plan; and
(5) Revision and update of operational plan" (p. 26)

Table 16.2 Coverage of forest area and HHs involvement in community forestry

S/No.	CFUG managed CF	Total unit	Remarks
1	Total forest area	5.83 Million hector	The area is 39.6 % of the total land coverage of the country
1.1	Dense forest area	4.27 Million hector	The forest area occupies 29 % of the total land
1.2	Shrubs area	1.56 Million hector	The shrubs area occupies 10.6 % of the total land
2	Forest area handed over to CFUGs	1.66 Million hector	28.47 % of total forest area
3	Total no. of CFUGs	17,810	Scattered all over the country
4	Total no. of HHs involved in CFUGs	2.21 Million	Which covers around 41 % of the total HHs (5.42 million) of the country

Source Ministry of Forest and Soil Conservation. Also see Ministry of Finance Economic Survey, 2013, www.mof.gov.np. Accessed on 4th May, 2014)

Usually, the community people should identify the users and form user group. For this, they have to identify potential community forest and its boundaries and the resources and the status of usages of the proposed community forests. The group should discuss the proposed community forests among the households. It has to prepare the draft list of users and categorise the users in different social strata to make final decision on user group members or users. Usually, the system is that the member of the community forestry group is composed on the basis of one-member-one house. Once the groups are formed, and the proposed forest is identified, the groups should make CFUG constitution specifying its objectives, working modalities with operational rules, and the formation of executive committees. The constitutions also should specify the inclusion of women and disadvantaged groups, etc. In addition, it requires to make the operational plan that guides to manage their forest and for benefits sharing schemes. Specifying all these, the group should apply to the DFO for the approval, and the District Forest Officer provides approval after inspection. It often monitors the CFUGs under their respective districts on the basis of operational plan.

16.4.2 Discussion of CFUG's Roles in Climate Change Adaptation in Nepal

During the past three decades of community forest implementation in Nepal, the size of community-managed forest has been increasing. In early 1980s, when the CFUGs were piloted in some selected districts, the total coverage of land by forest was only 29 %, which at present occupies nearly 40 %. In this period, the government has changed its legal and policy provisions giving emphasis to communities to manage the forest at the local level. On the other hand, a lot of enthusiasm from among the communities can easily be noticed as the number of CFUGs has dramatically increased over the period. The community forestry governance is based on win–win strategy as it provides benefits to the communities who take care of the forest and also to the national government for better protection of the forest. As a result, the end product is the community forestry conservation initiative which is contributing to maintaining the ecosystem and contributing to climate change.

16.4.3 Contribution to Household Benefits

With regard to understand the people's involvement to protect forest, some 102 general members from among the community members in Hekuli VDC and Markhu VDC of the selected fields were asked to rank about their perceived benefits for joining the CFUGs. The responses are ranked as per their perceptions. Accordingly, villagers, i.e. the household members, found their involvement as the first choice for involving in the community forestry users' group for fulfilling subsistence needs such as fuel wood and animal fodder (see Table 16.3). Similarly, the income generating opportunities to the people stands as second best option for joining the CFUGs. As 85 % of the people still depend on firewood for cooking for at homes and animals mostly depend on forest fodder, their choice for joining such CFUGs is obvious. Similarly, as given in Table 16.3 'income generation' stands as second rank in the list. This is mainly because farmers make 'halo' (wooden plough) from the forest tree which is one of the key instruments of agriculture farming system in Nepal. Forest is also considered very important resource to support and promote animal husbandry—which has also marketing value to the farmers ultimately help to contribute income generation. As people have found CFUGs as a common platform to be united at local level and have control over the forest resources at the local level, their joining such groups were found to be interest-based. Such reasons were also found interested among the community members who stand as 3rd and 4th ranking.

Normally, all the households within a village become the members in the CFUGs which also open the door to the women and the disadvantage people and

Table 16.3 Perception on the use of community forestry ($N = 30$)

Statement	Ranking
Fulfilling subsistence need such as fuel wood, animal fodder, etc.	1
Income generation	2
Uniting the people	3
Establishment of rights over resources	4
Participation of women and other minority groups	5

Source Field Survey

Question Based on the given statement in the table, please rank as per the importance for joining CFUGs (ranking 1 for the most important and 5 for the least important)

enhance a feeling of inclusion within the community. When the government had all responsibilities to govern the forest of the country, people often had limited access to obtain such facilities. The government's control over forest resources for basic uses at household level created apathy at community level to support forest management initiatives.

But the approach adopted for involving the people for managing the forest—both protection and utilisation—has created ownership feeling among the people. One of the reasons of people's attraction to joining CFUGs is that it helps to establish traditional rights of the local people over their resources. The legislation itself also specifies that that local people have rights and duties to manage community forests. Of the total CFUGs, around 6 % CFUGs are managed by women; however, the adoption of inclusive approach of bringing women and the disadvantaged groups in the management is still negligible.

16.4.4 Contribution to Forest Protection

With regard to analyse the CFUG's role to contribute protection and conservation of forest, the sample CFUGs representatives were asked to rank on the basis of the contribution of the CFUGs. Of the total five statements, the creation of CFUGs for protecting forest is ranked as number one by the respondents (see Table 16.4). Following this, the CFUG has become one of the important local entities to take care of forest in the community. As a result, development of networks stands as 2nd ranking in the eyes of the respondents. Such networks are established at various levels ranging from range post to district and national levels. FECOFUN is the umbrella organisation of CFUGs and stands as one of the strongest networks to operate as pressure groups for formulating favourable policies and promoting good governance within the community forestry programme (Kanel n.d.). Respondents stated that utilisation of forest products for community development as the 3rd rank provides home to the animals and protection of environment as the 4th and the 5th ranking, respectively.

Table 16.4 Usefulness of CFUGs for protecting community forest ($N = 30$)

Statement	Rank
Protection of forest	1
Development of networks	2
Utilisation of forest products for community development	3
Provide home for wild animals	4
Protection of environment	5

Source Field Survey

Question Based on the given statement in the table, please rank as per the usefulness of CFUGs for protecting community forest (ranking 1 for the most important and 5 for the least important)

Altogether 28 % of national forest has been handed over to the 17,810 local CFUGs. This shows that the forest area has been increasing and more involvement of people are also found for managing the community forest. The forest protection and judicious use of forest product can be taken as an appropriate choice to manage the forest. According to Kanel (n.d.), "the achievements of the community forestry can be seen in terms of better forest condition, better participation of the people and income generation for rural development and institutional building at grass root level" (p. 30).

16.5 Some Issues

Although community forestry has been considered a widely accepted programme at people's level, and the impact can visibly be seen both to the beneficiaries and to the expansion of forest and maintenance of greenery, many challenges also lie ahead of the programme. Any lack of addressing these challenges in time may create a number of problems for effectively managing these entities. Discussion with the respondents revealed a number of issues both caused by CFUG itself and by the government procedures.

16.5.1 Issues Created Within the CFUGs

The management of CFUGs depends mostly on trust of the people. In the absence of trust among the people and with the government, the management becomes difficult. A number of CFUGs particularly in the Terai-based CFUGs are found to be selling wood logs illegally which create difficulty to maintain the trust of the people.

In some cases such as in Terai, the distance of settlement from the community forest creates difficulty to get benefits from the community forest products and also monitor the forest resources. Such situation creates apathy among the members living longer distance from the forest. Including and involving distant users in community forestry has been a topic of long debate and also remains a contentious issue (Kanel n.d.).

Many CFUGs fail to develop operational plan, which is a mandatory requirement of forest inventory in the year 2000. Due to the inability of making such plan, they also face problems for the effective implementation of CFUGs activities. Often, the CFUGs are found to be involved in political activities and/or used for political purposes. Such situation impedes smooth operation of these groups, and, on many occasions, they depend on DFO for preparing the operational plans.

The access of benefit from the community forest is also not equally and judiciously managed. Compared to small size family, families having less number of livestock may get less benefit than the large families and those who have more number of livestock. CFUGs are still facing challenges in addressing the concerns of weaker section and disadvantage groups of the community. Thus, "making these institutions more accountable and responsible towards poor, disadvantaged groups and women is still considered as one of the main challenges in community forestry" (Kanel n.d., p. 32).

16.5.2 Issues Created from External Reasons

There is a legal issue to take responsibility of managing natural resources within the given VDC or DDC (HMGN 1999). The Local Self-Governance Act 1999 and its Regulation 2000 specifies the jurisdiction of forest belonging to the local government, and, on the other hand, Forest Act 1993 and Forest Rules 1995 authorise CFUGs to manage the forest plot within the boundary the specified VDC. Though both entities—VDC/DDC and the CFUGs–are legal entities, however, conflict of interest especially to get benefit from the given plot of forest often arises. The duplication of legal procedures often creates problems for managing community forest. In addition, there is another issue of revenue sharing among local people, community forest user's groups, local bodies such as VDC and DDC, and the government as the government also charges some amount (e.g. 40 % of the sale of Saal and Khayar wood of the Terai).

The implementation of CFUGs is mostly done by community members who are less educated. The development of forestry needs more research and training, which is lacking in any programme of community forestry management. Until now, the educated mass is still not interested for managing community forest, and the transformation of technical forestry into the forestry of local people has been lacking in Nepal. Due to the expansion of forest areas, the number wildlife has also been increasing. The fight between people and the wildlife grazing in these forests is also a major problem. Wildlife farming can be an additional benefit from the community forestry. But due to the lack of clarity in the legal procedure, the potential benefits are yet to be explored.

Finally, Kanel (n.d.) commented that "the community forestry process is facing problems of appropriate monitoring and evaluation systems in terms of its impact on poverty reduction. The linkages between results and impacts, and poor documentation or synthesis of CF lessons learned; and week non-participatory approaches are some key areas for consideration" (p. 34).

16.6 Conclusion

This chapter has discussed forest governance via community forestry. In Nepal, by 2013, around 28 % of total forest areas are managed under community forestry schemes where a huge number of people, i.e. 40 %, are involved for managing these CFUGs (Ministry of Forests and Soil Conservation 2009). The last three decades' efforts of community forestry show that people are interested to participation in order to manage the community forest which generates direct benefits to the CFUG households on one hand, and, on the other hand, the forest areas are increasing which gives a visible good result. In this context, the community forestry implementation in Nepal is found to be successful in managing forest in a sustainable manner and also achieving some objectives in the plan for the climate change adaptation. Community forestry, to some extent, provides some socio-economic benefits for local residents.

However, some issues related to land ownership, technical knowledge, legal framework, technical problems, clandestine loggers, market access, infrastructures development, managerial skills, etc. have been observed in this case (Pandit and Bevilacqua 2011). In addition, sufficient protection of community forests in a sustainable manner for the economic returns and promotion of tourism and conservation has also stood as question of debate (Pandit and Bevilacqua 2011).

Besides handing over forest resources to local communities as one of the best ways to conserve, manage, and utilise this resource, it is also important to establish cordial relations and collaborations between local governments and forest communities seems to be a key point for better community forest management. Then, only the win–win situation can be created which benefit both community members and also nationally and globally contributing the climate change adaptation. On the other hand, it helps in the conservation of forest which is also a national need. For this, a successful legal and institutional framework is needed to be incorporated and strengthened to enable the CFUGs to be more viable institutions. It is also important to disseminate and transfer locally appropriate practices for which conservation education need to impart at academic and training institutes.

References

Acharya KP (2002) Twenty-four years of community forestry in Nepal. Int For Rev 4(2):149–156

Anheier HK, Salamon LM (eds) (1998) The non-profit sector in the developing world. Manchester University Press, Manchester

Bartlett AG (1992) A review of community forestry advances in Nepal. Commonwealth Forestry Review, 71(2):95–100

DoF (2002) Community forestry program implementation guidelines (2000/2001). Kathmandu: Department of forest

HMGN (1956) Private forest nationalisation act 1956. Ministry of Law and Justice, His Majesty's Government of Nepal

HMGN (1961) Forest act 1961. Ministry of Law and Justice, His Majesty's Government of Nepal

HMGN (1967) Forest protection act 1967. Ministry of Law and Justice, His Majesty's Government of Nepal

HMGN (1973) National parks and wildlife conservation act 1973, fourth amendment. Ministry of HMGN (1976) National Forestry Plan 1976 (2033), Ministry of Forests and Soil Conservation

HMGN (1978) Panchayat forest and panchayat protected forest rules 1978. Ministry of Law and Justice, His Majesty's Government of Nepal

HMGN (1989) Master plan for the forestry sector, forest sector policy, 1989. Ministry of Forests and Soil Conservation and Law and Justice, His Majesty's Government of Nepal

HMGN (1996) National parks and wildlife conservation rules 2052. Ministry of Law and Justice, His Majesty's Government of Nepal

HMGN (1993) Forest act 2049. Ministry of Law and Justice, His Majesty's Government of Nepal

HMGN (1995) Forest rules 2051. Ministry of Law and Justice, His Majesty's Government of Nepal

HMGN (1999) Local self-governance act, 1999. Ministry of Local Development, Kathmandu, Nepal

HMGN (2000) Forest sector policy 2000. Ministry of Forests and Soil Conservation, Kathmandu, Nepal

Joshi AL (1993) Effects on administration of changed forest policies in Nepal. In: Proceedings of a workshop on policy and legislation in community forestry, 27–29 Jan 1993, RECOFTC, Bangkok

Kanel KR (n.d.) Current status of community forestry in Nepal. A paper submitted to Regional Community Forestry Training Centre for Asia and the Pacific Bangkok, Thailand

Kanel KR (1993) Community forestry and the 1993 forestry legislation: implications for policy and implementation. BankoJanakari 4(1):2–5

Kanel KR, Niraula DR (2004) Can rural livelihood be improved in Nepal through community forestry? BankoJankari 14(1):19–26

Kanel KR, Poudyal RP, Baral JC (2005) Current status of community forestry in Nepal. Regional Community Forestry Training Centre for Asia and the Pacific Bangkok, Thailand

Ministry of Forests and Soil Conservation (2009) Nepal forestry outlook study. Working paper no. Apfsos Ii/Wp/2009/05, Food And Agriculture Organisation Of The United Nations Regional Office For Asia And The Pacific, Bangkok, Thailand

Ministry of Forest and Soil Conservation (2013) The economic survey 2013. (www.mof.gov.org)

Pierre J, Peters BG (2000) Governance, politics and the state. Macmillan, Houndmills

Pandit R, Bevilacqua E (2011) Social heterogeneity and community forestry processes: reflections from forest users of Dhading District. Nepal Small-scale For 10(1):97–113

Chapter 17
Land Use and Disaster Governance: Implications and Policy Recommendations

Huong Ha and Tek Nath Dhakal

17.1 Introduction

There are three main themes which the authors of 16 chapters in this edited volume have endeavoured to draw the attention of readers, namely (i) land use and planning, (ii) disaster risk reduction and management and (iii) associated issues such as food security, public education, climate change, disability issues, geopolitics of disaster governance and peace. First of all, the way man uses land would have a decisive influence on the overall development prospects of societies, not only for this generation but also for future generations as well. This is consistent with the principles of sustainable development (Ha and Dhakal 2013). As explained by Regional Consultative Committee on Disaster Management (2011), "certain global development trends such as population growth and rapid urbanisation lead to land use conflicts, increased vulnerabilities and disaster risks" (p. 10).

The disaster risk management can be addressed in two phases—the pre-disaster phase and the post-disaster period. The pre-disaster phase includes four areas, such as "risk identification, risk mitigation, risk transfer and preparedness" (Espinoza and Peterson 2008, p. 341). On the other hand, the post-disaster phase comprises areas such as emergency response, rehabilitation and reconstruction. A comprehensive risk management has to address all these seven components. Yet, it is not a simple task to address all the components given resource and capacity constraint. Although many countries have developed, or are developing, national programs to

H. Ha (✉)
UON (University of Newcastle, Singapore Campus), Singapore, Singapore
e-mail: artp6075singapore@yahoo.com

T.N. Dhakal
Central Department of Public Administration, Tribhuvan University, Kathmandu, Nepal
e-mail: dhakaltn@gmail.com

H. Ha (ed.), *Land and Disaster Management Strategies in Asia*,
DOI: 10.1007/978-81-322-1976-7_17

partly or fully incorporate these elements of an integrated disaster risk management involving different sectors and actors, the outcome are limited. The search for effective and productive strategies to address the impact of disasters and reduce risk has remained a relentless mission.

It is realised that the interrogate policies of consumerism and commoditisation of nature destroys the carrying capacity of land and water resources which made the world more vulnerable to natural calamities which are labelled as "natural" but are mostly "man-made." To get relief from such issues decision making processes within countries should incorporate "sustainable lifestyle" and "consumer patterns" in their plans and strategies (Jackson 2008, p. 1). Actually "land use planning is instrumental in addressing the challenges posed by natural hazards on built environment … With its array of regulatory and non-regulatory techniques and mechanisms, land use planning can become an effective tool for disaster risk reduction" (Regional Consultative Committee on Disaster Management 2011, p. 13). The impact of disasters includes human and livelihood losses, disaster-related physical and mental health problems, losses of assets, productivity and services, which could occur to any community any country for a prolonged period of time. Thus, land planning and its proper use are very important which may have direct and indirect impacts to disasters and other associated issues, such as climate change. Countries like Nepal stands as a combination of topography and climate which means that it is very landslide prone. Thus, the impacts are exacerbated by the vulnerability of the population. A devastating loss of human lives and assets, and damaged infrastructure caused by the Uttarakhand flood in India in the 3rd week of June 2013 is a live example of failure of planning of land and the settlement (Rahman 2013). Hence, there is also a close connotation with land and water use and occurrence of disaster, though disaster is both manmade and nature made.

17.2 The Way Forward and Recommendations

Looking into the value of land, governments in many countries often launch a series of land policy reforms for improving land use efficiency, rationalising land allocation, enhancing land management, and coordinating urban and rural development. This is an important factor for fuelling production required from agriculture, industry and other economic activities. In fact, it also serves as a principal instrument in fostering "social justice, development, provision of decent dwellings and health conditions, and therefore should be used in the interest of the society as a whole" (United Nations 1976, p. 1). Yet, the report 2011–2013 of the Food and Agriculture Organisation of the United Nations indicated that still 842 million people, i.e. 12 % of the global population, were unable to meet their dietary energy requirements (FAO et al. 2013). Of them, a huge number of people live in developing regions, where the prevalence of undernourishment is estimated at 14.3 % (FAO et al. 2013). Therefore, food security has become a complex issue which is subject to many forces. In addition, food security particularly put pressure on land, water and the natural setting, which directly or indirectly become one of the causes of disasters and broadly climate change. In other words, food insecurity is a result of climate change, urban development and population growth, etc. These issues are interconnected beyond the boundaries of a country or a region. Therefore, the issues become global importance.

One of the issues arising from the discussion in this volume is how to preserve forest ecosystems and the services which enhance food production (Lambin and Meyfroid 2011). Many Asian countries have adopted a transitional approach to manage land use during the last several years which aim to simultaneously increase the coverage of their forest, and improve agricultural productivity by relying on various mixes of mechanisms. However, due to the lack of ability and opportunity to compete with economic globalisation, people in developing countries (especially in the rural areas) have illegally cut down trees in forests to seek cropland which triggers deforestation (Lambin and Meyfroid 2011). This has entailed many adverse effects, such as forest fires, haze caused by forest fires, food insecurity and environmental degradation. Therefore, land use and planning has been continuing a chronically subject for debate in many countries.

Another issue is disaster risk management. The disaster risk management system should comprise "formal or informal interaction between institutions, financial mechanisms, regulations and policies" (Freeman et al. 2003, p. i). It is commonly believed that governments must be active participant in its creation and implementation of national disaster systems. In addition, civil society organisations have also developed community-driven projects and programs for reducing natural hazard risk (Freeman et al. 2003). Though such approach to risk management may not be comprehensive, it helps directly to identifiable needs and the empowerment of local people. In this context, the Community Forest Users' Groups (CFUGs) (Chap. 16), for an example, in Nepal has stood as a successful program for protecting forest by mobilising the huge number of community people (Kanel n.d.).

It is plausible that all countries discussed in this volume have put much effort to set plans and launch initiatives to tackle various issues linked to land management and DRM. Government still plays the key role in all plans and activities. While civil society has been active in support of such initiatives, and embarked on their own program to alleviate the burdens caused by natural disasters, there has been insufficient information about activities by the private sector in risk reduction.

Efforts have been exerted, but different outcomes have been produced due to several reasons. Some reasons are pertinent to governance, such as administrative and institutional framework, structures and operations, and legal shortcomings, for example, land reforms in Kerala and in Pakistan. Other reasons include lack of resources or lack of collaboration or coordination as in the case of the people with disability (PWD) in Bangladesh (Chap. 8). Sometimes, the plans were too ambitious, the objectives are too wide or too challenging, and the implementation is not strict enough which drive the outcome to another direction (Chaps. 2 and 3).

There are strong linkages among the issues discussed in each chapter in this volume. For instance, lack of public education will lead to ignorant citizens who may not have the required knowledge to prepare for pre-, during and post-disasters (Chap. 12). This may reduce their abilities to timely respond to threats in case of emergencies.

From the conclusions of the chapters, we suggest the following recommendations which theoretically may help stakeholders address the challenges of land and disaster management in the context of Asia. Yet, more research with empirical tests and evidence is required for better governance of land and disasters.

Governance refers to complex systems and networks, including structure, institutional, legal and administrative arrangements, instruments and mechanisms, interaction among stakeholders, and connection among political, economic, socio-cultural and technological activities and the environment (Kooiman 1993, 1999; Koiman et al. 2008; Ha 2013). Thus, solutions to address land and disaster management should take into consideration all relevant elements of governance. In this instance, similar to the authors of many chapters in this book (Chaps. 6, 7, 12, 15 and 16), we firstly proposed a multi-sector governance approach for land and disaster management, and also for addressing issues with regard to disability, peace and climate change in the context of risk reduction. This approach should embrace both state and non-state sectors, and all groups of stakeholders in the process. Strong collaboration and coordination among sectors and actors in each sector is required to produce the desirable outcome in the governance process as highlighted in Chaps. 7 and 9. This has been demonstrated in the activities of civil society organisations, working with government bodies to contribute to risk reduction and post-disaster reconstruction efforts (Chaps. 6, 11, 12 and 16). In some isolated cases, local governments may distant themselves from working with other non-sectors, as described in Chap. 13, for some reasons, causing a crippling governance system where civil society can still function well given the strong support of local residents.

Secondly, institutional and administrative arrangements must be in place in order to ensure (i) a comprehensive and workable legal framework, (ii) an effective complaint system together with strict legal enforcement and implementation. A legal framework with many loopholes will ultimately lead to not only positive but also negative effects, or in many cases only negative outcome, as in the case of land reform in Kerala. The authors of Chap. 2 pointed out that land reform might benefit some target recipients (the poor, the landless) in the short term, but might made them distant from land in the long run. Misuse and mismanagement of land have led to food insecurity in Pakistan as highlighted in Chap. 4, whereas policy misalignment has also affected food supply and food security in Malaysia as discussed in Chap. 5.

Finally, governance approaches and processes have been and will be shaped and decided by political leaders. Hence, leaders in Asian countries must have the ability to develop a vision, and to share their vision with the followers in order to gain their support and mobilise the required resources. But what kind of leaders do we need? Apparently, transformational leaders having both the head (i.e. competency, ability, etc.) and the heart (servant leadership, caring, etc.) are sought. Above all, we need leaders with integrity and virtue who can make transformational change in the world for the benefits of the majority as explained by utilitarianism theory (Shaw et al. 2014). Genuine leaders will be able to deal with complex and increasingly political, economic and socio-cultural tensions in the future, including pressure of the Mother Nature as discussed in Chap. 14. The list of recommendations is not exhausted.

17.3 Limitations

Limitations are difficult to avoid in any scholarly research work, and the same applies to this edited volume. Due to time and resource constraints, this edited volume cannot cover all issues associated with land use and planning, risk reduction and disaster management. Also, this volume cannot include research studies in all countries in Asia. However, the selected chapters cover a wide range of important topics associated with land and disaster management in the context of Asia which can reflect the state of land use and planning, and disaster risk management in the selected countries.

17.4 Conclusion

In short, there is no simple or single blue print or a magic formula to address issues associated with land and disaster governance across countries. Thus, the concept of a contingency approach as discussed by Dunphy and Stace (1993) in a business context may be applied to tackle local, national and regional issues. Nevertheless, it is an urgent task for government to search for governance approaches and new ways to foster collaboration and coordination with the private sector, industry and professional associations, and civil society organisation, including international and national non-governmental organisations and community-based groups to manage land and disasters.

References

Dunphy D, Stace D (1993) The strategic management of corporate change. Hum Relat 46(8):905–918

Espinoza N, Peterson M (2008) Incomplete preferences in disaster risk management. Int J Technol Policy Manage 8(4):341–358

FAO, IFAD, WFP (2013) The state of food insecurity in the World 2013. The multiple dimensions of food security. FAO, Rome

Freeman PK, Martin LA, Linnerooth-Bayer J, Mechler R, Pflug G, Warner K (2003) Disaster risk management: national systems for the comprehensive management of disaster risk and financial strategies for natural disaster reconstruction. Inter-American Development Bank, Washington, D.C., USA

Ha H (2013) Climate change governance: Singapore case. In: Ha H, Dhakal TN (eds) Governance approaches for mitigation and adaptation of climate change in Asia. Palgrave McMillan, London, UK, pp 182–199

Ha H, Dhakal TN (2013) Governance approaches to mitigation of and adaptation to climate change in Asia. Palgrave McMillan, London, UK

Jackson T (2008) Chapter 4: The challenge of sustainable lifestyles. In: Worldwatch institute report state of the World 2008: innovations for a sustainable economy. Worldwatch Institute, Washington, D.C., USA, pp 1–2

Kanel KR (n.d.). Current status of community forestry in Nepal. A paper submitted to regional community forestry training centre for Asia and the Pacific Bangkok, Thailand. Retrieved from http://nepalpolicynet.com/images/documents/forest/research/Current%20Status%20of%20CF%20in%20Nepal.pdf (Accessed 2 July 2014)

Kooiman J (1993) Modern governance. London: Sage

Kooiman J (1999) Socio-political governance. Public Manag 1(1):67–92

Koiman J, Bavinck M, Chuenpagdee R, Mahon R, Pullin R (2008) Interactive governance and governability: an introduction. J Transdisciplinary Environ Stud 7(1):1–11

Lambin EF, Meyfroid P (2011) Global land use change, economic globalization, and the looming land scarcity. Proc Natl Acad Sci USA 108(9):3465–3472

Rahman M (2013) Indian floods leave tens of thousands stranded in Uttarakhand state. The Guardian, 20 June 2013

Regional Consultative Committee on Disaster Management (2011) Promoting use of disaster risk information in land-use planning. Regional Consultative Committee (RCC) Secretariat Asian Disaster Preparedness Centre, Bangkok, Thailand

Shaw W, Barry V, Issa T, Catley B (2014) Moral issues in business (The 2nd Asia Pacific Edition). Cengage Publishing, Melbourne, Australia

United Nations (1976) The Vancouver action plan. D. Land. United Nations, New York

Erratum to: Land and Disaster Management Strategies in Asia

Huong Ha

Erratum to:
H. Ha (ed.), *Land and Disaster Management Strategies in Asia*, DOI 10.1007/978-81-322-1976-7

Dr. Charlita Andales-Escano's name was published incorrectly as Dr. Charlatan Andales-Escano in the 'Notes on Reviewers' page (*page xxxvii*). The name should read Dr. Charlita Andales-Escano.

The online version of the original book can be found at http://dx.doi.org/10.1007/978-81-322-1976-7.

H. Ha (✉)
UON Singapore (University of Newcastle, Singapore Campus), Singapore, Singapore
e-mail: artp6075singapore@yahoo.com

H. Ha (ed.), *Land and Disaster Management Strategies in Asia*,
DOI: 10.1007/978-81-322-1976-7_18

FSC
www.fsc.org
MIX
Papier | Fördert
gute Waldnutzung
FSC® C083411

Kolibri 360 GmbH
Ferdinand-Jühlke-Straße 7
99095 Erfurt, Deutschland
produktsicherheit@kolibri360.de